EFFECTIVE ENVIRONMENTAL ASSESSMENTS

How to Manage and Prepare NEPA EAs

EFFECTIVE ENVIRONMENTAL ASSESSMENTS

How to Manage and Prepare NEPA EAs

Charles H. Eccleston

CRC Press
Taylor & Francis Group
Boca Raton London New York

CRC Press is an imprint of the
Taylor & Francis Group, an **informa** business
A TAYLOR & FRANCIS BOOK

CRC Press
Taylor & Francis Group
6000 Broken Sound Parkway NW, Suite 300
Boca Raton, FL 33487-2742

First issued in paperback 2019

ISBN-13: 978-1-56670-559-2 (hbk)
ISBN-13: 978-0-367-39764-7 (pbk)

Library of Congress Cataloging-in-Publication Data
Eccleston, Charles H.
Effective environmental assessments:how to manage and prepare NEPA EAs/Charles Eccleston.
p. cm.
Includes bibliographical references and index.
ISBN 1-56670-559-2 (alk. paper)
1. Environmental impact analysis--United States. 2. Environmental impact statements--United States. 3. United States. National Environmental Policy Act of 1969.
I. Title.
TD194.65 .E35 2000
658.4 '08—dc21   00-052035
CIP

Library of Congress Card Number 00-052035

Visit the Taylor & Francis Web site at
http://www.taylorandfrancis.com

and the CRC Press Web site at
http://www.crcpress.com

Foreword

A comprehensive guide to managing and preparing environmental assessments (EAs) has long been needed. Charles Eccleston, author of two previous books on The National Environmental Policy Act (NEPA) has now filled this void.

As one of the chief architects of NEPA, I bring a unique perspective to the historical development, benefits, and problems encountered in implementing the Act. For example, in drafting NEPA, the form and substance of an environmental impact statement (EIS) was incorporated as Section 102 of the Act. But the range of environmental impacts from minor to major varied greatly among proposals. In some cases, minor adjustments would satisfy the NEPA intent. For large and complex projects, a more in-depth inquiry was required. To determine the methods and extent of impact evaluation, a new tool — the EA — was necessary. The findings of such an assessment should indicate the nature and extent of the probable impact, from no significant impact (FONSI) to a fully developed environmental investigation and EIS.

The EA is a critical document in the implementation of NEPA. Eccleston's text, *Effective Environmental Assessments*, organizes and summarizes the experience of 30 years in the development of this aspect of the NEPA process. This comprehensive guide answers a need that was evident in A Study of Ways to Improve the Scientific Content and Methodology of Environmental Impact Analysis, funded by the National Science Foundation. This study found numerous reports of incomprehension and frustration in ascertaining and evaluating the environmental impacts of federal programs. The Council on Environmental Quality (CEQ), through its regulations, has undertaken to remedy this circumstance. Eccleston's comprehensive guide provides an operational tool for applying what has been learned over the decades. For those who need to know how the principles of NEPA are best implemented, it is a "dictionary" for EA procedures. It belongs on the desk of anyone involved in the NEPA process.

Because the EA is a critical step in the implementation of NEPA's goals, it is important that the responsible agency personnel understand the context that the EA is to serve. Evidence gathered in A Study of Ways to Improve the Scientific Content and Methodology of Environmental Impact Analysis pointed toward a conclusion that inadequacy in assessments was primarily conceptual rather than scientific or technical; this short-fall resulted as often from the mind-set or political bias of assessors as in failure to identify all relevant factors in the assessment. *Effective Environmental Assessments* provides specific direction and procedures for addressing this weakness in the process.

Previous texts by Charles Eccleston are *Environmental Impact Statements: A Comprehensive Guide to Project and Strategic Planning* (John Wiley & Sons, 2000), and *The NEPA Planning Process: A Comprehensive Guide with Emphasis on Efficiency* (John Wiley & Sons, 1999). Both books are very useful companions to *Effective Environmental Assessments.* I regard all three as valuable resources. All are comprehensive and detailed, but indexed so they are very accessible to users seeking clarification on the complexities of implementing NEPA.

Research on impact assessment is continuing and refinements and advances in procedure and criteria can be expected. However, for the present and for some time to come, Eccleston's guidebook is likely to be the leading source of information for students, practitioners, and decision makers.

Lynton K. Caldwell, Ph.D.
Professor Emeritus
Indiana University
(Consultant to the U.S. Senate Committee that drafted NEPA)

Preface

A man who carries a cat by the tail learns something he can learn in no other way.

Mark Twain

An act of historic proportions, the National Environmental Policy Act (NEPA) of 1969 pronounced the world's first environmental policy for protecting the state of the environment. At the time, few foresaw that NEPA would become a model for environmental policies adopted by nations the world over. With upwards of 50,000 NEPA Environmental Assessments (EAs) prepared annually, the influence of NEPA on federal planning is pervasive.

Problems

Recently, however, NEPA has come under scrutiny as federal funding has tightened. Emphasis has increasingly shifted toward doing more with less. This text is, in part, a response to this trend. One of the goals of this book is to present federal agencies and practitioners with an efficient and effective framework for planning federal actions.

Unlike an Environmental Impact Statement (EIS), an EA is normally more difficult to defend when challenged, because the agency bears the burden of proof in demonstrating that no significant impact would result. Thus, adversaries have increasingly focused efforts on attacking EAs. Agencies can defend against this strategy by devoting particular care to the preparation of their EAs. Alternatively, adversaries may increase their odds of success by directing efforts at flawed or poorly prepared EAs.

Unfortunately, the focus of the Council on Environmental Quality's (CEQ) NEPA Regulations is on providing regulatory direction for preparing EISs. Very little attention has been placed on propagating requirements governing the preparation of Environmental Assessments (EA). Consequently, a regulatory vacuum exists with respect to preparing EAs. Practitioners have struggled with the problem of preparing publicly defensible EAs given only minimal regulatory direction. Lack of definitive regulatory direction has resulted in many problems, not the least of which is that there is wide variation in the scope, content, and quality of EAs among federal agencies and even within the same agencies. This has further opened the avenue for litigation.

Goals of this Text

This text has been specifically designed to bridge this regulatory chasm; applying the "Rule of Reason," this text identifies and describes relevant EIS regulatory requirements that can be logically interpreted to also apply to preparation of EAs. This text also draws on the professional experiences and assimilates best professional practices from seasoned practitioners who have spent years preparing EAs. The goal is to provide the reader with a reasonable, definitive, consistent, and comprehensive methodology for managing, analyzing, and writing EAs.

To assist the practitioner, three representative EAs are presented in Appendices D–F that are typical of assessments produced in recent years by federal agencies. The three EAs illustrate principles inherent in a proficient analysis. They also illustrate some deficiencies common to many EAs. Critiques of each assessment are offered to assist the reader in improving preparation of future EAs.

The limited direction that has been propagated by the CEQ for preparing EAs is scattered throughout NEPA's implementing regulations, executive orders, guidance, and case law. To date, no text has collectively compiled and synthesized this information into a single source. This book is designed to provide the reader with a single, integrated, and comprehensive source of the relevant guidance and requirements. Diverse sources have been drawn from, including the CEQ NEPA Regulations and guidance documents, executive orders, professional papers, and experience of NEPA practitioners. Case law has also been integrated throughout this book to provide additional direction or clarification. Finally, the National Association of Environmental Professional's NEPA Working Group, has provided a valuable source of information, experience, and expertise from which to draw.

Objectives of this Text

This text is unique in that it

- Provides the user with the most comprehensive and thorough description of the EA process written to date. Ultimately, the goal is to provide the reader with a reasonable, definitive, consistent, and comprehensive methodology for managing, analyzing, and writing EAs.
- Comprehensively describes the step-by step process for managing and preparing an EA. An approach for evaluating environmental impacts is also described.

- Details all documentation requirements that the EA must meet. Additionally, recommendations are offered for promoting a more defensible assessment as well as improving the goal of excellent environmental planning. Lessons from case law are integrated with the relevant requirements.

- Addresses the regulatory vacuum under which practitioners have struggled with the paradoxical problem of preparing publicly defensible EAs given only minimal regulatory direction. The text identifies and describes relevant EIS regulatory requirements that can logically be interpreted to apply to preparation of EAs. The experiences of seasoned practitioners and best professional practices are assimilated to provide the reader with a comprehensive, definitive, and defensible methodology for preparing EAs.

- Addresses problems and dilemmas that have traditionally plagued preparation of EAs. Specific tools and approaches are suggested for resolving such problems. Emphasis is placed on introducing methods and procedures for streamlining the EA process.

Audience

This text is designed for beginners and experts alike. It begins with the fundamentals and advances into increasingly more complex subject matter. Experienced practitioners can use the book as a resource for quickly reviewing issues, or as a comprehensive textbook. While primarily aimed at professionals in government, consulting, and the private sector who prepare and review EAs, this book also lends itself to individuals who seek only an introduction to certain selected topics. Individuals and groups include decision-makers, analysts, scientists, planners, regulators, project engineers, and lawyers, to name just a few. Persons in advocacy or citizen groups who seek to challenge a NEPA compliance action will find the book equally useful. The book can be used by university students in environmental curricula and by instructors who teach professional courses.

While the book provides the reader with direction for complying with NEPA's requirements, the author stresses the importance of seeking the counsel of NEPA specialists, regulatory analysts, and legal experts, particularly in areas involving complex or controversial issues. The reader is advised to consult the actual regulatory provision for the details and precise wording.

Charles H. Eccleston
Chairman, Tools and Techniques (TNT) NEPA
Practice Committee
National Association of Environmental Professionals

Introduction

It is noble to teach oneself, but still nobler to teach others—and less trouble.

Mark Twain

While clearly seeing the need for a national environmental policy, the framers of the NEPA did not foresee the need for establishing the concept of an Environmental Assessment (EA). This was left to a later date, when the Council on Environmental Quality (CEQ) issued its NEPA Regulations, which formally introduced the concept of the EA.

The Environmental Impact Statement (EIS) was clearly envisioned by the architects of NEPA to be the principal tool for assessing impacts and planning actions. Instead, EAs have become the principal instrument used in NEPA for investigating impacts. Approximately 100 EAs are prepared for each EIS; the CEQ estimates that approximately 30,000–50,000 EAs are prepared annually.

FIGURE I.1
The EIS versus the EA.

Approximately 100 EAs are prepared for each EIS. From 30,000–50,000 EAs are prepared annually.

Much of the reliance on EAs can be attributed to agencies learning to incorporate NEPA's principles into the early stages of project planning, thus reducing the potential for significant impacts. This practice promotes the original intent and purpose of NEPA. EAs also provide an important instrument for planning actions, as evidenced in a CEQ survey that found that nearly all agencies have reported that at least some of their EAs have led to modification in project design or the incorporation of mitigation measures.

FIGURE I.2
EAs provide an effective tool for protecting the environment.

A CEQ survey found that nearly all agencies have reported at least some of their EAs have led to modification in project design or the incorporation of mitigation measures.

When questioned about their effectiveness, every agency surveyed in CEQ's study reported that EAs have succeeded in fulfilling the purposes for which

FIGURE I.3
Importance of EAs in agency planning.

In a CEQ survey, every agency reported that EAs have succeeded in fulfilling the specific purposes for which they were used.

they have been used. Over 60 percent of the respondents indicated that their EA process has been very successful in meeting these purposes.[1]

I.1 Sliding Scale and Rule of Reason

Each EA is generally prepared under a unique set of circumstances. Guidance provided in this text is not unequivocal; common sense and professional experience must be exercised in determining the scope and depth of analysis appropriate for a given investigation. It is not possible to provide a cookbook approach governing all circumstances that might arise. However, the following tools are provided to assist practitioners in preparing a cost-effective analysis. Properly used, these tools provide instruments for performing an analysis that is more concise and can be prepared in a shorter period of time. These principles are discussed in more detail in subsequent chapters. The user is directed to the author's companion book, *The NEPA Planning Process: A Comprehensive Guide with Emphasis on Efficiency*, for a more detailed description of the basis and application of these tools.[2]

I.1.1 The Sliding-Scale Approach

A *sliding-scale approach* can prove particularly useful in preparing an EA. The sliding-scale approach recognizes that the standards that an EA and Environmental Impact Statement (EIS) are expected to meet depend on the particular circumstances surrounding the action. Under a sliding-scale approach, impacts, issues, and related regulatory requirements are investigated and addressed with a degree of effort commensurate with their importance. The amount of consideration devoted to specific issues and impacts varies with respect to its potential for significant impacts. Investigation of impacts that are clearly small or nonsignificant requires less analysis than actions with a correspondingly larger potential for impact.

I.1.2 Rule of Reason

An overstrict or unreasonable application of a regulatory requirement can lead to decisions, a course of action, or a level of effort that is wasteful or unreasonable. The Rule of Reason is a mechanism used by the courts for injecting reason and commonsense into the NEPA process. Common sense is applied in determining the scope and detail of the issues, alternatives, and impacts that are analyzed.

References

1. A. Blaug, Use of the Environmental Assessment by Federal Agencies in NEPA Implementation, *The Environmental Professional*, Volume 15 pp. 57–65, 1993.
2. C. H. Eccleston, *The NEPA Planning Process: A Comprehensive Guide with Emphasis on Efficiency*, John Wiley & Sons, Inc., New York, N. Y., 1999.

About the Author

Charles H. Eccleston is the author of the successful text, *The NEPA Planning Process: A Comprehensive Guide with Emphasis on Efficiency* (John Wiley & Sons, 1999), and *Environmental Impact Statements: A Comprehensive Guide to Project and Strategic Planning* (John Wiley & Sons, 2000).[1] He has lectured and taught, and authored more than two dozen professional publications on National Environmental Policy Act (NEPA) and environmental impact assessment.
With nearly 20 years of diverse engineering and scientific experience, he has managed a diverse array of environmental analysis and planning efforts. As a Principal Environmental Scientist at the U. S. Department of Energy's Hanford site in Richland, Washington, he has developed innovative tools, techniques, and strategies for effectively integrating NEPA with site-wide planning and other environmental processes such as ISO-14000 and pollution prevention. In this position, he has developed numerous methodologies that have received national attention for their ability to streamline NEPA compliance while reducing project cost and delays.

Mr. Eccleston has chaired two national committees chartered with responsibility for establishing nationally Accepted Methods of Professional Practice (AMPPs) for addressing problems that have traditionally hindered NEPA and environmental planning. Currently, he chairs the National Association of Environmental Professional's Tools and Techniques (TNT) NEPA Practice Committee.

As a member of *Who's Who in America,* he has participated in two White House-sponsored environmental workshops held to develop approaches for improving NEPA effectiveness and for spearheading a national environmental/industrial coalition.

Prior to working at the Department of Energy's Hanford site, he held a position with the GTE Corporation's Defense Electronics Division in Sunnyvale, California, where he contributed to the development of advanced strategic weapon systems that helped to bring an end to the Cold War. Prior to this, he was a Senior Engineer in the Advanced Design Branch (ADB) at the Texas Instruments Corporation in Dallas, Texas.

Mr. Eccleston holds a Master's of Science degree in environmental geology/geophysics and Bachelor of Science degrees in environmental geology and computer science. As a Certified Environmental Professional (CEP), he consults on NEPA and planning problems and can be contacted by e-mail at: ecclestonc@msn.com.

Acknowledgments

I am indebted to many people who reviewed and provided comments on this book. While it does not endorse or necessarily represent the formal view or position of the National Association of Environmental Professionals (NAEP), numerous individuals from within the NAEP's Tools and Techniques (TNT) Committee were particularly helpful.

Although it is not feasible to mention all individuals by name, I would like to call attention to the following three professionals who are members of NAEP's TNT Committee: Mr. Christopher D. Damour (U.S. Army) and Mr. Craig T. Casper (Wilbur Smith Associates) provided numerous suggestions for enhancing the effectiveness of the Environmental Assessment (EA) process; I am particularly thankful to Mr. J. Peyton Doub (Halliburton NUS), who provided comments, contributed a section on collecting data and interpreting significance, and provided three example EAs that are presented in the appendices.

Special thanks are also extended to Ms. B. D. Williamson (Office of General Counsel, U.S. Department of Energy, Richland Operations Office), who reviewed and contributed to various legal concepts. Finally, I would like to thank Mr. M. C. McMillen (Energetics, Inc.), who performed an in-depth review and contributed important comments on the section involving cumulative impact analyses

DEDICATION

To Jean, Alice, and Brandie

List of Acronyms

CATX	Categorical Exclusion
CEQ	Council on Environmental Quality
CFR	Code of Federal Regulations
DBS	Decision-Based Scoping
DIT	Decision-Identification Tree
EA	Environmental Assessment
EIS	Environmental Impact Statement
EJ	Environmental Justice
EO	Executive Order
EPA	Environmental Protection Agency
FONSI	Finding of No Significant Impact
FR	Federal Register
IDT	Interdisciplinary Team
ISO	International Organization for Standardization
MAP	Mitigation Action Plan
NAAQS	National Ambient Air Quality Standards
NAEP	National Association of Environmental Professionals
NEPA	National Environmental Policy Act
NOI	Notice of Intent
SHPO	State Historic Preservation Officer
SIP	State Implementation Plan
TNT	Tools and Techniques
USC	United States Code
USFWS	United States Fish and Wildlife Service

List of Acronyms

Contents

1

An Overview of NEPA

> Do not tell fish stories where the people know you; but particularly, don't tell them where they know the fish.
>
> Mark Twain

In encouraging growth and expanding American economic power, the United States government has played a preeminent role in the development and exploitation of our nation's natural resources. Directly or indirectly, the federal government has sponsored an unfathomable number of actions that have exacted a significant toll on environmental quality.

Historically, most federal decisions were based primarily on economic and technical considerations. All this was to change in the late 1960s. In elevating the importance of environmental considerations, the U.S. Congress enacted the National Environmental Policy Act (NEPA) of 1969, establishing a national policy for protecting our nation's environment.[1] The passage of NEPA paved the way for elevating environmental considerations to the forefront of federal decisionmaking. A copy of the NEPA Act is provided in Appendix A.

NEPA's profound importance is perhaps due less in part to the national policy it establishes, than to the rigorous procedural requirements that have been imposed on all federal agencies. NEPA was enacted as a supplement to the traditional mission of federal agencies. In reaching a final decision regarding a proposal, federal agencies must now balance potential environmental consequences with other more established considerations.

1.1 NEPA's PURPOSE

In enacting NEPA, the U.S. Congress established the world's first national policy for protecting the quality of the human environment for future generations. The Act is composed of two titles. Title 1 declares a national environmental policy and sets forth procedural requirements that must be followed in reaching a decision to pursue a federal action, while Title II creates a Council on Environmental Quality (CEQ).

1

NEPA's Purpose

The purposes of this Act are: To decare a national policy which will encourage productive and enjoyable harmony between man and his environment; to promote efforts which will prevent or eliminate damage to the environment and biosphere and stimulate the health and welfare of man; to enrich the understanding of the ecological systems and natural resources important to the Nation; and to establish a Council on Environmental Quality.

(Sec. 2 of NEPA [42 U.S.C. § 4321])

The statute is most notable for three principal elements:

1. Declaration of the world's first national policy for protecting and preserving the environment

2. Establishment of an "action-forcing" mechanism (i.e., environmental impact statement) for implementing this policy

3. Creation of a CEQ for overseeing NEPA and elevating environmental concerns directly to the presidential level

1.1.1 Title I - Declaration of a National Environmental Policy Act

---

Title 1, Section 101 of NEPA

Section 101 declares a policy to use all practical means, including financial and technical means, to promote the general well being of the environment. Specifically, all practicable means will be used to

1. *fulfill the responsibilities of each generation as trustee of the environment for succeeding generations;*

2. *assure for all Americans, safe, healthful, productive, esthetically and culturally pleasing surroundings;*

3. *attain the widest range of beneficial uses of the environment, without degradation, risk to health or safety, or other undesirable or unintended consequences;*

4. *preserve important historic, cultural, and natural aspects or our national heritage, and maintain, wherever possible, an environment which supports diversity, and variety of individual choice;*

5. *achieve a balance between population and resource use which will permit high standards of living, and a wide sharing of life's amenities; and*

6. *enhance the quality of renewable resources and approach the maximum attainable recycling of depletable resources.*

Section 101 of Title I is the heart of NEPA. It announces the world's first national environmental policy and creates goals that supplement each federal agency's mission.

Section 102 of Title I establishes the procedural mechanism for carrying out the policy established in Section 101. Under Section 102, all agencies of the federal government shall "... *include in every recommendation or report on proposals for legislation and other major federal actions significantly affecting the quality of the human environment, a detailed statement by the responsible official ...*"[2]

The term "detailed statement" has since become more commonly known as the Environmental Impact Statement (EIS). The reader is referred to the author's companion text, *The NEPA Planning Process*, for an overview of the NEPA Act.[3]

The NEPA regulations — Formal regulations (Regulations) for implementing NEPA were issued by the CEQ in 1978.[4] To promote efficiency, the Regulations provide direction for reducing paperwork and delay, use of "scoping" to help agencies focus on key issues, incorporation of material by reference, and also stresses the need to focus on important issues while de-emphasizing nonsignificant ones. A copy of CEQ's NEPA regulations is provided in Appendix B.

Regulatory Nomenclature Throughout this text, the term "NEPA" is sometimes shortened to "Act." Similarly, the phrase "CEQ NEPA Regulations" is shortened to simply "Regulations." For brevity, references to a particular section in the Regulations (40 Code of Federal Regulations [CFR] 1500–1508) are abbreviated so that they simply reference the specific section number in the Regulations. For instance, a reference to "40 CFR 1500.3" is shortened to the more convenient expression "§1500.3."

1.1.2 Complying to the Fullest Extent Possible

Federal agencies are directed to interpret and administer policies, regulations, and public laws in accordance with the policies and requirements of NEPA. Moreover, agencies are instructed to: "integrate the requirements of NEPA with other planning and environmental review procedures required by law or by agency practice so that all such procedures run concurrently rather than consecutively" (§1500.2[c]).

> **Agency Responsibility**
>
> *... to the fullest extent possible the policies, regulations, and public laws of the United States shall be interpreted and administered in accordance with the policies set forth in this Act*
> **(Sec. 102 [42 U.S.C. § 4332])**

1.2 Overview of the NEPA Compliance Process

This section provides a general overview of the NEPA process with emphasis on describing the three levels of NEPA compliance. A detailed accounting of every aspect and intricacy inherent to the NEPA process is beyond the scope of this text. For a detailed review, the reader is referred to the companion text, *The NEPA Planning Process*.[5]

Details governing the actual implementation of NEPA can vary, particularly with respect to the way individual agencies choose to implement specific aspects of their respective processes. For requirements governing specific circumstances, the reader is referred to the regulations, and the agency's internal orders and NEPA implementation procedures.

1.2.1   When Must NEPA Begin?

> **Definition of a Proposal**
>
> *With respect to NEPA, a "Proposal" exists when the following three conditions are met:*
>
> 1. *an agency has a goal*
>
> 2. *the agency is actively preparing to make a decision on one or more alternative means of accomplishing the goal.*
>
> 3. *the effects can be meaningfully evaluated.*
> (1508.23)

The EIS process must be started so that it "can be completed in time for the final statement to be included in any recommendation or report on the proposal" (§1502.5); thus, the NEPA process is to begin as close as possible to the time in which an agency is developing or is presented with a proposal. This provision is interpreted to apply to preparation of Environmental Assessments (EA) as well as an EIS, as all federal actions are considered potentially subject to the requirements of an EIS until it can be demonstrated otherwise.

1.2.2   When Must NEPA be Completed?

> **Timing**
>
> *"An agency shall commence preparation of an environmental impact statement as close as possible to the time the agency is developing or is presented with a proposal ... so that preparation can be completed in time for the final statement to be included in any recommendation or report on the proposal. The statement shall be prepared early enough so that it can serve practically as an important contribution to the decisionmaking process and will not be used to rationalize or justify decisions already made..."*
> (§1502.5)

The NEPA process must be initiated early enough so that it can contribute to the decision-making process and will not be used to rationalize or justify decisions already made. This timing requirement is met when a NEPA analysis has been prepared in time to meet the decision deadline, but not so early that it cannot meaningfully contribute to the decisionmaking process (§1502.5).

1.2.3   Three Levels of NEPA Compliance

The NEPA process can be viewed as consisting of three levels of planning and environmental compliance. These three levels, defined from the least to the most demanding, are

- **Categorical Exclusion (CATX)** — Some federal actions might qualify for a CATX, thus excluding them from further NEPA review and documentation requirements.
- **Environmental Assessment (EA)** — If a federal action does not qualify for a CATX, an EA may be prepared to determine whether

a federal action qualifies for a Finding of No Significant Impact (FONSI), thus exempting it from the requirement to prepare an EIS.

* **Environmental Impact Statement (EIS)** — In general, an EIS must be prepared for proposed federal actions that do not qualify for either a CATX or a FONSI.

1.2.4 Initiating the NEPA Process

Care must be exercised in selecting a manager who can exercise good judgment and who possesses experience in preparing NEPA analyses. Inexperienced management and/or analysts have been one of the principal causes responsible for cost overruns, poor planning, and project delays.

In one instance, a NEPA technical support document was prepared for a waste management project at a federal installation operated by the Department of Energy. The group manager lacked experience preparing environmental impact studies and shied away from making key decisions. When decisions were made, they were frequently formulated with little regard to future ramifications. While this manager eventually left her position, the quality of this technical planning process and other similar environmental compliance efforts was questionable. If management lacks experience, it is vital that the counsel of experienced practitioners be sought.

Examining proposals for existing NEPA coverage — The NEPA planning process normally begins when a need for taking action has been identified (See oval-shaped rectangle in the upper left hand corner of Figure 1.1). Figure 1.1 provides a simplified overview of the entire NEPA process (including preparation of the EIS).

A new action might fall within the scope of an existing EIS or EA. If the proposal is sufficiently covered under existing NEPA documentation, the agency can proceed with the action (with respect to NEPA's requirements). Existing NEPA documentation should be examined to determine whether the proposal has been subject to a previous NEPA review (first and second decision diamond, Figure 1.1). Reaching such a determination often requires exercising a substantial degree of professional judgment. Under certain conditions, the proposal might require supplementing an existing EIS (§1502.9[c]).

Categorically excluding actions — A proposal might be excluded from further NEPA review if it falls within an existing class of actions (i.e., CATXs) that has been previously determined to result in no significant impact

Categorical Exclusion
A category of actions which do not individually or cumulatively have a significant effect on the human environment ... and for which, therefore, neither an environmental assessment nor an environmental impact statement is required.
(§1508.4)

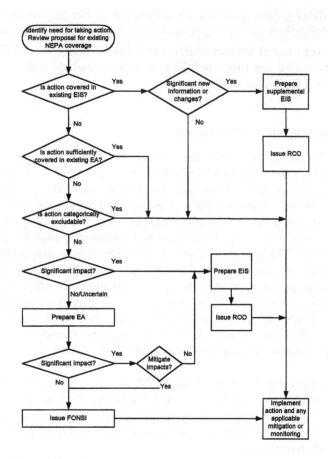

FIGURE 1.1
Overview of typical NEPA process.

(cumulative or otherwise) and for which preparation of an EA/EIS is therefore not required (see third decision diamond, Figure 1.1). Each agency is required to prepare a list of CATXs as part of its individual NEPA Implementation Procedures. If a CATX is applicable, the agency can proceed with the action (with respect to NEPA requirements).

It is important to note that, under §1508.4, agencies are also required to "... provide for *extraordinary circumstances* in which a normally excluded action may have significant environmental impacts." For more information on the application of CATXs, the reader is referred to *The NEPA Planning Process.*[6]

If the proposal has not been excluded, the agency should review its NEPA Implementation Procedures for guidance in determining the appropriate level of NEPA compliance.

CATXs and Case Law   Care must be exercised in applying a CATX. An agency's review process and administrative record should support its decision to apply the CATX. Witness a recent case, in which the Department of Interior was sued for allowing a company to bioprospect for microbes in Yellowstone National Park. The Department of Interior argued that the bioprospecting activities fell under its CATX for "day-to-day resource management and research activities." According to the court, it is inappropriate to claim that an action is covered by a CATX when the agency lacks "evidence in the administrative record or elsewhere that such a determination was made at the appropriate time." The court went on to state that use of this CATX was questionable because (1) the commercial exploitation of natural resources is probably not equivalent to "day-to-day resource management and research activities," and (2) such activities involved extraordinary circumstances associated with "unique geographic characteristics" and "ecologically significant or critical areas," thus making the proposed action ineligible for a CATX under the agency's own rules.[7]

1.2.5 The Environmental Assessment

If the environmental review determines that the action is not eligible for a CATX, the agency can choose to prepare an EA to determine whether the proposal could significantly affect quality of the human environment (see box labeled "Prepare EA" Figure 1.1).

Definition of an Environmental Assessment
A concise public document for which a federal agency is responsible that serves to: (1) briefly provide sufficient evidence and analysis for determining whether to prepare an environmental impact statement or a finding of no significance impact...
(§1508.9)

If the agency believes that the proposal would result in a significant impact, it can choose to prepare an EIS without first writing an EA (see fourth decision diamond labeled "Significant impact?" Figure 1.1). To facilitate future analysis, an agency can also prepare an EA even if it plans to eventually prepare an EIS for the action.

With respect to NEPA's requirements, the agency is free to pursue any course of action as long as it does not result in a significant impact. A Finding of No Significant Impact (FONSI) is prepared if, based on the EA, the decisionmaker concludes that no significant impact would occur. (see box labeled "Issue FONSI?" Figure 1.1). If the EA leads to a FONSI, an EIS is not required, and the agency can proceed with the action (with respect to NEPA's requirements).

Once a FONSI has been issued, the agency is generally free to proceed with the action (with respect to NEPA's requirements), in accordance with any applicable mitigation or monitoring measures. This process is described in detail in Chapter 5.

When should an assessment be prepared? — An EA may be prepared on any action at any time to foster agency planning objectives, or to provide sufficient evidence for determining whether an EIS is required (§1501.3[b]).

When an Environmental Assessment Should be Prepared

"Agencies shall prepare an environmental assessment ... when necessary under the procedures adopted ... to supplement these regulations...An assessment is not necessary if the agency has decided to prepare an environmental impact statement."

(§1501.3[a])

"Agencies may prepare an environmental assessment on any action at any time in order to assist agency planning and decisionmaking."

(§1501.3[b])

"Briefly provide sufficient evidence and analysis for determining whether to prepare an environmental impact statement or a finding of no significant impact."

(§1508.9[a][1])

Three Purposes of an Environmental Assessment

1. *Briefly provide sufficient evidence and analysis for determining whether to prepare an environmental impact statement or a finding of no significant impact.*

2. *Aid an agency's compliance with the Act when no environmental impact statement is necessary.*

3. *Facilitate preparation of a statement when one is necessary.*

(§1508.9[a])

Three purposes of an assessment — Preparation of an EA can serve three basic purposes (§1508.9[a]). The most important of these, is to briefly provide sufficient evidence and analysis for determining whether to prepare an EIS or a FONSI.

The assessment and significant impacts — Two options exist if the proposal could result in a significant impact (see fifth decision diamond labeled "Significant impact" Figure 1.1): (1) mitigate the potentially significant impacts to the point of nonsignificance (see box labeled "Mitigate impacts?" Figure 1.1), or (2) prepare an EIS.

1.2.6   The Environmental Impact Statement

Definition of an Environmental Impact Statement

... a detailed written statement as required by section 102(2)(C) of the Act.

(§1508.11)

An EIS is prepared to investigate environmental consequences and alternatives for pursuing a proposal. An EIS must be prepared (with few exceptions) if a federal proposal cannot be categorically excluded and is not eligible for a FONSI (see box labeled "Prepare EIS" Figure 1.1).

The EIS must be used by decisionmakers in reaching a final decision regarding the course of action to be taken (§1502.1). A Record of Decision (ROD) is issued, publicly documenting the course of action that the agency has chosen to pursue. On completing this process, the agency is free to pursue the course of action described in the ROD. Once an ROD has been issued, the agency is free to proceed with the action (with respect to NEPA's requirements), in accordance with any applicable mitigation or monitoring measures. The reader is referred to the companion text, *Environmental Impact Statements*, for a detailed discussion of EISs.[8]

NEPA's Requirements for Preparing an EIS

...include in every recommendation or report on proposals for legislation and other major federal actions significantly affecting the quality of the human environment, a detailed statement by the responsible official ... on

i. *The environmental impact of the proposed action*

ii. *Any adverse environmental effects which cannot be avoided should the proposal be implemented*

iii. *Alternatives to the proposed action*

iv. *The relationship between local short-term uses of man's environment and the maintenance and enhancement of long-term productivity*

v. *Any irreversible and irretrievable commitments of resources which would be involved in the proposed action should it be implemented*

(Section 102[2][c] of NEPA)

Supplementing an EIS — As indicated in Figure 1.1 (see diamond labeled "Significant new information or changes"), under certain circumstances, an agency must supplement a draft or final EIS (§1502.9[c]). Once the supplemental EIS process has been completed, the agency issues an ROD documenting its decision with respect to the course of action that will be taken. The agency is then free to proceed with its decision, in accordance with any applicable mitigation or monitoring measures.

Circumstances Under which an EIS must be Supplemented

1. *The agency makes substantial changes in the proposed action, relevant to environmental effects or,*

2. *Significant new circumstances or information becomes available, relevant to environmental concerns that bear on the proposed action or its impacts.*

§1502.9[c]

References

1. Public Law 91-190, 42 U.S.C. 4321–4347, January 1, 1970.
2. Title 1, Section 102[c] of NEPA, emphasis added.
3. Eccleston C. H., *The NEPA Planning Process: A Comprehensive Guide with Emphasis on Efficiency*, John Wiley & Sons Inc, New York, N.Y., 1999.
4. CEQ, *Regulations for Implementing the Procedural Provisions of the National Environmental Policy Act*, 40 Code of Federal Regulations, Pts. 1500–1508, 1978.
5. Eccleston C. H., *The NEPA Planning Process: A Comprehensive Guide with Emphasis on Efficiency*, John Wiley & Sons, Inc., New York, N.Y., 1999.
6. Eccleston C. H., *The NEPA Planning Process: A Comprehensive Guide with Emphasis on Efficiency*, John Wiley & Sons Inc, New York, N.Y., 1999.
7. *Edmonds Institute v. Department of the Interior*, No. 42 F. Supp. 2d, 1999; US Dist. LEXIS 4168 (D.D.C. March 24, 1999).
8. Eccleston C. H., *Environmental Impact Statements: A Comprehensive Guide to Project and Strategic Planning*, John Wiley & Sons Inc, New York, N.Y., 2000; Eccleston C. H., *The NEPA Planning Process: A Comprehensive Guide with Emphasis on Efficiency*, John Wiley & Sons Inc, New York, N.Y., 1999.

2

General Concepts and Requirements

> I have been complimented many times and they always embarrass me;
> I always feel that they have not said enough.
>
> Mark Twain

The National Environmental Policy Act (NEPA) regulations (Regulations) were primarily written with the objective of specifying regulatory requirements governing preparation of Environmental Impact Statements (EISs). The focus was clearly on preparing EISs. Given this focus, only limited attention was devoted to describing regulatory requirements which also apply to the preparation of Environmental Assessments (EAs). In fact, the concept of an EA appears to have been more an afterthought than a thoroughly defined procedural mechanism. Hence a vacuum has existed with respect to determining which requirements specified in the Regulations also apply to preparation of EAs. One of the objectives of this text is to bridge this gap.

Accordingly, this chapter identifies and describes those regulatory provisions that are interpreted to apply to the preparation of EAs as well as EISs. A thorough understanding of these basic requirements and concepts is essential in effectively planning agency actions and in complying with NEPA's requirements, and sets the stage for application of these principles in subsequent chapters. For a detailed discussion of NEPA's essential concepts and efficiency requirements, the reader is referred to Chapters 3 and 4 of *The NEPA Planning Process*.[1]

The following sections contain basic requirements that are interpreted to apply to the entire NEPA process, including preparation of EAs. The reader is encouraged to read the actual regulatory provisions that are cited.

2.1 NEPA is a Planning and Decisionmaking Process

NEPA is often incorrectly viewed as a document preparation process. Nothing could be further from the truth. While preparation of environmental documents is an integral component, it is not why NEPA was enacted. The central purpose of NEPA is to provide decisionmakers with information on which to base decisions.

11

> ### NEPA's Purpose
>
> *Ultimately, of course, it is not better documents but better decisions that count. NEPA's purpose is not to generate paperwork — even excellent paperwork — but to foster excellent action. The NEPA process is intended to help public officials make decisions that are based on understanding of environmental consequences, and take actions that protect, restore, and enhance the environment.*
>
> **(1500.1[c])**

2.1.1   Reasonable Alternatives

Traditionally, the primary reason for preparing an EA has been to determine whether a proposed action would significantly impact environmental quality. Yet, an EA can also be used as a tool for planning actions and determining an optimum course of action. For example, in one recent case, an agency had already made a decision to take some kind of action to avoid an environmental fine. An EA was prepared to evaluate "all reasonable alternatives." The agency reported that this EA provided an effective planning tool for specifically determining the best course of action to take.[2]

2.2   Interim Actions

> **FIGURE 2.1**
> Restriction on actions that may be taken prior to completing the requirements of NEPA.
>
> *"NEPA procedures must ensure that environmental information is available to public officials and citizens before decisions are made and before actions are taken."*
>
> **§1500.1[b])**
>
> *"Until an agency issues a record of decision ... (except as provided in paragraph (c) of this section), no action concerning the proposal shall be taken which would ..."*
>
> **(§1506.1[a])**
>
> *"While work on a required program environmental impact statement is in progress and the action is not covered by an existing program statement, agencies shall not undertake in the interim any major Federal action covered by the program which may significantly affect the quality of the human environment unless ..."*
>
> **(§1506.1[c])**

Periodically, an agency might need to implement individual project or program element actions that fall within the scope of an ongoing NEPA analysis. As depicted in Figure 2.1, pursuing an action under such circumstances would normally constitute a violation of NEPA's regulatory provisions. Actions that might legitimately proceed in advance of completing an ongoing NEPA review process are referred to as *interim actions*. Chapter 10 of *The NEPA Planning Process* provides an in-depth discussion of the requirements governing interim actions.[3]

2.2.1 Eligibility for Interim Action Status

Section 1506.1 of the Regulations places specific limitations and requirements on actions permitted to take place prior to completing NEPA. As depicted in Figure 2.2, §1506.1 separates interim action requirements into two categories: (1) Non-programmatic EIS and (2) Programmatic EIS requirements.

Before an interim action can be implemented, §1506.1[c][2] requires that the proposal must first have been adequately investigated in an EIS. Strictly interpreted, this requirement leads to a paradox with potential repercussions in terms of cost, schedules, and resource requirements, because it appears to preclude actions that qualify for a CATX or FONSI, but have not been the subject of an EIS. Thus, even if an interim action can be shown to have no significant impact, a strict interpretation leads to the unreasonable conclusion that it would still have to be reviewed in an EIS before the agency could pursue the action. The text, *The NEPA Planning Process,* provides a mechanism for resolving this paradox.[4]

FIGURE 2.2
Requirements for proceeding with an action in advance of completing the EIS process.

Non-Programmatic EIS

Until an agency issues an ROD, no action concerning the proposal shall be taken that would

1. *Have an adverse environmental impact,*

2. *Limit the choice of reasonable alternatives*
   (§1506.1[a])

Programmatic EIS

While work on a required program environmental impact statement is in progress and the action is not covered by an existing program statement, agencies shall not undertake in the interim any major federal action covered by the program that might significantly affect the quality of the human environment unless such action

1. *Is justified independently of the program*

2. *Is itself accompanied by an adequate environmental impact statement*

3. *Will not prejudice the ultimate decision on the program. (An interim action prejudices the ultimate decision on the program when it tends to determine subsequent development or limit alternatives.)*
   (§1506.1[c])

2.3 Integrating NEPA with Other Requirements

Agencies are instructed to integrate NEPA with other environmental planning and review efforts (e.g., regulatory requirements, permits, agreements, studies, project planning) so that procedures run concurrently rather than consecutively. Such practice can

1. avoid duplication of effort
2. reduce delays

3. minimize environmental compliance costs

4. result in more effective decisionmaking

Where possible, agencies are instructed to cooperate with state and local agencies to eliminate duplication.

Integrating NEPA

Integrating the NEPA process into early planning
                    (§1500.5[a], §1501.1[a], emphasis added)

Integrate the requirements of NEPA with other planning and environmental review procedures...so that all such procedures run concurrently rather than consecutively
                                    (§1500.2[c], emphasis added)

Agencies shall integrate the NEPA process with other planning at the earliest possible time...
                                            (§1501.2, emphasis added)

Identify other environmental review and consultation requirements ... prepare other required analyses and studies concurrently with, and integrated with ...
                                        (§1501.7[a][6], emphasis added)

Any environmental document in compliance with NEPA may be combined with any other agency document...
                                            (§1506.4, emphasis added)

Agencies shall cooperate with state and local agencies to the fullest extent possible to reduce duplication between NEPA and comparable state and local requirements ...
                                        (1506.2[c], emphasis added)

2.3.1   Integrating Environmental Design Arts

Integrating Environmental Design Arts

All agencies of the Federal Government shall:

(A) utilize (and) ... insure the integrated use of ... environmental design arts in planning and in decisionmaking which may have an impact on man's environment;

(Sec. 102 [2][A], of NEPA [42 U.S.C. § 4332])

Under NEPA, agencies are required to "integrate environmental design arts into their planning and in decisionmaking..."[5] This requirement is interpreted to mean that disciplines such as architecture and urban planning (e.g., environmental design arts) must be integrated with the agency's planning process so that federal actions are blended more naturally into the environment. Compliance with this requirement implies that an EA should be used to assist the agency in planning actions as well as determining whether an EIS is required.

2.4 Conducting an Early and Open Process

Agencies are required to conduct an early and open process. NEPA documents are required to be prepared and publicly issued at the same time as other planning documents (§1501.2[b]). Fulfilling this requirement is crucial if an agency is to truly use NEPA as a planning and decisionmaking tool, which informs decisionmakers and the public about the consequences of potential actions.

> **An Early and Open Process**
>
> *There shall be an early and open process for determining the scope of issues to be addressed and for identifying the significant issues...*
> (§1501.7, emphasis added)
>
> *...insure that environmental information is available to public officials and citizens before decisions are made and before actions are taken.*
> (§1500.1[b], emphasis added)
>
> *... [a NEPA document is to be] prepared early enough so that it can serve practically as an important contribution to the decisionmaking process and will not be used to rationalize or justify decisions already made.*
> (§1502.5, emphasis added)

2.5 Public Involvement

In the past, some agencies have operated under the mistaken belief that the EA process does not require public involvement; an EA is, in fact, a "public document" (§1508.9). Moreover, agencies are instructed to make diligent efforts to involve the public in preparing and executing their NEPA implementation procedures (§1506.6[b]).[6]

> **NEPA is a Public Process**
>
> *Copies of such statement and the comments and views of the appropriate federal, state and local agencies, ... shall be made available to the President, the Council on Environmental Quality and the public...*
> (Section 102[2][C] of the NEPA Act, emphasis added)
>
> *There shall be an early and open process for determining the scope of issues to be addressed and for identifying the significant issues...*
> (§1501.7, emphasis added)

2.5.1 Public Notice

Agencies are required to "provide *public notice* of NEPA-related *hearings, public meetings,* and the *availability of environmental documents* so as to inform those persons and agencies who may be interested or affected (§1506.6[a], emphasis added)." The term "environmental documents" is defined to include EAs (§1508.10).

2.6   Scoping

Scoping
There shall be an __early__ and __open__ process for determining the __scope of issues__ to be addressed and for identifying the significant issues related to a proposed action. This process shall be termed scoping.
(§1501.7, emphasis added)

Agencies are required to conduct an "early and open" process in determining the *scope* of an EIS. The regulations are silent with respect to the application of scoping during the EA process. Nevertheless, application of scoping (whether internal or public) during the EA process is crucial to the objective of providing relevant information to the decisionmaker. Chapter 11 of the text, *The NEPA Planning Process*, provides a detailed discussion of NEPA's concept of scope.[7]

2.7   Systematic and Interdisciplinary Planning

Systematic and Interdisciplinary Process
... utilize a __systematic, interdisciplinary approach__ which will insure the integrated use of the natural and social sciences and the environmental design arts in planning and in decisionmaking which may have an impact on man's environment.
(§1501.2[a], emphasis added)

Agencies are required to use a *systematic and interdisciplinary* approach in implementing their NEPA process. This requirement has been interpreted to extend to the preparation of EAs as well as to EISs.[8]

2.7.1   Systematic

The term systematic places a mandate on agencies to utilize a logical, ordered, and methodological approach in which each stage of the EA process builds upon previous stages.

2.7.2   Interdisciplinary

Multidisciplinary Versus Interdisciplinary Approach
Multidisciplinary approach: Denotes a process in which specialists representing various disciplines perform their assigned tasks with little or no interaction.
Interdisciplinary approach: Indicates a process in which specialists interface and work together on common issues.

The requirement to perform an interdisciplinary approach places a burden on agencies to ensure that the analysis is performed by knowledgeable specialists who possess expertise in the disciplines for which they have been assigned responsibility.

The terms multidisciplinary and interdisciplinary are not equivalent. A multidisciplinary approach refers to a process in which specialists perform their assigned tasks with little or no interaction. In contrast, an interdisciplinary approach acknowledges that environmental analysis involves a multitude of interconnected disciplines that can only be understood if specialists from these diverse fields interface and work together on common issues.

Specialists representing those resource area(s) most critical to the analysis should either directly or indirectly participate in, or should be consulted in, preparation of the EA.

2.8 Writing Documents in Plain English

The Regulations require agencies to "employ writers of clear prose" preparing analyses that are written in "plain language" so that they can be clearly understood by decisionmakers and the public (§1502.8).

In the words of one court, a NEPA document is to be "...orga-

> **Preparing Understandable Documents**
>
> *Environmental impact statements shall be written in plain language and may use appropriate graphics so that decisionmakers and the public can readily understand them. Agencies should employ writers of clear prose or editors to write, review, or edit statements...*
>
> **(§1502.8, emphasis added)**

nized and written so as to be readily understandable by governmental decisionmakers and by interested non-professional laypersons likely to be affected by actions ..."[9]

2.9 Incorporation by Reference

The Regulations encourage use of incorporating information by reference as a means of reducing the size of an EIS. For example, agencies are *required* to incorporate existing material into an EIS by reference if it reduces the length of the statement without impeding either the agency's or public's ability to review the document (§1506.3). This is also an appropriate method for reducing the length of EAs.[10]

> **Incorporation by Reference**
>
> *Agencies shall incorporate material ... by reference when the effect will be to cut down on bulk without impeding agency and public review of the action. The incorporated material shall be cited ... and its content briefly described. No material may be incorporated by reference unless it is reasonably available for inspection by potentially interested persons within the time allowed for comment...*
>
> **(§1502.21)**

When material is incorporated by reference, the assessment must reference this material and provide a brief description of its content. All referenced material must be reasonably available for inspection by interested persons. Material not publicly available can be added as an appendix to the EA.

2.10 Adopting Another Agency's EA

In an effort to promote efficiency, CEQ encourages agencies to adopt, where appropriate, EISs prepared by other federal agencies (§1500.4[n], §1500.5[h], and §1506.3). However, the regulations are silent concerning the question of whether another agency's EA can be adopted.[11]

2.10.1 Department of Energy's Adoption Process

The Department of Energy (DOE) has taken the position that the concept of adoption also extends to EAs as well. The DOE has provided the following guidance with respect to this question:

> Any federal agency may adopt any other federal or state agency's EA and is encouraged to do so when such adoption would save time and money. In deciding that adoption is the appropriate course of action, DOE ... must conclude that the EA adequately describes DOE's proposed action and in all other respects is satisfactory for DOE's purposes...[12]

Once DOE determines that the originating agency's document is adequate (possibly after adding additional information), DOE is responsible for transmitting the EA to the state(s), Indian tribes, and, as appropriate, the public for preapproval review and comment (unless the originating agency has already conducted an equivalent public involvement process). After considering all comments and complying with the previously mentioned requirements, DOE may issue its own FONSI.

Because the adopting agency is responsible for verifying the adequacy of the analysis and conclusions, it must perform an independent review of the document to be adopted. The EA checklists in Appendix C provide a useful tool for assisting practitioners in performing this review.

2.11 Methodology

The EA analysis must be accurate, of high quality, and scientifically credible. Completion of the preliminary analysis should be followed by a rigorous peer review. Comments (internal or external) should be maintained as part of the agency's administrative record.

2.11.1 A Rigorous, Accurate, and Scientific Analysis

Emphasis is placed on conducting a rigorous, accurate, and scientific analysis that thoroughly investigates potential environmental issues and impacts.

Performing a Rigorous, Accurate, and Scientific Analysis

The information must be of high quality. Accurate scientific analysis... [is] essential to implementing NEPA.

(§1500.1[b], emphasis added)

Agencies shall insure the professional integrity, including scientific integrity...They shall identify any methodologies used...

(§1502.24, emphasis added)

...the analysis is supported by credible scientific evidence...

(§1502.22 [b][4], emphasis added)

...supported by evidence that the agency has made the necessary environmental analyses.

(§1502.1, emphasis added)

Environmental impact statements shall be analytic rather than encyclopedic...

(§1502.2[a], emphasis added)

2.12 A Fair and Objective Analysis

The EA must provide the public and decisionmaker with a fair, objective, and impartial analysis. Practitioners should strive to avoid even the slightest perception that the analysis may be less than impartial.

Providing an Objective Analysis

... shall provide full and fair discussion...

(§1502.1, emphasis added)

2.12.1 Disclosing Opposing Points of View

In preparing an EIS, agencies must make a concerted effort to "...disclose and discuss...all major points of view..." (§1502.9[a]). This requirement is interpreted to also apply to the preparation of EAs. It is therefore recommended that the EA address any reasonable opposing views that have been publicly voiced.

2.13 Dealing with Incomplete and Unavailable Information

Special procedures have been established for responding to circumstances involving incomplete or unavailable information. The author's companion text, *Environmental Impact Statements,* provides a detailed discussion on the requirements for dealing with circumstances involving incomplete and unavailable information.[13]

2.13.1 Incomplete Information

If the "incomplete information" is necessary for making an informed choice between alternatives, and the overall cost is not exorbitant, the information must be obtained and included in the analysis (§1502.22[a]).

2.13.2 Unavailable Information

The agency must clearly indicate when information, relevant to "reasonably foreseeable significant adverse impacts" cannot be obtained either because (1) the overall costs of obtaining it are exorbitant, or (2) the means of obtaining it are not known.(§1502.22[b])

As indicated in Figure 2.3, four principal requirements must be satisfied when faced with circumstances involving unavailable information. Where the agency has indicated that information is either incomplete or unavailable, it should, at a minimum, prepare a qualitative description of the most relevant impacts. However, the reader is cautioned that the inability to satisfactorily quantify an important impact may well render the EA ineffective in supporting a Finding Of No Significant Impact (FONSI).[14]

It is recommended that the "rule of reason," in conjunction with a sliding-scale approach, be applied in determining the degree of effort most appropriate for addressing incomplete or unavailable information. Thus, the cost and level of effort expended should be commensurate with the potential for significance and the value that this information would contribute to decisionmaking.

FIGURE 2.3
Requirements for dealing with incomplete or unavailable information.

1. *A statement that such information is incomplete or unavailable;*

2. *A statement of the relevance of the incomplete or unavailable information to evaluating reasonably foreseeable significant adverse impacts on the human environment;*

3. *A summary of existing credible scientific evidence which is relevant to evaluating the reasonably foreseeable significant adverse impacts on the human environment; and*

4. *The agency's evaluation of such impacts based upon theoretical approaches or research methods generally accepted in the scientific community.*

(§1502.22[a])

References

1. Eccleston C. H., *The NEPA Planning Process: A Comprehensive Guide with Emphasis on Efficiency,* John Wiley & Sons Inc, New York, N.Y., 1999.
2. DOE, *NEPA Lessons Learned,* Issue No. 18, March 1, 1999.
3. Eccleston C. H., *The NEPA Planning Process: A Comprehensive Guide with Emphasis on Efficiency,* John Wiley & Sons Inc, New York, N.Y., 1999.
4. Eccleston C. H., *The NEPA Planning Process: A Comprehensive Guide with Emphasis on Efficiency,* pp258–259, John Wiley & Sons Inc, New York, N.Y., 1999.
5. §102[2][A] of NEPA, 42 U.S.C. §4332.
6. E. A. Blaug, Use of the Environmental Assessment by Federal Agencies in NEPA Implementation, *The Environmental Professional,* Volume 15 pp. 57–65, 1993.
7. Eccleston C. H., *The NEPA Planning Process: A Comprehensive Guide with Emphasis on Efficiency,* John Wiley & Sons Inc, New York, N.Y., 1999.
8. Government Institutes, Inc., *Environmental Law Handbook,* Chapter 10, 10 edition, 1989.
9. *Oregon Environmental. Council v. Kunzman,* 636 F. Supp 632, (U.S.D.C for Oregon 1986).
10. CEQ, *Council on Environmental Quality — Forty Most Asked Questions Concerning CEQ's National Environmental Policy Act Regulations (40 CFR 1500–1508), Federal Register* Vol. 46, No. 55, 18026–18038, March, 23, 1981, Question number 36a.
11. U.S. Department of Energy, *NEPA Lessons Learned,* pp14–15, Issue No. 23, June 1, 2000.
12. U.S. Department of Energy, *Frequently Asked Questions on the Department of Energy's National Environmental Policy Act Regulations,* revised August 1998, Question #15.
13. Eccleston C. H., *Environmental Impact Statements: A Comprehensive Guide to Project and Strategic Planning,* John Wiley & Sons Inc, New York, N.Y., 2000; Eccleston C. H., *The NEPA Planning Process: A Comprehensive Guide with Emphasis on Efficiency,* John Wiley & Sons Inc, New York, N.Y., 1999.
14. DOE, *NEPA Lessons Learned,* Issue No. 18, p6, March 1, 1999

3

NEPA's Concept of Environmental Impact Analysis

Predictions are notoriously difficult to make — especially when they concern the future

Mark Twain

Despite intense effort, the prediction of environmental impacts remains a challenging and inexact science. In *Life on the Mississippi*, Mark Twain humorously expressed how scientific observations can be misused to produce bizarre conclusions:

> In the space of one hundred and seventy-six years the Lower Mississippi has shortened itself two hundred and forty two miles. That is an average of a trifle over one mile and a third per year. Therefore, any calm person, who is not blind or idiotic, can see that in the Old Oolitic Silurian Period, just a million years ago next November, The Lowest Mississippi River was upward of one million three hundred thousand miles long, and stuck out over the Gulf of Mexico like a fishing rod. And by the same token any person can see that seven hundred and forty-two years from now the Lower Mississippi will be only a mile and three quarters long, and Cairo and New Orleans will have joined their streets together and be plodding comfortably along under a single mayor and a mutual board of aldermen. There is something fascinating about science. One gets such wholesale returns of conjecture out of such a trifling investment of fact.

It is for reasons such as this that this chapter has been written. Specifically, this section describes a systematic, general-purpose approach for analyzing environmental impacts. This Action-Impact Model is depicted in Figure 3.1. The reader is referred to Chapter 2 in the companion text, *Environmental Impact Statements*, for a more thorough discussion of environmental impact analysis.[1]

3.1 Actions

Potential actions can be viewed as having several stages of development, one of which is a proposal: "*Proposal* exists at that stage in the development of an

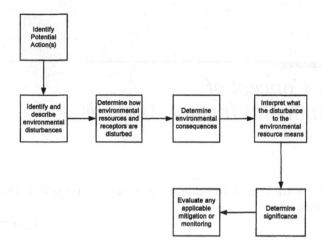

FIGURE 3.1
The Action-Impact Model. A general-purpose approach for analyzing environmental impacts.

action when an agency subject to the Act has a goal and is actively preparing to make a decision on one or more alternative means of accomplishing that goal and the effects can be meaningfully evaluated." (§1508.23)

> **NEPA's Definition of an Action**
>
> *"Actions include new and continuing activities including projects and programs entirely or partly financed, assisted, conducted, regulated, or approved by federal agencies; new or revised agency rules, regulations, plans, policies, or procedures and legislative proposals."*
> (§1508.18[a])

With respect to a NEPA analysis, a proposal typically consists of a set of discrete component actions (e.g., site preparation, construction of an access road and infrastructure, construction of a facility, operation of a facility). All actions related to the proposal must be identified and adequately evaluated in a section of the EA that describes the proposed action (Figure 3.1).

3.2 Environmental Disturbances

Actions produce environmental disturbances (e.g., air emission, effluents, disruption of flora or fauna, or waste products) that must be identified and described in detail sufficient to support a subsequent analysis of their effect on environmental resources (see Figure 3.1). The environmental disturbances must be identified and characterized in either the proposed action and/or environmental impact section of the EA.

3.3 Receptors and Resources

Environmental distur-bances change or perturb one or more receptors (i.e., air or water quality, cul-tural resources, wildlife, habitat, or human health) that are often referred to as environmental resources (see Figure 3.1). It is con-sidered good practice to describe the affected envi-ronment (receptors/envi-ronmental resources) in an EA (i.e., Affected Environ-ment section) before they have been affected by the proposed action.

> ### Affected Environment
>
> *Affected Environment: The NEPA Regulations require that an environmental impact statement include a chapter on the affected environment (§1502.15). The Regulations do not require that such a chapter be included as part of an environmental assessment. However, the inclusion of this section in an EA is considered good professional practice. Some agencies' NEPA implementation regulations even require that such a section be included in their EAs.*

> ### Definition of the Human Environment
>
> *Human environment shall be interpreted comprehensively to include the natural and physical environment and the relationship of people with that environment.*
>
> **(§1508.14)**

3.4 Impact Analysis

The EA must determine the impact on the human environment. To this end, the environmental analysis is directed at determining how a given environmen-tal disturbance would affect receptors/environ-mental resources. This analysis is conducted on a resource-by-resource basis in the environmental impact section of the EA (see Figure 3.1).

The result is a set of con-sequences (i.e., environ-mental effects or impacts). The terms "impacts" and "effects" are synonymous (§1508.8[b]). The Council

> ### The Three Types of Impacts
>
> *Direct Effects: are caused by the action and occur at the same time and place.*
>
> **(§1508.8)**
>
> *Indirect Effects: are caused by the action and are later in time or farther removed in distance, but are still reasonably foreseeable. Indirect effects may include growth-inducing effects and other effects related to induced changes in the pattern of land use, population density, or growth rate, and related effects on air, water, and other natural systems, including ecosystems.*
>
> **(§1508.8, emphasis added)**
>
> *Cumulative Impact: is the impact on the environment which results from the incremental impact of the action when added to other past, present, and reasonably foreseeable future actions regardless of what agency (federal or non-federal) or person undertakes such other actions. Cumulative actions can result from individually minor but collectively significant actions taking place over a period of time.*
>
> **(§1508.7)**

on Environmental Quality (CEQ) recognizes three distinct types of impacts (§1508.25[c]): (1) direct, (2) indirect, and (3) cumulative.

The term "effects" is defined to include: "... ecological (such as the effects on natural resources and on the components, structures, and functioning of affected ecosystems), aesthetic, historic, cultural, economic, social, or health, whether direct, indirect, or cumulative" (§1508.8).

3.5   Significance

The environmental consequences can be considered either significant or non-significant, according to specific factors presented in the Regulations (§1508.27). The EA describes the environmental disturbances in a manner that allows the decisionmaker to reach a decision regarding their *significance* or *nonsignificance* (*see* Figure 3.1). The reader is directed to Section 4.6 of this text for a more detailed description of the concept of significance. Additional information on significance and its interpretation can be found in Chapter 8 of *The NEPA Planning Process*.[2]

3.6   Mitigation and Monitoring

The agency can choose to mitigate potentially significant impacts (see Figure 3.1). A monitoring program is also an integral element of a well planned environmental process, particularly where there is a chance that the impact projections could be exceeded. The reader is referred to Chapter 4 in the companion text, *Environmental Impact Statements,* for a more detailed treatment of this subject.[3]

3.6.1   Mitigation

If the EA concludes that one or more significant impacts would result from the proposed action, the agency might elect to implement mitigation measures that might render a potentially *significant* impact *nonsignificant*. An EIS is not required if the potentially significant impacts can be mitigated to the point of nonsignificance.

Methods of mitigation can include avoiding or minimizing the impacts of an action, repairing the effects of impacts that do occur, and compensating for impacts by replacing or substituting resources that have been damaged (§1508.20). If a decision is made to mitigate the impacts to the point of nonsignificance, it is recommended that a Mitigation Action Plan (MAP) be prepared as an integral part of the action.

Mitigation Methods

Mitigation methods may include

a. avoiding the impact altogether by not taking a certain action or parts of an action;

b. minimizing impacts by limiting the degree or magnitude of the action and its implementation;

c. rectifying the impact by repairing, rehabilitating, or restoring the affected environment;

d. reducing or eliminating the impact over time by preservation and maintenance operations during the life of the action;

e. compensating for the impact by replacing or providing substitute resources or environments.

(§1508.20)

3.6.2 Monitoring

Post-monitoring is an important step in ensuring that environmental predictions are not exceeded and that commitments made in the EA/Finding of No Significant Impact (FONSI) are not lost in the haste and confusion of project implementation. Post-monitoring is useful in ensuring that:

1. Environmental standards are met.
2. Mitigation measures are adequately implemented.
3. No impacts are encountered that are substantially different from those originally forecast. A monitoring and enforcement plan (as part of the mitigation action plan) should be adopted and summarized for any mitigation measures that are chosen (§1505.2[c]). Agencies are also responsible for making the results of relevant monitoring available to the public.

It is important to note that specific performance standards against which mitigation measures can be assessed need to be established.

References

1. Eccleston C.H., *Environmental Impact Statements: A Comprehensive Guide to Project and Strategic Planning*, John Wiley & Sons Inc, 2000; Eccleston C. H., *The NEPA Planning Process: A Comprehensive Guide with Emphasis on Efficiency*, John Wiley & Sons Inc, 1999.

2. Eccleston C. H., *The NEPA Planning Process: A Comprehensive Guide with Emphasis on Efficiency,* John Wiley & Sons Inc, 1999.
3. Eccleston C.H., *Environmental Impact Statements: A Comprehensive Guide to Project and Strategic Planning,* John Wiley & Sons Inc, 2000; Eccleston C. H., *The NEPA Planning Process: A Comprehensive Guide with Emphasis on Efficiency,* John Wiley & Sons Inc, 1999.

4

Determining Whether
an EA or EIS is Required

The Threshold Question

> Every time I reform in one direction, I go overboard in another.
>
> Mark Twain

Section 102 of NEPA, requires all agencies of the federal government to prepare and include an Environmental Impact Statement (EIS) in every recommendation or report on ... **proposals** for **legislation** and other **major federal actions significantly affecting** the quality of the **human environment** ...[1]

Often referred to as the threshold question of significance, this mandate provides the linchpin for determining whether an Environmental Assessment (EA) is sufficient or whether an EIS must be prepared for the proposal. Once the EA has been completed, the decisionmaker is left with what is often a formidable task of determining whether the action could result in a significant environmental impact, requiring preparation of a much more rigorous EIS. The outcome of this determination can have profound ramifications in terms of cost, schedule, and potential litigation, not to mention environmental quality.

NEPA's Threshold Criteria
1. *proposals*
2. *legislation*
3. *major*
4. *federal*
5. *actions*
6. *significantly*
7. *affecting*
8. *human*
9. *environment*
42 U.S.C. §4332 [102][2][c])

The issue of significance has been the subject of substantial litigation. The threshold question can essentially be dissected into nine discrete criteria that must be satisfied before the EIS requirement as a whole is triggered. Each of these threshold criteria is briefly described in the following sections. An in-depth description of the threshold question is beyond the scope of this book. The reader is referred to the author's companion book, *The NEPA Planning Process: A Comprehensive Guide with Emphasis on Efficiency*, for a comprehensive overview of this subject.[2]

29

4.1   Proposals

A proposal might exist in actual fact, even though the agency has not officially declared one to exist. As described in Chapter 1, a proposal is considered to exist either officially or unofficially when (§1508.23)

1. a federal agency has a goal;
2. the agency is actively preparing to make a decision on one or more alternative means of accomplishing the goal;
3. the effects can be meaningfully evaluated.

4.2   Legislation

Proposals also include submittals for congressional legislation. Specifically, a *legislative proposal* involves ... "a bill or legislative proposal to Congress developed by or with the significant cooperation and support of a federal agency, but does not include requests for appropriations." (§1508.17)

For the purposes of NEPA, a legislative proposal must be developed by or with "significant cooperation" of a federal agency. The test for "significant cooperation" hinges on whether a proposal is developed predominantly by the federal agency. Drafting legislation does not, by itself, necessarily constitute significant cooperation.

4.3   Major

> **The Most Common Interpretation of "Major"**
>
> *"Major" reinforces, but does not have a meaning independent from, the term "significantly."*

The courts have recognized two distinct interpretations for the term "major." A few courts have interpreted "major" to be a separate criterion, independent of the term "significantly." Under this interpretation, "major" has generally been interpreted as an indicator of the size of a potential action.

In contrast, most courts, and also the Council on Environmental Quality (CEQ), have simply interpreted "major" to reinforce, but not to have a meaning independent from, the term "significantly." Thus, under this interpretation, the size of a proposal has little bearing on whether the action requires preparation of an EIS. This interpretation avoids dilemmas where a federal action could be deemed major in terms of size, yet would not significantly affect the quality of the human environment; it also avoids dilemmas where a relatively small action could result in a significant impact.

4.4 Federal

Under NEPA, the term "federal" includes all agencies of the federal government; a federal agency does not include " ...the Congress, the Judiciary, or the President, including the performance of staff functions for the President in his Executive Office" (§1508.12).

In certain instances, the courts have determined that actions undertaken by a nonfederal agency can be "federalized" for the purposes of NEPA. The amount of federal involvement necessary to "federalize" a nonfederal activity, triggering NEPA,

> **Federal Involvement that Might Federalize an Otherwise Nonfederal Action**
>
> 1. *federal permits or approvals.*
> 2. *federal control.*
> 3. *federal funding.*

can be a particularly difficult issue to assess. A general-purpose tool for determining when NEPA applies to nonfederal actions is provided in the author's companion book, *Environmental Impact Statements — A Comprehensive Guide to Project and Strategic Planning.*[3]

4.5 Action

For the purposes of NEPA, the concept of an action is pervasive. Actions include both new and continuing activities. Actions also include circumstances where the responsible officials fail to act and that failure to act is reviewable by courts or administrative tribunals under the Administrative Procedure Act or other applicable law.

> **Definition of an Action**
>
> *...new and continuing activities, including projects and programs entirely or partly financed, assisted, conducted, regulated, or approved by federal agencies; new or revised agency rules, regulations, plans, policies, or procedures; and legislative proposals.*
> (§1508.18[a])

4.6 Significantly

> **Context and Intensity**
>
> *The Regulations require that both the context and intensity be considered in reaching a determination regarding significance.*
>
> **(§1508.18[a])**

When deciding whether to prepare an EIS, the concept of "significance" is perhaps both the most important and elusive of the nine threshold criteria. Determining the potential significance of an impact can be a daunting task. For this reason, Chapter 8 provides the reader with a rigorous and systematic procedure for assisting decisionmakers and practitioners in reaching a defensible determination regarding potential significance.

As specified in the Regulations, both the intensity and the context in which an impact would occur must be considered. For additional information on significance and its interpretation, the reader is directed to Chapter 8 of the companion book, *The NEPA Planning Process*.[4]

4.6.1 Context

> **Context**
>
> *"...the significance of an action must be analyzed in several contexts such as society as a whole (human, national), the affected region, the affected interests, and the locality. Significance varies with the setting of the proposed action. For instance, in the case of a site-specific action, significance would usually depend upon the effects in the locale rather than in the world as a whole."*
>
> **(§1508.27[a])**

Significance is a function of the setting (i.e., context) in which an impact would occur. The term "context" recognizes potentially affected resources, as well as the location and setting in which an environmental impact would occur.

Four distinct contexts are explicitly identified in the Regulations: (1) society as a whole, (2) the affected region, (3) the affected interests, and (4) the locality.

Table 4.1 presents examples of categories of environmental effects and the corresponding context that might be appropriate for the assessment of significance.

4.6.2 Intensity

Intensity is a measure of the degree or severity of an impact. The Regulations define ten factors (i.e., significance factors) that are to be used in assessing intensity (§1508.27[b]). These significance factors are indicated in Figure 4.1.

An impact cannot necessarily be deemed nonsignificant simply because the action is temporary. Moreover, agencies are cautioned against *segmenting*

TABLE 4.1

Examples of Effects and the Corresponding Context that Might be Most Appropriate for Assessing Significance

Specific Effect	Corresponding Context
Soil impacts	Site and adjoining properties
Wetlands impacts	Site and remainder of sub-watershed
Visual impacts	Viewsheds that include the site
Socioeconomic impacts	Political jurisdictions such as countries and municipalities

FIGURE 4.1
Ten significance factors used in assessing the intensity of an environmental impact.

1. *Impacts that may be both beneficial and adverse. A significant effect may exist even if the federal agency believes that on balance the effect will be beneficial.*

2. *The degree to which the proposed action affects public health or safety.*

3. *Unique characteristics of the geographic area such as proximity to historic or cultural resources, park lands, prime farmlands, wetlands, wild and scenic rivers, or ecologically critical areas.*

4. *The degree to which the effects on the quality of the human environment are likely to be highly controversial.*

5. *The degree to which the possible effects on the human environment are highly uncertain or involve unique or unknown risks.*

6. *The degree to which the action may establish a precedent for future actions with significant effects or represents a decision in principle about a future consideration.*

7. *Whether the action is related to other actions with individually insignificant but cumulatively significant impacts. Significance exists if it is reasonable to anticipate a cumulatively significant impact on the environment. Significance cannot be avoided by terming an action temporary or by breaking it down into small component parts.*

8. *The degree to which the action may adversely affect districts, sites, highways, structures, or objects listed in or eligible for listing in the National Register of Historic Places or may cause loss or destruction of significant scientific, cultural, or historical resources.*

9. *The degree to which the action may adversely affect an endangered or threatened species or its habitat that has been determined to be critical under the Endangered Species Act of 1973.*

10. *Whether the action threatens a violation of federal, state, or local law or requirements imposed for the protection of the environment.*
§1508.27[b]

or *piecemealing* an action by "breaking a project down into smaller component parts" that are individually nonsignificant (§1508.27[b][7]). An example of segmentation involves dissecting a powerplant project into a separate analysis of the plant, electric transmission line, gas pipeline, support buildings, and effluent cooling towers.

4.7　Affecting

Effects on the Human Environment
The "effects" on the human environment includes attributes as diverse as: ecological, and natural resources (including the components, structures, and functions of ecosystems), aesthetic, historic, cultural, economic, and social resources and health issues.
(§1508.8)

An action is said to affect the quality of the human environment. The Regulations define "affecting" to mean "will or *may* have an effect on" (§1508.3).

With respect to NEPA, an action affects the environment if it produces a change in one or more environmental resources. A reasonably close connection must exist between a disturbance and its resulting effect on the environment.

4.8　Human Environment

Human Environment
"Human Environment" shall be interpreted comprehensively to include the natural and physical environment and the relationship of people with that environment... This means that economic or social effects are not intended by themselves to require preparation of an environmental impact statement. When an environmental impact statement is prepared and economic or social and natural or physical environmental effects are interrelated, then the environmental impact statement will discuss all of these effects on the human environment.
(§1508.14)

To result in a finding of significance, an action must significantly affect the quality of the "human environment." As some relationship exists between humans and virtually every aspect of the physical and natural environment, the courts have viewed this term broadly. From a practical standpoint, there is little distinction between the terms "environment" and "human environment."

The "environment" can be divided into two distinct categories: (1) natural and physical environs and (2) man-made or built environs (§1502.16[g], §1508.8[b], §1508.14).

References

1. 42 U.S.C. §4332 (102)(2)(c).
2. Eccleston C. H., *The NEPA Planning Process: A Comprehensive Guide with Emphasis on Efficiency*, Chapter 8, John Wiley & Sons Inc., New York, N.Y., 1999.
3. Eccleston C. H., *Environmental Impact Statements: A Comprehensive Guide to Project and Strategic Planning*, Chapter 1, John Wiley & Sons, Inc., New York, N.Y., 2000.
4. Eccleston C. H., *The NEPA Planning Process: A Comprehensive Guide with Emphasis on Efficiency*, John Wiley & Sons Inc, 1999; *Environmental Impact Statements: A Comprehensive Guide to Project and Strategic Planning*, John Wiley & Sons Inc, New York, N.Y. 2000.

5

The Environmental Assessment Process: the Procedural Steps

Always acknowledge a fault frankly. This will throw those in authority off their guard and give you opportunity to commit more.

Mark Twain

To justify issuing a Finding of No Significant Impact (FONSI), an Environmental Assessment (EA) must provide clear and convincing evidence that the proposal would not result in any significant impacts, or that any significant impacts can be mitigated to the point of nonsignificance. In contrast, an Environmental Impact Statement (EIS) does not assume this evidentiary burden, as it essentially acknowledges the presence of significant impacts. Consequently, if challenged, an EIS is often easier to defend.

Potential adversaries have taken note of this fact and have refocused efforts on challenging EAs, which are considered more vulnerable. Agencies can counter the trend toward increased scrutiny of their EAs by ensuring that a rigorous investigation has been performed and thoroughly documented.

5.1 Preparing the Environmental Assessment

Figure 5.1 depicts a generalized procedure for preparing an EA. A National Environmental Policy Act (NEPA) review is typically initiated when a need for taking action has been identified (*see* oval-shaped box, Figure 5.1). If the action is not eligible for a Categorical Exclusion (CATX), the agency's NEPA implementation procedures should be consulted for guidance in determining whether the action is one that normally requires preparation of either an EA or EIS.

If the agency is uncertain that an action would result in a significant impact, it may choose to prepare an EA to determine whether the action qualifies for a Finding of No Significant Impact (FONSI); this is the course normally taken, as preparation of an EIS requires a substantially larger amount of effort than does an EA.

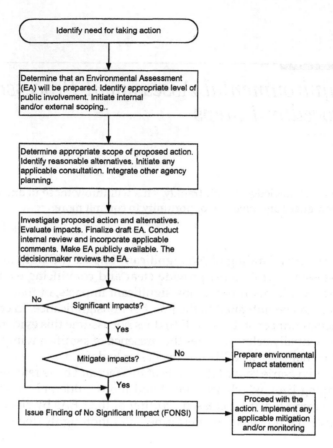

FIGURE 5.1
Typical environmental assessment process.

5.1.1   Public Involvement

Once a decision to prepare an EA is made, the level of public involvement most appropriate for this proposal must be determined. The stage is now set for initiating internal (and, if applicable, public) scoping (see first rectangle, Figure 5.1).

As warranted, consultations with outside authorities and agencies are also initiated. The importance of this step is witnessed by the fact that one agency reported that the NEPA process was particularly useful in helping the state and tribe resolve their differences regarding the proposed action.[1] Chapter 2 of the author's companion text, *Environmental Impact Statements* provides additional information that may be of use in promoting public involvement.[2] The EA process is integrated with other planning studies or analyses (see second rectangle, Figure 5.1).

Public comments — The Regulations do not specifically require an agency to incorporate or respond to public comments in an EA. However, as NEPA is intended to be an open public process, such practice is recommended.

One agency reported that such practice contributed significantly to the success of its EA process. Comments received on the proposed action were placed in the beginning of the EA. References were added to steer the reader to corresponding sections of the EA where the reader could see how these comments were addressed.[3]

5.1.2　Preparing the EA

A systematic process must be performed to identify all reasonable alternatives. Analysts investigate both the proposed action and reasonable alternatives (see third rectangle, Figure 5.1). Alternatives deemed to be unreasonable are dismissed from further study.

TABLE 5.1

Attributes of a High Quality Decision Process

The Decision Process

Accurately describes the problem and the criteria for solving it.

Uses available information effectively.

Collects new information wisely.

Generates and chooses from a wide range of alternatives.

Distinguishes between facts, myths, values, and unknowns.

Describes consequences associated with alternative solutions.

Leads to choices that are consistent with organizational, stakeholder, personal, or other important values.

Potentially significant impacts are rigorously investigated. Once the EA has been finalized, it is circulated for internal (and possible public) review; relevant comments are incorporated. The finalized EA is made publicly available, and the assessment is carefully reviewed by the decisionmaker.

Decision Protocol process — Picture a "typical" Interdisciplinary Team (IDT). While one person is describing the problem, another has already generated the solution, a third is wondering how stakeholders will react, the fourth is questioning how the project will be financed, while the fifth is busy evaluating the environmental impacts. It is not uncommon to find that the planning/decisionmaking process lacks a common structured direction.[4]

To address such problems, the U.S. Forest Service has developed a *Decision Protocol* process. After obtaining input and developing a pilot process, the Forest Service tested its Decision Protocol on approximately 20 proposed projects across the country. Decision Protocol is based on the observation that a high quality decision is based on the attributes listed in Table 5.1.

In April 1999, the U.S. Forest Service issued its Decision Protocol 2.0.[5] The protocol is a question-based administrative aid that the Forest Service claims can help an IDT improve preparation of EAs/EISs. Decision Protocol 2.0 is based on the following five-cycle procedure:

1. *Process* The team determines potential decisions that need to be considered, how they can be implemented, and potential constraints (see section later in this chapter titled "Decision-Based Scoping").

2. *Problem* The context of the problem is defined through verbal and graphic depictions of the situation, the purpose and need, and definition of the existing information base (including any gaps and uncertainties).

3. *Design* This cycle results in a description of the proposal. The team identifies activities to accomplish the objectives, defines alternatives, and identifies cause-and-effect relationships and potential mitigation measures.

4. *Consequences* The set of alternatives is refined. Potential impacts are defined. The IDT considers interactions among other proposed activities, uncertainties, and how design changes could affect key consequences.

5. *Action* The IDT compares alternatives in terms of meeting predefined objectives, cost and reasonability, and avoiding adverse effects. A preferred design is chosen and a defensible rationale is developed and documented. The IDT examines the sensitivity of this choice in terms of changes in the assumptions. An implementation plan is prepared. As necessary, a mitigation/monitoring plan is also developed.

5.1.3 Determining the Potential for Significance

As indicated by the first decision diamond in Figure 5.1, two options exist if an agency cannot demonstrate that the potential environmental impacts are clearly nonsignificant: (1) mitigate impacts to the point of nonsignificance or (2) prepare an EIS. The scope of any mitigation measures must address the range of significant impacts that could occur.

If based on review of the EA, the decisionmaker concludes that, no significant impacts would result (or they can be mitigated to the point of nonsignificance), the agency issues a FONSI; once the FONSI is issued, the agency is free (with respect to NEPA) to pursue the action. The reader should consult the agency's implementation procedures for any variations or additional requirements that must be met.

5.1.4 The Three Purposes That an EA May Serve

Agencies may prepare an EA on any action at any time to further agency planning and decisionmaking (§1501.3[b], §1508.8[a][2], and §1508.8[a][3]).

As depicted in Table 5.2, an EA can serve any one of the three objectives (§1508.9). An EA's most important function is as a screening device for determining whether an action would result in a significant impact requiring preparation of an EIS.

TABLE 5.2

Three Purposes of an EA

1. Briefly provide sufficient evidence and analysis for determining whether to prepare an EIS or a FONSI;

2. Aid an agency's compliance with the Act when no EIS is necessary;

3. Facilitate preparation of an EIS, when one is necessary.

5.1.5 Timing

Preparation of an EA must be timed so that it can be circulated with other planning documents (§1501.2[b]).

According to the Council on Environmental Quality (CEQ), the EA process should normally require a maximum of three months to complete; typically, the process should be completed in less than three months.[6] Such guidance may well be unrealistic, as many agencies report that their EAs typically require more than three months to prepare and issue.

> **Recommended Time Needed to Complete an EA**
>
> *The Council on Environmental Quality advises that an EA process should normally require a maximum of three months to complete.*

5.1.6 EAs are Public Documents

Some agencies have operated under the mistaken belief that preparation of an EA does not require public involvement. An EA is in fact a "public document" (§1508.9). Moreover, as defined in the Regulations, an EA and FONSI are both included under the definition of an "environmental document," (§1508.10). As an "environmental document," agencies are required to make provisions for public involvement (§1506.6[b]).[7] Agencies are also required to solicit appropriate information from the public (§1506.6[d]).

> **An EA is a Public Document**
>
> *An EA means "Means a concise public document for which a Federal agency is responsible..."*
>
> **(§1508.9[a])**

> **An EA is an Environmental Document**
>
> *"Environmental document" includes ... environmental assessment... finding of no significant impact.*
>
> **(§1508.10)**

While EAs must be made available, no public-comment review and incorporation period is required by the Regulations. Agencies are, therefore, not specifically required to respond to public comments as they are in EISs. While an agency is not required to publicly respond to such comments, it is recommended that the agency at least have some manner of response recorded in its administrative record.

Successfully managing public involvement is one of the keys to effective NEPA practice. A study performed by Reinke and Robitaille sheds light on which public involvement problems have led the courts to find that an EA is inadequate.[8] Problems with EAs typically occurred because the documents had not been made available to the public or had unknowingly excluded interested parties. It is the burden of the preparer to seek out interested or impacted peoples.

> **Involving the Public and Other Agencies**
>
> *If an EA is being prepared to determine whether an EIS will be required, other environmental agencies, applicants, and the public must be included in its preparation to the maximum practical extent.*
>
> **(§1501.4[b]).**

If an agency is preparing an EA to determine whether an EIS is required, other environmental agencies, applicants, and the public must be included to the maximum practical extent in the process (§1501.4[b]).

A scoping process should follow the decision to prepare the EA; this process may take many forms, including internal and/or public scoping meetings. The following sections discuss the agency's responsibility to include the public in the EA process.

Public Notification — Agencies are required to make "... diligent efforts to involve the public in preparing and implementing their NEPA procedures (§1506.6[b])." Moreover, agencies are required to provide "public notice" of "...the availability of environmental documents so as to inform those persons and agencies who may be interested or affected (§1506.6[b])." Table 5.3 provides direction on how to notify the public of the availability of either an EA or FONSI (§1506.6[b]).

Although EAs and FONSIs are public documents, they are not required to be filed with the EPA or stored in a national repository, as are EISs.

Consultation — Federal agencies are required to consult with other agencies, as appropriate, in preparing their NEPA documents. Consultation facilitates a more thorough analysis (§1501.1, §1501.2[d], §1502.25).

TABLE 5.3

Method for Notifying the Public of the Availability of an EA or FONSI

All Cases

In all cases, the agency shall mail notice to those who have requested it on an individual action (§1506.6[b][1]).

Where an Action is of National Concern

In the case of an action with effects of national concern, notice shall include publication in the Federal Register and notice by mail to national organizations reasonably expected to be interested in the matter, and may include listing in the 102 Monitor. An agency engaged in rulemaking may provide notice by mail to national organizations who have requested that notice regularly be provided. Agencies shall maintain a list of such organizations (§1506.6[b][2]).

Where An Action Is Of Local Concern

In the case of an action with effects primarily of local concern, the notice may include (§1506.6[b][3]):

(i) Notice to state and areawide clearinghouses pursuant to OMB Circular A-95 (Revised)

(ii) Notice to Indian tribes when effects may occur on reservations

(iii) Following the affected state's public notice procedures for comparable actions

(iv) Publication in local newspapers (in papers of general circulation rather than legal papers)

(v) Notice through other local media

(vi) Notice to potentially interested community organizations including small business associations

(vii) Publication in newsletters that may be expected to reach potentially interested persons

(viii) Direct mailing to owners and occupants of nearby or affected property

(ix) Posting of notice on and off site in the area where the action is to be located

---

Regulatory direction provided in §1502.25 is directed at preparation of EIS but is interpreted to be equally applicable to EAs as well; this is evidenced by the fact that an EA is required to list "... agencies and persons consulted. (1508.9[b])." As a planning vehicle,

Environmental Review and Consultation Requirements

To the fullest extent possible, agencies shall prepare draft environmental impact statements concurrently with and integrated with environmental impact analyses and related surveys and studies required by the Fish and Wildlife Coordination Act ... the National Historic Preservation ... the Endangered Species Act ... and other environmental review laws and executive orders.

(§1502.25[a])

such a step is particularly useful because it can identify related permitting requirements early in the federal planning cycle.

Scoping and public meetings — The CEQ strongly encourages use of a public scoping process to assist agencies in determining the scope of an EA. As indicated in Table 5.4, under some circumstances, a public hearing or meeting may be necessary (§1506.6[c]).

> **TABLE 5.4**
>
> Public Hearings and Meetings
>
> ---
>
> A public hearing or meeting is to be held for circumstances involving
>
> (1) substantial environmental controversy concerning the proposed action or substantial interest in holding the hearing,
>
> (2) a request for a hearing by another agency with jurisdiction over the action supported by reasons why a hearing will be helpful.

5.1.7   Applicants and Environmental Assessment Contractors

If an applicant is applying for a federal permit, licence, or approval, preparation of the EA should begin "no later than immediately after the application is received" by the agency (§1502.5[b]).

> **EA Contractors**
>
> *If an outside contractor is hired to prepare an EA, the agency is still responsible for evaluating the environmental issues and must take full responsibility for its scope and content.*

EA contractors — If an agency obtains the services of a contractor to prepare an EIS, the contractor must sign a disclosure statement indicating that it has "no financial or other interest in the outcome of the project" (§1506.5[b]). This regulatory requirement does not apply to a contractor hired to prepare an EA; however, if a contractor is used in preparing an EA, the agency is still responsible for evaluating the environmental issues and must take full responsibility for its scope and content.

5.2   The Analysis

The Regulations focus on defining the requirements for performing an EIS analysis. Only limited direction is provided concerning these requirements when an EA is prepared.

5.2.1 Decision-Based Scoping

To prevent potential disconnects andfacilitate more comprehensiveplanning, analysts should carefully consider the scope of future decisionmaking before initiating the analysis. Specifically, an effort should be mounted to address the following question: "What types of decisions might need to be considered by the decisionmaker?" The response to this seemingly simple question is essential in shaping the scope and bounds of the analysis.

Chapter 1 of the companion text, *Environmental Impact Statements*, provides a methodology referred to as **Decision-Based Scoping (DBS),** which is designed to assist practitioners addressing this problem.[9] Under DBS, emphasis is placed on first identifying potential decisions that eventually may need to be considered. Once potential decisions are identified, the scope of actions and alternatives naturally follows. A DBS approach underscores the fact that an EA can be as useful in making decisions regarding implementation of a proposal as it can in its more traditional role of assessing the potentially significant impacts.

The Decision-Identification Tree — To facilitate the DBS task, a tool, referred to as a **Decision-Identification Tree (DIT),** is also introduced in Chapter 1 of the text, *Environmental Impact Statements*, to assist practitioners in identifying potential decision points. A DIT can be especially useful in scoping complex proposals that may involve numerous engineering, technical, socioeconomic, or regulatory decisions, or where the potential decision points are not clearly obvious.

5.2.2 Analysis of Impacts

As appropriate, potential impacts should be quantified whenever feasible. While the Regulations do not specifically state that an EA must consider impacts of connected, similar, and cumulative actions, a determination of nonsignificance cannot be reached without considering such impacts (§1508.27[b][7]).

A FONSI that fails to conclusively demonstrate nonsignificance is highly vulnerable to a successful legal challenge. To this end, the analysis must be rigorous and presented so that a decisionmaker can clearly reach a conclusion regarding the potential for significance. No stone should be left unturned. The EA should provide specific evidence addressing the significance factors indicated in §1508.27.

5.2.3 Attention Centered on Proposed Action

An EIS must devote "substantial consideration" to each of the analyzed alternatives (§1502.4[b]). In contrast, the EA analysis is normally centered on the proposed action.

The Proposed Action
The EA analysis is commonly focused on the proposed action. Reasonable alternatives are only briefly described and then dismissed. Reasons for dismissing the reasonable alternatives should be clearly explained.

Typically, reasonable alternatives are only briefly described and then dismissed from more detailed study. Discussion is limited to topics that are necessary to give the reader an adequate understanding of what the alternative covers and to allow the reader to discriminate between the alternatives.

Care should be exercised in clearly explaining to the reader why each alternative has been dismissed. Alternatives that have been considered but either not analyzed or rejected are often placed in a separate section labeled "Alternatives Considered but not Carried Forward."

Such practice can be justified, as the purpose of an EA is generally focused on determining whether an action could result in significant impacts. As long as the proposed action would not significantly affect the environment, there is correspondingly less justification for exploring other alternatives in detail.

Planning Future Action
Besides its primary purpose, an EA also provides an effective tool for assessing reasonable actions and planning future actions.

While the primary function of EAs is determining the potential significance of a proposed action, it is important to remember that an assessment can also be used to plan actions or determine an optimum course of action. In a recent case, for example, a decision had already been made to take action to avoid an environmental fine; an EA was prepared that investigated "all reasonable alternatives." The agency reported that the EA provided an effective planning tool for determining specifically what action to take.[10]

Describing Alternatives
A sliding-scale approach may be useful in determining how many alternatives to address and the degree to which each alternative is described.

Describing Alternatives — A sliding-scale approach may be useful in determining the range of alternatives and the degree to which such alternatives are described; thus, projects that are more complex or controversial may necessitate a more thorough investigation.

5.2.4   Significance Determinations are Reserved for the FONSI

Significance Determinations
A determination regarding nonsignificance should not be made in the EA. Such determinations should be made by the decisionmaker based on the analysis provided in the EA; the actual determination of nonsignificance is reserved for the FONSI.

While there is not complete agreement, some agencies purport that an EA should be limited to providing decisionmakers with objective information that they can use in reaching a conclusion regarding significance. Thus, the actual determination of significance is made based upon the decisionmaker's review of the EA. Analysts are therefore confronted with the difficult challenge of preparing an EA that contains information that will clearly lead the reader to a conclusion, yet does not prejudge the potential significance of the impacts.

Accordingly, this text takes the position that the analysis should avoid even the appearance of partiality or the perception of predetermining significance. However, potential impacts that are clearly and unequivacally nonsignificant may be dismissed during the scoping process from more detailed review. If, based

> **Avoid Use of Judgmental Terms**
>
> *With respect to describing impacts, judgmental terms such as "acceptable," "tolerable," or "nonsignificant," should be avoided in describing the impacts. Less-judgmental terms such as "substantial," "consequential," "inconsequential," "large," or "small" may be used instead.*

on review of the EA, the decisionmaker reaches a determination that all impacts are nonsignificant (or can be mitigated to such a point), this determination is documented in the FONSI.

Consistent with this approach, judgmental terms such as "nonsignificant," or "acceptable," should be avoided when describing impacts. Less-judgmental terms such as "consequential,"

> **Example of a Nonjudgmental Statement**
>
> *Suppose an impact analysis concludes that a proposed action would result in a very small loss of critical habitat. The analysis should avoid making statements such as, "The action would result in a nonsignificant loss of critical habitat." Instead, the analysis might more appropriately quantify the impact and state that, "The action would result in a relatively small loss of critical habitat, approximately 50 square meters in size out of a total area of 5 square kilometers."*

"inconsequential," "large," or "small" are more preferable. Better yet, the description of an impact should be quantified whenever feasible.

5.3 Streamlining the EA Compliance Process

The CEQ encourages use of EAs as a means of streamlining the NEPA process (§1500.4[p] and §1500.5[k]). Some agencies have experienced exceptionally long review and approval cycles, resulting in project delays. The EA review and approval cycle should be examined periodically for inefficiencies. Where practical, reviews involving more than one entity should be conducted in parallel. Agencies should consider delegating approval to the lowest competent decisionmaking level within an organization. Additional methods are described in the following sections for streamlining the EA process.

5.3.1   Reducing the Length of an EA

Many EAs are excessively lengthy and detailed. Incorporating data by reference is a particularly useful method for reducing document length. Focusing on potentially significant issues while reducing attention devoted to nonsignificant issues is another.

Avoid Setting Future Precedents

If a decision is made to add material not normally required in an EA, a disclaimer should be included to prevent setting a future precedent.

A sliding-scale approach may be useful in determining the detail and complexity of analysis that is appropriate. The amount of attention devoted to a given impact increases with the complexity of the proposed action and the potential for significance.

If a decision is made to add material not normally required in an EA, a disclaimer should be included to prevent setting a future precedent. The disclaimer should indicate the reason why inclusion of the material was deemed necessary.

5.3.2   Cooperating with Other Agencies

Reducing Duplication of Efforts

To reduce duplication with other environmental requirements, agencies are mandated to cooperate with state and local agencies in the development of EAs.

To reduce duplication with other environmental requirements, agencies are mandated to cooperate with state and local agencies in preparing their EAs (§1506.2[b][4], §1502.5[b]). Significant savings have been reported when efforts were made to identify and coordinate NEPA with other environmental studies.

5.3.3   Is Time Money?

In preparing NEPA documents, there was a widely held belief that "time is money." In other words, documents that take an inordinate period to prepare generally cost more than those consuming shorter periods. This premise has recently been challenged. A study of cost versus preparation time has been performed by the U.S. Department of Energy.[11] This study involved 177 EAs prepared between August 1992 and June 1999.

As indicated in Figure 5.2, there is essentially no correlation between document cost and preparation time for EAs. This finding, however, does not suggest that the goal of reducing preparation time is unimportant. Rather, the study suggests that, in striving to reduce cost, one should focus on factors other than preparation time. A related study of EISs reached a similar conclusion.

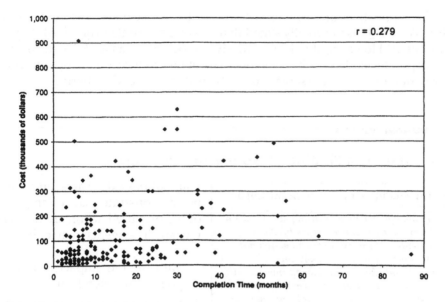

FIGURE 5.2
Document costs versus preparation time for environmental assessments.

Instead of concentrating on preparation time for controlling costs, it is recommended that practitioners focus on efficient use of

- existing environmental data (e.g., affected environment, designs, accident analyses)
- in-house resources in preparing all or portions of the document

5.3.4 Tiering

The CEQ strongly advocates a concept referred to as tiering for expediting NEPA compliance. Under tiering, an EA prepared for an action within the scope of an EIS needs only to summarize the issues discussed in the EIS and state where a copy of the statement can be obtained. Discussions presented in the EIS can then be incorporated by reference (§1502.20).

Tiering an EA

"Agencies are encouraged to tier their environmental impact statements to eliminate repetitive discussions of the same issues and to focus on the actual issues ripe for decision at each level of environmental review... the subsequent ... environmental assessment need only summarize the issues discussed in the broader statement and incorporate discussions from the broader statement by reference and shall concentrate on the issues specific to the subsequent action. The subsequent document shall state where the earlier document is available."

(§1502.20)

The Regulations are silent on whether an EA can be tiered from an existing EA. The CEQ has informally stated that it is inappropriate to tier one EA from another.[12] However, this is essentially a moot point as an EA can simply incorporate another EA by reference. The ability to incorporate by reference provides essentially the same capability as that provided by tiering.

5.4   Issuing a Finding of No Significant Impact

> **The Finding Of No Significant Impact.**
>
> *"Finding of No Significant Impact" means a document by a federal agency briefly presenting the reasons why an action not otherwise excluded...will not have a significant effect on the human environment, and for which an environmental impact statement therefore will not be prepared. It shall include the environmental assessment or a summary of it, and shall note any other environmental documents related to it.... If the assessment is included, the finding need not repeat any of the discussion in the assessment but may incorporate it by reference."*
>
> **(§1508.13)**

Four options exist if an agency cannot demonstrate that the impacts of the proposed action are nonsignificant. The agency may either

1. Mitigate the impacts to the point of nonsignificance

2. Modify the action to render the impacts nonsignificant (a subset of #1 above)

3. Select another alternative (such as no action) that was adequately analyzed in the EA

4. Prepare an EIS

5.4.1   FONSI is a Public Document

Like an EA, a FONSI is a "public document," and must be made publicly available. While the Regulations state that the FONSI must be made available to the "affected public" (§1501.4[e][1]), little direction is provided as to how this should accomplished. The regulations do state that, when an action affects a wetland or floodplain, special effort should be made to make the FONSI publicly available.[13] For more information, see Section 5.1.6.

5.4.2   Waiting Period

Typically, the proposed action may be initiated as soon as the FONSI is issued.

However, under some restricted circumstances (see Table 5.5), a FONSI must be made publicly available for a minimum review period of 30 days before the agency makes a final determination regarding preparation of an EIS and before the action may begin (§1501.4[e][2]). This waiting period is intended to give the public an opportunity to review the FONSI and, if appropriate, take action.

TABLE 5.5

Circumstances Where a 30-Day Waiting Period is Required

Under certain circumstances, an agency shall make the FONSI available to the public for a period of 30 days before determining whether an EIS must be prepared and before the action may be taken. These circumstances are

1. a proposed action is similar to one normally requiring preparation of an EIS under the agency's implementation procedures, or

2. the nature of the proposed action is one without precedent.

(§1501.4[e][2])

As indicated in Table 5.6, the CEQ also indicates that a 30-day public review period is warranted for circumstances involving:[14]

- Borderline actions
- Unusual, new types of actions or precedent-setting actions
- Controversial proposals

TABLE 5.6

Additional CEQ Guidance for Determining Whether a 30-Day Waiting Period is Applicable

1. Borderline case, such that there is a reasonable argument for preparation of an EIS;

2. Unusual case, a new kind of action, or a precedent-setting case such as a first intrusion of even minor development into a pristine area;

3. When scientific or public controversy exists over the proposal.

5.5 The Administrative Record

An agency should not depend solely on environmentally sound decisions as a means of demonstrating compliance with NEPA's requirements. If the agency is challenged, a court may require the agency to turn over files and documentation related to the suit.

What Constitutes the Agency's Administrative Record?

For the purposes of NEPA, the agency's administrative record can be considered to include the entire record that existed at the time the decision was made and not simply the portion of the record read by the decisionmaker.

Many professionals are typically involved in shaping plans, assumptions, and internal decisions before an EA reaches the decisionmaker. For the purposes of NEPA, the agency's *administrative record* can be considered to include the entire record that existed at the time the decision was made, and not simply the portion of the record read by the decisionmaker.[15] The reader is directed to the companion book, *Environmental Impact Statements: A Comprehensive Guide to Project and Strategic Planning,* for additional information on developing a thorough administrative record.[16]

5.5.1 Administrative Record and Case Law

Care must be taken to ensure that the determination and resulting FONSI is consistent with the facts and the agency's administrative record. A recent court case illustrates the need for performing a thorough analysis and preparing a comprehensive administrative record. This case involved a state-sponsored action that required federal authorization. Both NEPA and the state's environmental policy act were triggered. The action was later challenged in court. The state was asked to produce the administrative record detailing how the analysis had been performed; it was unable to produce a well-defined administrative record detailing the assumptions used in the analysis and how various decisions had been made. Sensing that it was in a weak position, the state agency agreed to settle the case out of court.

In another case, an EA was prepared for construction of a parking lot in Glacier National Park. The proposed action involved removal of vegetation and 500-year-old cedar trees that the agency's administrative record characterized as "significant in light of the cumulative impacts that have occurred and the extreme rarity of the habitat involved." The EA also contained statements implying that an EIS was needed to consider the project's impacts on unique resources. Further, while other National Park Service analyses identified approximately 200 types of important trees, the original FONSI identified only nine important types of trees that would need to be removed. The court found the FONSI inadequate.[17]

5.6   Serving as an Expert Witness

Whether you are a client being sued or an environmental practitioner, you may find yourself in the position of serving as an expert witness. Expert witnesses often serve in different capacities. An attorney, for example, may ask an expert witness or an environmental consultant to evaluate the merits of a particular case. An expert in NEPA might also be called upon in a court of law to help prove or disprove a point, or the entire case.

If you are asked to serve in such a capacity, the first thing you will learn is that there is very little literature written on this subject for the non-lawyer. The following sections are based on an article by M. J. Rogoff that is intended to fill this void.[18]

5.6.1 Are You an Expert Witness?

An expert witness is generally regarded as a person whose education, skill, or expertise goes beyond the experience of ordinary citizens. If you work for a company and are asked to serve as an expert witness, you should first check to ensure that there would be no conflict of interest.

The first formal notice of a lawsuit or *pleading* will typically allege that harm has been done to the plaintiff (the party bringing the suit). The plaintiff bears the burden of proving this allegation. Based on a number of different factors, a judge may deny the plaintiff's claim. An expert can help the lawyer in developing a defensible allegation.

5.6.2 Discovery

Prior to the actual trial, a legal process known as discovery normally allows the plaintiff and defendant to request, obtain, and review information (e.g., business plans, memorandums, e-mails, and technical data) accumulated by the opposing party. Discovery is

> **Discovery**
>
> *Discovery is the legal process that allows a plaintiff and defendant to request, obtain, and review information accumulated by the opposing party*

designed to eliminate surprises by allowing one to mount a defense against charges or evidence that may be entered into trial by the opposing party. A consultant or expert can be a valuable player in helping a lawyer identify and evaluate the types of information that need to be requested from the opposing party.

Experts can be indispensable in helping a judge or jury understand the facts of a case. They can also be particularly effective in sifting through highly technical data and in interpreting the facts of the case.

5.6.3 Depositions and Reports

Depositions provide a means of gathering evidence from anyone possessing relevant knowledge about the case. The expert witness is sworn to tell the truth by a court reporter. Expert witnesses might be questioned by the opposing party to identify biases or weaknesses in their testimony. A transcript is prepared documenting the expert witness' statements and responses to questions.

Experts might also be asked to prepare an expert report detailing their findings.

5.6.4   Trial

The expert witness can be indispensable in deciphering the technical mysteries surrounding a case. During cross-examination, the opposing party will attempt to destroy the credibility of the expert. Experts may be confronted with apparent contradictions or errors in their reports of deposition. Experts might even find that their credentials have been brought into question. Table 5.7 provides suggestions to follow in testifying as an expert witness.[18]

TABLE 5.7

Suggestions to Follow when Testifying as an Expert Witness

- Cases and careers have been ended when falsifications were presented in a deposition or at trial. Think before answering a question.

- Take time to think out your answer. At the very least, this will give your attorney time to object to the question or line of reasoning.

- Answer only the question asked. Even if you believe the opposition party's question to be irrelevant, don't follow-up with the question you think the examiner should have asked.

- Don't volunteer information. Answer only the question asked and then stop. You are not responsible for educating the examiner. Once you have answered, remain quiet. The examiner may use this pause and stare at you in a manner designed to intimidate or make you feel uncomfortable; in doing so, the examiner hopes to elicit more information that might be used against your position.

- Don't answer a question you don't understand. It is not your responsibility to ask the question. If you don't understand a question, tell the examiner that you don't understand and ask him to rephrase the question.

- Don't guess. Be as specific as you can, but never guess. If you can't recall, simply tell the examiner that you can't remember.

When testifying, expert witnesses must exercise care in the way they answer the opposing party's questions. This includes their facial expressions and overall demeanor.

References

1. U.S. DOE, *NEPA Lessons Learned*, Issue No. 20, p17, September 1, 1999.
2. Eccleston C. H., *Environmental Impact Statements: A Comprehensive Guide to Project and Strategic Planning,* John Wiley & Sons Inc, New York, N.Y., 2000; and Eccleston C. H., *The NEPA Planning Process: A Comprehensive Guide with Emphasis on Efficiency,* John Wiley & Sons Inc, 1999; *Environmental Impact Statements: A Comprehensive Guide to Project and Strategic Planning,* John Wiley & Sons Inc, New York, N.Y. 2000.

3. U.S. DOE, *NEPA Lessons Learned,* Issue No. 20, p16, September 1, 1999.

4. Berg J. E., U.S. DOE's *NEPA Lessons Learned,* September 1, 1999, Issue No.20.

5. U.S. Forest Service, Decision Protocol 2.0, http://www.fs.fed.us/forum/nepa/dp2roadmap.htm

6. Council on Environmental Quality, Forty Most Asked Questions Concerning CEQ's National Environmental Policy Act Regulations (40 CFR 1500–1508), Federal Register Vol. 46, No.55, 18026–18038, Question Number 35, March 23, 1981.

7. Blaug E.A., Use of the Environmental Assessment by Federal Agencies in NEPA Implementation, *The Environmental Professwional,* Volume 15 pp. 57–65, 1993.

8. Reinke D. C. and P. Robitaille, NEPA litigation 1988–1995: A Detailed Statistical Analysis, *Proceedings of the 22nd Annual Conference of the National Association of Environmental Professionals,* pp 759–765, 1997.

9. Eccleston C. H., *Environmental Impact Statements: A Comprehensive Guide to Project and Strategic Planning,* John Wiley & Sons Inc, New York, N.Y., 2000.

10. DOE, *NEPA Lessons Learned,* Issue No. 18, March 1, 1999.

11. U.S. DOE, *NEPA Lessons Learned,* Issue No. 20, pp 19–20, September 1, 1999.

12. Personal communications with the CEQ.

13. Executive Order 11988, 42 FR 26951, May 24, 1977; and Executive Order 11990, 42 FR 26961, May 24, 1977.

14. CEQ, Council on Environmental Quality - Forty Most Asked Questions Concerning CEQ's National Environmental Policy Act Regulations (40 CFR 1500–1508), Federal Register Vol. 46, No. 55, 18026-18038, March, 23, 1981, Question number 37b.

15. *Haynes* v. *United States,* 891 F.2d 235 (9th Cir. 1989).

16. Eccleston C. H., *Environmental Impact Statements: A Comprehensive Guide to Project and Strategic Planning,* John Wiley & Sons, Inc., New York, N.Y., 2000.

17. *Coalition for Canyon Preservation and Wildlands Center for Preventing Roads v Department of Transportation,* No. CV 98-84-M-DWM, 1999, U.S. District., LEXIS 835 (D. Mont. January 19, 1999).

18. Rogoff M J., ...So you Want to be an Expert Witness, *Environmental Practice Journal,* National Association of Environmental Professionals, Vol. 2, Number 1, pp 5–6, March 2000, ISSN 1466–0466.

6

Writing the
Environmental Assessment

> When in doubt, tell the truth.
>
> Mark Twain

Agencies are ultimately responsible for both the preparation and content of their National Environmental Policy Act (NEPA) documents. This chapter presents the reader with the documentation requirements that must be satisfied in writing an Environmental Assessment (EA). Because most agencies have internal requirements and nuances, readers are encouraged to consult their agency's NEPA implementing procedures and internal guidance for any requirements that may supplement the regulatory direction provided in the Council on Environmental Quality's (CEQ) NEPA regulations (Regulations). Mr. Peyton J. Doubs (Tetra Tech NUS, Inc.) has graciously provided representative examples of EAs prepared for three diverse projects, including a site-wide assessment; these EAs are found in Appendices D through F.

6.1 General Direction for Preparing an EA

The direction provided in Section 6.1 applies to preparation of the entire EA.

6.1.1 Documenting Assumptions

Future uncertainties are often dealt with by making reasonable assumptions. One of the most common criticisms of NEPA documents is inadequate referencing of source material, including assumptions. The EA must clearly identify any uncertainties and

> **Documenting Assumptions**
>
> *All significant assumptions used in performing the analysis need to be documented in the EA.*

the assumptions used to bridge them. The rationale on which each assumption is based should be clearly documented. Such practice can be critical in defending the validity of the analysis, if challenged.

6.1.2   "Will" Versus "Would"

To clearly demonstrate that no decision has yet been made, discussions of potential actions should be written as if the action *might* take place. Words such as "proposed" and "would" should be used to clearly indicate that a final decision has not been made.

> **Example of "Will" Versus "Would"**
>
> *Instead of stating that "The action will result in the loss of 8,000 square meters of critical habitat," the assessment should indicate, "The proposed action would result in the loss of 8,000 square meters of critical habitat."*

6.1.3   Readability

If citizens cannot understand the material in the assessment, they cannot effectively participate in the NEPA process; this can lead to distrust, resentment, hostility, and even lawsuits. Reader-friendly documents tend to foster public understanding and cooperation, and lessen the risk of litigation. For instance, technical supporting data can be briefly summarized in simple terms in the EA with the technical detail provided in the appendix or incorporated by reference from publicly available material.

To increase readability, practitioners should make good use of graphic aids (maps, tables, figures, graphs, and flowcharts) to enhance a reader's comprehension.

Graphic aids increase comprehension — How can environmental documents be improved to more effectively convey information to the target audience? This question was recently the topic of two published articles in *Environmental Impact Assessment Review*.[1] Three researchers from the University of Illinois conducted tests on students to measure their ability in understanding and recalling the project description and environmental consequences of a local flood-control plan in an Environmental Impact Statement (EIS). While the study was directed at EISs, the results are generally applicable to an EA as well.

Comprehension   In the first study, students read portions of the EIS and then answered questions about the project and its environmental effects.[2] The study concluded that the student's understanding of the material was "atrocious," even among the best readers. The students' performance was generally below 70 percent — a measure the authors of the study considered adequate for comprehension (the equivalent of an academic "C").

Photosimulation   The second study by the same researchers offers cost-effective suggestions for improving readability.[3] The first of these — "photosimulation," involves a series of before and after pictures of a project area; the "after" picture is created using photograph manipulation software such as Adobe Photoshop®, to show possible changes in the landscape. Prudence must be exercised in providing the reader with an "after" picture that accurately depicts how the affected environment would be changed.

TABLE 6.1

Photosimulation

Photosimulation: The use of "before" and "after" pictures of a project study area to depict how an action would change the affected environment can markedly improve reading understanding and comprehension.

In the Illinois example, pictures of a local creek were used to show how the creek would look if flood control measures were implemented (see Figure 6.1). The study concluded that reading comprehension substantially improved through the use of photosimulation. Specifically, comprehension of the proposal and the understanding of environmental impacts improved to a level significantly greater than 70 percent.

Other techniques for improving readability — The Illinois researchers' second suggestion for improving comprehension involves better editing. Practitioners can improve the effectiveness of their documents by following seven simple techniques: provide an overview, provide clear headings, state

FIGURE 6.1
Potential flood control features suggested for Hickory Creek involved three different changes of the creek banks. The banks were to be changed from their existing condition (top left) to either a fabric-formed concrete embankment (top right), a vertical concrete wall (lower left), or an earthen embankment (lower right). Photos from Sullivan, W. C., Kuo, F.E., and Prabhu, M., Communicating with Citizens: The Power of Photosimulations and Simple Editing, *Environmental Impact Assessment Review* 17(4):295–310, July 1997. Reprinted by permission of Elsevier Science, NY.[4]

TABLE 6.2
Seven Simple Techniques for Improving Readability

1. Provide an overview.
2. Provide clear headings.
3. State headings as questions.
4. Make headings distinct.
5. Use locally recognizable landmarks to identify the location(s) of future actions.
6. Explain technical terms as they come up (rather than in a glossary).
7. Use text bullets.

some headings as questions, make headings distinct, use locally recognizable landmarks to identify locations of project work, explain technical terms as they come up (in addition to a glossary), and use text bullets. When combined with photosimulation, these seven simple techniques increased students' comprehension to over 80 percent.

Use Ordinary Words and Phrases	
Instead of	*Try using*
Adjacent to	*Next to*
Due to the fact that	*Because*
Initiate	*Start, begin*
In the event that	*If*
Prior to	*Before*

Where possible, use simple, ordinary words and phrases.

Glossaries An easy and effective way to make NEPA reader-friendly is to mark in bold or italics the first occurrence of terms that are defined in the glossary. This will effectively signal the reader to consult the glossary. This system should be explained in a footnote or text box at the beginning of the NEPA document and also in the glossary.

Use Active Voice Over Passive Voice	
Passive Voice	*Active Voice*
"An EIS will be prepared"	*"The DOE will prepare an EIS"*
"Comments may be submitted to"	*"You may submit comments to"*

Using plain language — Write to express — not to impress. The following suggestions can enhance the readability of the EA:

- Promote use of active voice over passive voice.

Example of Minimizing the Use of Highly Technical Terms
Instead of using a term such as "nonelutable resin," try "resin from which adsorbed material cannot be separated."

- Use ordinary words and phrases;
- Minimize or explain use of technical terms. Instead of using a term such as "nonelutable resin," try "resin from which adsorbed material cannot be separated."

- Reduce use of abbreviations, including acronyms;
- Divide long sentences into shorter ones.

Emphasizing use of active voice (see first bullet) should not be misconstrued to mean that passive voice should never be used. For example, the statement, "The forest is dominated by chestnut oak and pignut hickory,"is preferable to the active voice "Chestnut oak and pignut hickory dominate the forest."

6.1.4   How Long Should the EA be?

The CEQ views an EA as a "concise public document." It should not contain long descriptions or overdetailed data. Rather, CEQ states that it is to be a concise document that provides a brief discussion of the need for the proposal, alternatives, and environmental impacts (§1508.9).

> **Recommended Length of an EA**
>
> *The CEQ advises that an EA should be no more than 10 to 15 pages in length. In reality, most EAs exceed this recommended page limitation.*

While the regulations do not contain page limits for an EA, the CEQ has advised agencies to keep the length of an EA to a maximum of 10 to 15 pages.[5] To avoid undue length, the EA may incorporate by reference background data to support its concise discussion of the proposal and relevant issues.

In reality, the vast majority of EAs exceed CEQ's suggested page limit. The length of an EA varies with the size and complexity of the proposal, environmental issues, and the potential for significant impacts. For this reason, rigid adherence to a preconceived document length may not be prac-

> **The Length of an EA**
>
> *Rigid adherence to a preconceived document length may not be practical. Instead, the following advice is recommended: the length of an EA should be limited to that information that will concisely explain the proposal, and address important issues and the potential for significant environmental impacts.*

tical. Instead, the following advice is recommended: the length of an EA should be limited to that information that will concisely explain the proposal, and address important issues and the potential for significant environmental impacts.

6.2   Required Outline for an Environmental Assessment

As indicated in Table 6.3, the Regulations specify only four documentation requirements that an EA must satisfy (§1508.9[b]). The first three items indicated

TABLE 6.3

CEQ Documentation Requirements for an EA

1. Need for the Proposal.
2. Alternatives (as required by section 102[2][E] of NEPA).
3. Environmental Impacts of the Proposed Action and Alternatives.
4. Listing of Agencies and Persons Consulted.

(§1508.9[b])

in Table 6.3 are common to both the EA and EIS. The fourth item is required only in EAs.

The agency's internal guidance and NEPA implementation procedures should be consulted for any additional documentation requirements.

6.2.1 Suggested EA Outline

To improve decisionmaking and promote a more open public process, it may be advantageous to include information beyond CEQ's minimal requirements. However, it is important to strike a balance between the additional information and the potential for increased work scope.

A generalized outline for balancing these competing objectives is presented in Table 6.4. This outline should be tailored to meet the agency's specific mission and circumstances, including (as appropriate) any specific requirements cited in the agency's orders and NEPA implementation procedures.

TABLE 6.4

Suggested Outline for an Environmental Assessment

Title Page

Glossary

Executive Summary

1.0 Purpose and need for the proposed action

2.0 Description of the affected environment

3.0 Description of the proposed action and reasonable alternatives

4.0 Analysis of environmental impacts of the proposed action and reasonable alternatives

5.0 Applicable environmental permits and regulatory requirements that would need to be obtained

6.0 List of agencies and persons consulted

7.0 List of preparers

8.0 References

Including a title page, glossary, and executive summary is consistent with the Seven Methods for Increasing Readability (Table 6.2), while adding only marginally to the document size.

The reader should note that an EIS requires that the document define both a purpose and need for taking action (see first item in Table 6.3). Discussing the purpose in addition to the need may help explain the project's rationale to the public and decisionmaker.[7]

Adding a description of the affected environment can be an element essential in helping the decisionmaker to reach a defensible determination regarding significance.

Listing applicable permits and regulatory requirements helps the agency in identifying and planning for future actions, as well as in assessing potentially significant impacts (i.e., §1508.27[b[10]]).

Citing agencies and persons consulted is helpful in demonstrating that appropriate experts were consulted and that an interdisciplinary analysis was prepared. It can also demonstrate that the assessment was coordinated with other agencies having jurisdiction by law, or special expertise with respect to the environmental issues.

Listing the preparers can help the agency in demonstrating that it performed an interdisciplinary analysis (§1501.2[a]). Finally, it can help the agency in substantiating its analysis and in documenting its administrative record.

6.3 Performing the Analysis

Consensus must be reached regarding the appropriate use of any analytical methodologies or computer models that are employed. Where possible, analysts should use methods and computer models endorsed by the U.S. Environmental Protection Agency (EPA), other governmental agencies, or professional societies.

> **Analytical Methodologies**
>
> *Analysts should use analytical methodologies endorsed by the U.S. Environmental Protection Agency (EPA), other governmental agencies, or professional societies.*

6.3.1 A Five-Step Methodology

Practitioners are mandated to "utilize a systematic, interdisciplinary approach ... in planning and in decisionmaking which may have an impact on man's environment."[8] To this end, a five-step methodology for investigating environmental impacts is offered in Table 6.5. Professional discretion must be exercised in determining how these steps are most practically

TABLE 6.5
Systematic Methodology for Performing an EA Analysis

Step 1 - *Characterize Potentially Affected Resources:* Data is collected on the potentially affected resources. Temporal and spatial boundaries are delineated for the subsequent analysis. The potentially affected environmental resources are described.

Step 2 - *Identify Component Actions*: All component actions that may occur over the reasonable foreseeable life-cycle of the proposal are identified.

Step 3 - *Identify Environmental Disturbances:* Environmental disturbances (i.e., noise, effluents, emissions, ground disturbances) resulting from the component actions are described. Cause-and-effect relationships are described.

Step 4 - *Screen and Analyze Environmental Impacts:* The environmental disturbances are investigated to determine how they would *impact* (change or affect) environmental resources. The environmental disturbances are screened to determine their potential for significance. Disturbances deemed to be clearly and unequivocally nonsignificant are eliminated from further study. A more detailed analysis is then performed to determine how the remaining environmental impacts might significantly impact environmental resources.

Step 5 - *Assess Potential Monitoring and Mitigation Measures:* If appropriate, potential mitigation measures are evaluated to determine their effectiveness in mitigating potentially significant impacts. A monitoring plan can also be prepared.

applied based on the complexity of the analysis and the particular circumstances. The reader is referred to Section 2.4 of the companion text, *Environmental Impact Statements*, for additional information.

Step 1: Characterize potentially affected resources — The term "potentially affected environmental resources" refers to the existing physical, biological, cultural, and socioeconomic resources before they are altered by the agency's proposal. These resources should be identified and characterized prior to beginning the environmental impact analysis.

Commonly Consulted Sources of Information   The affected environment descriptions in many EAs utilize technical information from standardized sources available for most locales throughout the United States. Most of these sources are published by federal, state, and municipal agencies. Normally, this data can be rapidly obtained from the publishing agency or from local offices at no charge or for a nominal charge. Some sources are available online through the World Wide Web. Other sources can be requested on the telephone or in writing. Analysts often visit local government offices to obtain information sources while on a trip to visit the proposed site.

Table 6.6 lists some of the technical sources that are routinely consulted in preparing EAs. The list is not exhaustive. Readers are urged to consider other sources of technical information that might also be available and appropriate for the specific action in question.

A variety of site-specific environmental documents often provide useful technical information for analyzing actions taking place within existing

TABLE 6.6

Routinely Consulted Technical Information Sources

Source	Available From	Relevant to
Soil surveys	Natural Resources Conservation Service	Soils geology
National wetland inventory maps	U.S. Fish & Wildlife Service; often available from municipal offices	Wetlands, biological, and water resources
Flood insurance rate maps	Federal Emergency Management Agency; usually available from municipal offices	Floodplains and water resources
Topographic maps (7.5-minute)	U.S. Geological Survey	Topography water resources
Topographic maps (1- or 2- foot contours)	These more detailed topographic maps are sometimes available from municipal offices	Topography water resources
Comprehensive plan (municipal or county)	Municipal offices	Land use; socioeconomics; biological resources
Zoning maps	Municipal offices	Land use
Aerial photographs	Municipal offices; National Aerial Photography Program	Land use; biological resources
STORET (storage & retrieval) water quality database	U.S. Environmental Protection Agency (can access free online)	Water resources
Water-data reports	U.S. Geological Survey	Water resources
Environmental record database searches	Sold at nominal cost from private-sector database search companies	Hazardous materials/waste management
Census data (population, employment, etc.)	Bureau of the Census (can access free online)	Socioeconomics
State natural heritage databases	State natural resource management agency; must access in writing and might include a nominal charge	Threatened and endangered species; biological resources
State Historic Preservation Officer	State cultural resource agency; access in writing as part of Section 106 consultation process	Cultural resources

federal installation boundaries. These may include previous EAs and EISs, site environmental reports, natural resource management plans, environmental baseline surveys, and geographical information system layers. Readers are also advised that at least one site visit is recommended, regardless of how much published technical data is available.

Several commercial companies now offer environmental data from computerized record sources. These companies were primarily formed to provide rapid database searches to facilitate preparation of Phase I environmental site assessments addressing the potential for environmental contamination of property. Recently, however, they have expanded their database capabilities to offer information useful in preparing NEPA analyses. Sources of data include

- data on sites listed in the National Register of Historic Places,
- wetlands mapped on National Wetland Inventory maps,
- floodplains mapped by the Federal Emergency Management Agency, and
- threatened and endangered species.

The services offered by these companies are typically very fast (on the order of one to two days) and relatively inexpensive (typically less than $200).

The reader is cautioned that some of the information obtained from these database searches may be incomplete or inadequate. For example, sometimes only information on federally-recognized historic and archaeological sites, or on federally listed threatened or endangered species, is available. Information from state or local lists is often not provided.

Furthermore, the database searches may provide inadequate descriptive detail concerning the features noted. For example, a review of National Wetland Inventory maps does not constitute an adequate determination of whether wetlands exist on a given site. Perhaps most important, merely reviewing computerized database information fails to engage environmental agency personnel in the early planning process required in preparing a NEPA analysis.

While EA preparers should not be discouraged from using these environmental database search services, they should understand the limitations of the searches. The searches do not replace the need for an independent and thorough investigation by analysts preparing the NEPA analyses. Moreover, such searches do not meet the agency consultation requirements under NEPA.

Practitioners are cautioned to use the searches as a supplemental research tool; the practitioner is responsible for cross-checking the accuracy and supplementing the data as necessary to adequately support the impact assessment. As appropriate, regulatory agencies must still be contacted to determine whether additional relevant information is available.

Step 2: Identify component actions — NEPA recognizes three types of action: (1) connected, (2) cumulative, and (3) similar.[9] A final course of action will invariably consist of a set of individual or component actions. To adequately evaluate environmental impacts, all reasonably foreseeable component actions must be identified and described. For example, a proposal involving construction of a small office building might involve an array of *component actions* such as site clearing and excavation, construction of access roads, construction of power and communications cables, construction of a water and sanitary line, a parking lot, and drilling a water supply well.

Step 3: Identify environmental disturbances — The terms "environmental disturbances" and "environmental impacts" should not be confused. An environmental disturbance can be viewed as some type of stressor produced by an action that might change or affect an environmental resource. For example, a new aircraft might release

> **Environmental Disturbance Versus Environmental Impact**
>
> *The terms "environmental disturbances" and "environmental impacts" should not be confused. Environmental disturbances can be viewed as a stressor (e.g., emission or effluent) produced by an action that might change or affect an environmental resource. The environmental impact denotes how such a disturbance would affect or change an environmental resource.*

high-altitude carbon dioxide emissions (i.e., environmental disturbance). Such a release in itself is not an environmental impact. The environmental impact denotes how such a release would affect or change an environmental resource (e.g., contribute to global warming). Thus, an environmental impact depicts how a change in an environmental resource, caused by an environmental disturbance, would affect humans and environmental quality.

Each potential action may produce environmental disturbances. The disturbances must be identified and described in detail sufficient to allow analysts to investigate how they could affect or change environmental resources. Pathways and cause-and-effect relationships should also be identified.

Step 4: Screen and analyze environmental impacts — The environmental disturbances are screened (e.g. scoping procedure) to determine how they would affect environmental resources. The potential disturbances can be compared with quantitative and qualitative metrics (significance criteria) to determine whether they are clearly inconsequential. Disturbances deemed to be clearly and unequivocally nonsignificant are eliminated from more detailed study. A more detailed analysis is then performed to determine how the remaining environmental impacts would affect environmental resources.

A no-action alternative provides an effective environmental baseline for performing the analysis. Thus, analysts have a basis for comparing the effects of pursuing a particular action against the impacts of taking no action. Metrics can be used to determine whether a certain level/degree of change from the baseline level is deemed significant. Once environmental impacts of taking no-action are determined, an effort is mounted to determine the impacts of the proposed action.

Step 5: Assess potential monitoring and mitigation measures — If deemed appropriate, potential mitigation measures are also evaluated. An agency might also choose to implement impact reduction methods even though all impacts are found to be nonsignificant. The reader should note that a nonsignificant impact does not necessarily imply that there would be no impact — only that the impact is below the threshold of significance. A monitoring plan may also need to be prepared.

6.4   Specific Documentation Requirements and Guidance

> **Sufficiency of Information**
>
> *A tool (Sufficiency-Test) is offered in the companion text, Environmental Impact Statements, for assisting practitioners in the task of determining the amount of detail, discussion, and analysis that is sufficient to adequately cover a proposed action*

Some aspects of a proposal might be considered sufficiently covered by merely mentioning that the specific action will take place. In others cases, an extensive analysis might be considered necessary. Since NEPA's inception, no definitive direction has been established for determining the amount of detail, discussion, and analysis that is sufficient to adequately cover a proposed action. Yet, agency decisionmakers are routinely called upon to do just that. Inevitably, such determinations tend to be subjective. A tool (Sufficiency-Test) is offered in Chapter 3 of the companion text, *Environmental Impact Statements,* for assisting practitioners in dealing with this issue.[10]

Specific documentation requirements and guidance governing preparation of an EA are provided in the following sections.

6.4.1   Need Section

As indicated in Table 6.3, the Regulations require that the EA indicate the need for taking action. Interestingly, the Regulations do not require that an EA indicate the *purpose* for taking action, as is required for an EIS.

> **Determining the Underlying Need**
>
> *A simple technique that practitioners can use in determining the underlying need for a proposed action simply involves asking the following question: Why are we considering the proposed action?*

In the past, a recurring problem has involved discussing the need for preparing the EA rather than the need for taking action (i.e., proposal). For example, some EAs have mistakenly explained that the need was to comply with NEPA. A simple technique that practitioners can use in determining the underlying need for a proposed action, involves asking the following question: *Why are we considering the proposed action?*

> **The Need for Taking Action**
>
> *A need is something lacking or desired.*

"Need," as defined in Webster's Dictionary, is "a lack of something useful, required, or desired."[11] Put succinctly, a need is something *lacking* or *desired*. As espoused by Schmidt, the statement of the purpose and need is critical in successfully identifying the range of reasonable alternatives for later analysis.[7,12] Properly defined, the statement of need provides an effective tool for screening out what might otherwise be a diverse or even unbounded range of alternatives.

Defining a range of reasonable alternatives — A broadly written statement of need can facilitate agency planning, as it tends to compel consideration of more diverse approaches for satisfying the agency's need. The disadvantage is that it can increase the scope of analysis, because a greater number of alternatives might need to be investigated. If the number of alternatives is unwieldy, this range can be reduced by narrowing the statement of need. However, the reader is cautioned that it may be inappropriate (and possibly illegal) to define the statement of need so narrowly that it essentially eliminates all but the proposed action.

6.4.2 Affected Environment Section

As indicated earlier (see Table 6.3), the Regulations do not require a description of the affected environment in an EA. However, inclusion of an Affected Environment Section can be very useful, as it provides a baseline of the existing environment, against which both the intensity and context of potential impacts can be compared. Lacking such a baseline, it may be difficult or impossible to conclusively demonstrate that the potential impacts are truly nonsignificant.

> **Affected Environment**
>
> *"The environmental impact statement shall succinctly describe the environment of the area(s) to be affected or created by the alternatives under consideration. The descriptions shall be no longer than is necessary to understand the effects of the alternatives. Data and analyses in a statement shall be commensurate with the importance of the impact, with less important material summarized, consolidated, or simply referenced. Agencies shall avoid useless bulk in statements and shall concentrate effort and attention on important issues. Verbose descriptions of the affected environment are themselves no measure of the adequacy of an environmental impact statement."*
>
> **(§1502.15)**

The EA should succinctly describe potentially affected environmental resources. Each resource should be described in only enough detail to allow the reader to understand how it could be affected. Consistent with a sliding-scale approach, such discussions should be commensurate with the potential for sustaining a significant impact. As warranted, the discussion and data may be incorporated by reference (§1502.15).

Preparing the affected environment section — The introduction to the Affected Environment Section should indicate the purpose for which this section has been prepared (i.e., to provide a baseline for gauging impacts). The location or area that would be affected should be clearly delineated on a map.

> **Avoiding a Common Problem**
>
> *A problem commonly encountered in EAs involves mixing descriptions of the affected environment with discussions of potential impacts. The description of the affected environment should be limited to describing the baseline environment before the proposal is implemented — not as it would be affected if the proposal were implemented.*

The discussion should be presented on a resource-by-resource basis, in the same order as resources are evaluated in the section describing the environmental impacts of the proposal.

Interpreting baseline measurements — Where possible, indicators should be provided that can be used in gauging the environmental impacts that will be described later. For example, Table 6.7 indicates hypothetical baseline noise levels (measurements are in equivalent sound level [L_{eq}]) that have been measured at four alternative sitting locations. To assist the reader in interpreting these measurements, Table 6.8 indicates typical noise levels (measured in dBA) encountered in daily life and industry.

TABLE 6.7

Hypothetical Background Noise Levels at Four Alternative Locations

Location	L_{eq}-24 (dBA)
Proposed Location	39
Alternative Location #1	45
Alternative Location #2	34
Alternative Location #3	63

6.4.3   The Proposed Action and Alternatives Section

A generalized outline for describing the section on the proposed action and reasonable alternatives is offered in Table 6.9 (an expansion of Section 2.0 shown in Table 6.3). As necessary, this generic outline should be tailored to meet the agency's specific mission, circumstances, and requirements.

TABLE 6.8

Typical decibel (dBA) Values[13]

Sound	dBA
Rustling leaves	20
Room in a quiet dwelling at midnight	32
Conversational speech	60
Busy restaurant	65
Vacuum cleaner in private residence (at 3.1 meters)	69
Beginning of hearing damage for prolonged exposure	85
Heavy city traffic. Also a heavy diesel-propelled vehicle (at approximately 7.6 meters)	92
Home lawn mower	98
Air hammer	107
Jet airliner (153 meters overhead)	115

TABLE 6.9

General-Purpose Outline for Describing Alternatives

2.0 Description of Proposed Action and Alternatives

2.1 Brief Introduction
    2.1.1 Briefly explain process used to identify alternatives
    2.1.2 Briefly describe alternatives considered but not analyzed

2.2 No-action Alternative

2.3 Proposed Action
    2.3.1 Location
    2.3.2 Cost and schedule
    2.3.3 Construction activities
    2.3.4 Operational activities
    2.3.5 Support facilities and activities
    2.3.6 Routine maintenance and upgrades
    2.3.7 Project termination and decommissioning

2.4 Alternative A
    (Similar to that shown for Section 2.3, although alternatives in an EA are usually
    covered in less detail)

2.5 Alternative B
    •
    •
    •

While a description of the no-action alternative is not specifically required by the Regulations, its inclusion is considered good practice, and it might be required by the courts (as a reasonable alternative).

To the extent practical, the proposed action is described throughout its entire life-cycle.

Emphasis is normally placed on describing the proposed action. Correspondingly, less attention is devoted to describing the reasonable alternatives.

Pollution prevention — The CEQ has issued guidance instructing agencies to take every opportunity to incorporate **pollution prevention** considerations into their planning

> **Incorporating Pollution Prevention**
>
> *An EA should be used as a tool for identifying methods for reducing or preventing pollution.*

process. As appropriate, an EA should be used as a tool for identifying methods for reducing or preventing pollution.[14]

Comparing alternatives — While not required, a comparison of alternatives could be advantageous as it can help foster more informed decisionmaking. The results of such an investigation can provide the decisionmaker with factors for better decisionmaking. Table 6.10 provides a hypothetical comparison of three potential courses of action for obtaining a source of steam and electric power (e.g., co-generation):

1. Construct a small co-generation plant.
2. Renovate an existing plant.
3. Take no action with respect to replacing an existing co-generation plant.

Instead of using simple responses such as "Yes" or "No," many practitioners find it more informative to use quantitative or short descriptive phrases. For example, instead of indicating "Yes" to the question of land disturbance, a response might state, "Grading of .75 acres." Instead of indicating "Yes" to the question of demolition, a response might state "Limited to three office buildings constructed prior to 1950."

TABLE 6.10

Comparison of Plant Alternatives

Succinct Attributes	Construct New Plant	Renovation Alternative	No Action Alternative
Steam production	Yes	Yes	Yes
Electric production	Yes (50 megawatts)	No	No
Fuel type	Gas	Coal	Coal
Construction of fuel pipeline	Yes (.6 miles)	No	No
Highway fuel transport	No	Yes (55 miles)	Yes (55 miles)
New plant site	Yes (5 acres)	No	No
Ash sludge disposal	No	Yes (.5 tons/day)	Yes (.8 tons/day)
Requires new construction	Yes, a new plant	Yes, refurbishing an existing plant	No
Water use	50% reduction	No change	No change
Closure of existing plant	Yes	No	No
Land disturbance	Yes, grading 25 acres of habitat	No	No
Demolition of existing structures	Yes	Yes	No

6.4.4 Environmental Impact Section

It is recommended that the analysis begin with an evaluation of the potentially most important issues or impacts, ending with those of least importance. The order used for describing the impacts to environmental resources should follow the same sequence as that used in describing the resources in the Affected Environment Section.

The discussion of direct and indirect impacts should not be segregated into separate sections. However, because of impact complexity and the special analytical methodologies that might be employed, the cumulative impact analysis is sometimes delegated to a separate section.

In recent years, environmental justice and the analysis of accident scenarios have become issues of special interest. For this reason, these issues are covered in depth in Chapter 7.

Analysis of cumulative impacts — Cumulative impacts are to be considered along with the direct and indirect effects in reaching a determination regarding significance (§1508.27[b][7]). As cumulative impact assessments have increasingly become the focus of litigation in

> **Cumulative Impact Analyses Typically do not Support Decisionmaking**
>
> *Neither the discussion of cumulative impact considerations nor the thoroughness of the analysis is generally sufficient to support informed decisionmaking.*

recent years, the importance of performing an adequate analysis has correspondingly risen. Notwithstanding, the analysis of cumulative impacts is often either neglected or is not afforded adequate attention.

For example, one study reviewed cumulative impact considerations in 30 different EAs prepared for a variety of different projects throughout the U.S. Out of the 30 EAs, the term "cumulative impact" was mentioned in only 14 (47%) of the assessments. Moreover, only 7 (23%) of the EAs discussed cumulative impact in the "Environmental Consequences" section of the EA. When cumulative impacts were mentioned, they were typically addressed in a qualitative manner and lacked a clear delineation of the spatial and temporal study boundaries. The study's authors concluded that neither the discussion of cumulative impact considerations nor the thoroughness of the analysis was generally sufficient to support informed decisionmaking.[15]

The findings of this study are consistent with the conclusions reached in a similar study of 89 EAs. Out of the 89 EAs, only 35 (39%) mentioned the term "cumulative impact," with eight of these assessments stating that there were no cumulative impacts but providing no evidence to support such a conclusion. Five EAs identified potential cumulative impacts but concluded they were nonsignificant without presenting an analysis to support such a conclusion.[16]

In some EAs, the issue of cumulative impacts may be the single most important factor in determining whether an EIS must be prepared. If EAs are to be credibly used in reaching significance determinations, the issue of cumulative

> **The Importance of Evaluating Cumulative Impacts**
>
> *In some EAs, the issue of cumulative impacts can be the single most important factor in determining whether an EIS must be prepared.*

impacts must be addressed in a more rigorous and systematic manner.

Analysis of cumulative impacts is covered in detail in Chapter 7.

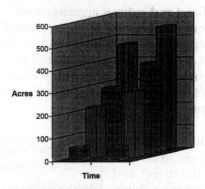

■ No Action	60	60
■ Proposed Action	220	290
■ Alternative 1	290	400
■ Alternative2	470	550

FIGURE 6.2
Acres of disturbed habitat.

Describing impacts — Agencies must make a concerted effort to "...disclose and discuss...all major points of view..." (§1502.9[a]). Graphics can be particularly useful in conveying pertinent information to the decisionmaker and public. For example, the bar chart shown in Figure 6.2 shows the amount of habitat that would be disturbed with respect to four different courses of action. The first row of bars shows the habitat disturbance that would result during the construction phase of a proposed military training facility. The second row of bars depicts the disturbance that would result during the operational phase.

> **Describing Impacts**
>
> *Instead of simply stating that a resource "would be impacted," the analysis must indicate* how *environmental resources would be affected.*

Instead of simply stating that a resource "would be impacted," the analysis must indicate *how* and to what extent environmental resources would be affected. For instance, it is much more informative to describe how much and to what degree a critical habitat would be perturbed than to simply state (as is sometimes the case) that the "...habitat would be impacted." Where possible, impacts should be quantified using an appropriate unit of measurement. Potentially affected resources or populations should be clearly described. The *time and period* over which the impact would occur should be indicated, as should its *likelihood* or *probability*. It is usually not necessary to categorize each impact as "short-term" or "long-term," or "direct" or "indirect." Selected topics that necessitate special elaboration are described in the following sections.

Site-specific Analyses: Case law indicates that the impact analysis must be site-specific. Lack of data on sensitive or endangered species at a site does not imply that the site has been studied and found to lack such species. Almost any EA involving some type of

> **Avoiding Common Problems**
>
> *In lieu of performing a site specific field study, some EAs have attempted to dismiss potential impacts or resources with a casual statement such as "Sensitive or endangered species are not known to exist in the location of the proposed site." Lacking a specific survey, such statements, by themselves, do not provide evidence sufficient to demonstrate that a significant impact would not occur.*

construction that could affect naturally vegetated areas should include a site visit by a biologist. Similar investigations might be required where the proposal could affect other environmental resources as well.

Conformance with Regulatory Standards: Conformance with regulatory standards does not necessarily ensure that an action would not significantly affect an environmental resource. A statement such

> **Conformance with Laws and Regulations Does Not Prove Nonsignificance**
>
> *Conformance with all applicable laws and regulations does not prove that impacts would be nonsignificant.*

as, "The activity would be conducted in accordance with all applicable regulatory requirements ...," should not be used as the sole or even primary source of evidence of nonsignificance. An action could comply with all applicable laws and regulations and still result in a significant environmental impact

For example, an impact to a wetlands that qualifies for an exemption or authorization under a general permit to Section 404 of the Clean Water Act is not necessarily nonsignificant in the context of NEPA. Certain impacts to wetlands, such as atmospheric emissions, noise, and even some types of draining, might not

> **Regulatory Compliance does not Necessarily Prove Nonsignificance**
>
> *A statement such as, "The activity would be conducted in accordance with all applicable regulatory requirements ...," should not be used as the sole or primary source of evidence of nonsignificance.*

even be regulated under Clean Water Permitting Programs.

As another example, shipment of radioactive waste might comply with all applicable laws and regulations. Yet, the fact that such a shipment would comply with all regulatory standards does not necessarily guarantee that a significant impact could not occur (e.g., the transportation truck has an accident, or the cargo ship strikes another ship and sinks).

Conformity Determinations with Air Quality Standards: Under the NEPA Regulations, a determination of significance requires considering "whether the action threatens a violation of federal, state, or local law, or requirements imposed for the protection of the environment" (§1508.27[b][10]). A new directive by the Environmental Protection Agency (EPA) establishes specific direction for reviewing potential violations of existing State Implementation Plan (SIP) conformity criteria under the National Ambient Air Quality Standards (NAAQS).[17] Specifically, this directive provides a mechanism for reviewing and enforcing criteria pollutant standards (i.e., carbon monoxide, lead, nitrogen dioxide, ozone, particulate matter, and sulfur dioxide).

Pursuant to EPA's direction, conformity determinations for proposed actions are to be integrated with the NEPA process. EPA's directive defines exemptions and the specific circumstances under which conformity must be addressed in a NEPA document. Where applicable, an EA might need to consider direct and indirect emissions in evaluating conformity with SIP criteria pollutant standards.

Mitigation measures — A description of proposed mitigation measures is frequently addressed in the environmental impact section of an EA. Alternatively, mitigation measures can be described and incorporated as an integral element of the proposed action; analysis of such measures and their effectiveness in mitigating potential impacts should still be relegated to the environmental impact section of the EA.

It may also be advantageous to summarize the description of the mitigation measures and their effectiveness in reducing impacts in a separate stand-alone section of the EA. Such a section provides the reader with an easily assessible source of information on the measures and their effectiveness, and facilitates incorporation of mitigation commitments into the Finding of No Significant Impact (FONSI).

Statements such as, "A mitigation plan will be developed later in cooperation with the EPA," do not provide sufficient evidence for a decisionmaker to determine whether significant impacts can be mitigated to the point of non-significance. A well planned and defensible mitigation plan should contain five essential ingredients depicted in Table 6.11.

TABLE 6.11

Five Elements Essential in a Mitigation Plan

1. Specific and detailed measures
2. Specific schedule
3. Appropriate funding
4. Measurable performance criteria
5. Assignment of responsibility for implementing such measures

---

The Effectiveness of Mitigation Measures Must be Analyzed

Where mitigation measures are incorporated, the EA analysis must evaluate their effectiveness in reducing impacts to the point of nonsignificance.

Analyzing Effectiveness of Mitigation Measures: Without an analysis of the effectiveness of mitigation measures, it is virtually impossible to determine whether the potential impacts would be reduced to the point of nonsignificance. Failure to conclusively demonstrate the effectiveness of potential mitigation measures in reducing impacts below the level of significance can leave an agency wide open to a legal challenge.

Comparing impacts among alternatives — While not required, a comparison of impacts that would result from pursuing various courses of action can be advantageous as it can assist the agency in determining an optimum course of action. The results of such an investigation can provide the decisionmaker with factors that can foster more informed decisionmaking. Table 6.12 provides a similar example in which cultural resource impacts associated with different alternatives have been compared for a hypothetical proposal involving construction of a pipeline.

Table 6.13 provides an example in which the incremental change in pollutant concentrations that would result from four hypothetical cogeneration plants are compared against one another. As indicated earlier, many practitioners find it more informative and less judgmental to use quantitative or short descriptive phrases where feasible.

6.4.5   Listing Permits, Approvals and Regulatory Requirements

A common method for determining significance involves assessing an action for conformance with applicable environmental requirements (§1508.27[b][10]). A section briefly (*see* Section 5.0 in Table 6.4) listing applicable federal, state, local, and tribal regulatory approvals and requirements can assist the decisionmaker in assessing the potential for significant impacts.

Such a list can also facilitate agency planning by identifying future requirements during the early planning stage. As an example, Table 6.14 compares federal, state, local, and tribal permits, licenses, and approvals required for three different alternatives.

6.4.6   Listing of Agencies and Persons Consulted

As indicated by the fourth item in Table 6.3, an EA is required to list agencies and persons consulted in preparing the analysis (e.g., U.S. EPA, local air pollution control boards). One of the purposes of this requirement is to publicly indicate the degree to which agencies have conducted a thorough and interdisciplinary process. Consultation is also an important regulatory mandate that facilitates a more thorough analysis and is useful in identifying related permitting requirements.

TABLE 6.12

Comparison of Potential Cultural Resource Sensitivities for a Hypothetical Proposal Involving Four Alternative Pipeline Corridors

Alternatives	Amount previously surveyed	Documented historic resources	Documented prehistoric resources	Degree disturbed	Amount of sensitive landforms present	Sensitivity rating
Proposed action	1.2 acres	1	1	Minimal	Moderate	Moderate
Alternative 1	1.3 acres	0	1	Low to moderate	Minimal	Low
Alternative 2	2.5 acres	2	1	Moderate	High	High
No action	3.4 acres	2	0	Low to moderate	Minimal	Low

TABLE 6.13

Incremental Pollutant Concentrations (ug/m^3) that Would Result From Four Alternative Cogeneration Plants

Pollutant	Averaging Period	Alternative 1 (small gas-fired plant)	Alternative 2 (large gas-fired plant)	Alternative 3 (small coal-fired plant)	Alternative 4 (large coal-fired plant)
NO2	Annual	0.02	0.08	0.03	0.13
SO2	1-hour	0.12	0.33	17.52	73.80
	3-hour	0.04	0.15	5.50	26.9
	24-hour	0.01	0.02	0.45	1.80
	Annual	0.01	0.01	0.03	0.13
CO	1-hour	3.1	15.70	26.44	106.3
	8-hour	0.02	1.01	1.81	6.8
Particulate matter PM$_{10}$ (10 microns)	24-hour	0.01	0.09	1.97	7.9
	Annual	0.01	0.01	0.03	0.11
Total Suspended Particles (TSP)	24-hour	0.01	0.09	3.85	16.3
	Annual	0.00	0.01	0.05	0.23

TABLE 6.14

Example of a Matrix Containing Permits, Licenses, and Approvals That Would Need to be Obtained for the Evaluated Alternatives

Proposed Action	Alternative #1	Alternative #2
RCRA Part B Permit	RCRA Part B Permit	RCRA Part B Permit
State Materials License No. 10170-A & 10303-B	State Hazardous Waste Facility Siting Permit	State Hazardous Waste Facility Siting Permit
State Broker's User Permit No. C1011-2	State Hazardous Waste Management Act Permit	State Air Pollution Control Permit
State Generator's User Permit No. H1043-2B	Health Department Construction and Operation Permit Implementing the CAA	State New Contaminant Source Construction Permit
State approval for PCB thermal treatment (TSCA)	State approval for PCB thermal treatment (TSCA)	State approval for PCB thermal treatment (TSCA)
Transporter Permit No. 02-000-1139-1C	State transportation permit, license, and certificate	N/A
County Temporary Air-Quality Permit	State transportation permit	State transportation permit
Transportation Department certification	Transportation Department certification	Transportation Department certification

References

1. DOE, *NEPA Lessons Learned*, Improving EIS Readability, Issue No. 12, p 14, September 2, 1997.
2. Sullivan W. C., F. E. Kuo, and M. Prabhu, Assessing the impact of environmental impact statements on citizens, *Environmental Impact Assessment Review*, 16(3):171–182, May 1996.
3. Sullivan, W. C., F. E. Kuo, and M. Prabhu, Communicating with citizens: the power of. photosimulations and simple editing. *Environmental Impact Assessment Review*, 17(4):295–310, July 1997.
4. Sullivan, W C., FE. Kuo and M. Prahu, *Environmental Impact Assessment Review*, Vol. 17(4), Communicating with Citizens: The Power of Photosimulations and Simple Editing. pp. 295–310, Elsevier Science Inc, July 1997.
5. CEQ, Council on Environmental Quality - Forty Most Asked Questions Concerning CEQ's National Environmental Policy Act Regulations (40 CFR 1500–1508), *Federal Register* Vol. 46, No. 55, 18026–18038, March, 23, 1981, Question number 36a.
7. Eccleston C.H., *Environmental Impact Statements: A Comprehensive Guide to Project and Strategic Planning*, John Wiley & Sons, New York, N.Y., pp 92–97, 2000.
8. Section 102(2)(a) of NEPA.
9. §1508.25[a].

10. Eccleston C. H., *Environmental Impact Statements: A Comprehensive Guide to Project and Strategic Planning,* John Wiley & Sons Inc, New York, N.Y., 2000; Eccleston C. H., *The NEPA Planning Process: A Comprehensive Guide with Emphasis on Efficiency,* John Wiley & Sons Inc, 1999.
11. *Webster's New Twentieth Century Dictionary,* unabridged second edition, Simon and Schuster, 1983.
12. Schmidt O. L., The Statement of Underlying Need Determines the Range of Alternatives in an Environmental Document, The Scientific Challenges of NEPA: Future Directions Based on 20 Years of Experience, Session 13- The NEPA Process, Knoxville Tennessee, October 25–27, 1989; Schmidt O. L., The Statement of Underlying Need Defines the Range of Alternatives in Environmental Documents, 18 *Environmental Law* 371–81, 1988.
13. Newman and Beattie, 1985.
14. CEQ, Guidance on Pollution Prevention and the National Environmental Policy Act, 58 *Federal Register,* 6478, January 29, 1993.
15. Burris R. K. and L. W. Canter, Cumulative Impacts are not Properly Addressed in Environmental Assessments, *Environmental Impact Assessment Review,* Elsevier Sciences Inc., 1997;17, pp 5–18.
16. Mcold L., and Holman J., 1995. Cumulative Impacts in Environmental Assessments: How Well are they Considered?, *The Environmental Professional,* 17(1):2–8.
17. EPA, Final rule for Determining Conformity of General Federal Actions to State or Federal Implementation Plans, 58 *Federal Register* 63214, No. 228, November 30, 1993, and took effect on January 31, 1994 (40 CFR parts 6, 51, and 93)

7

Special Topics
in Environmental Analysis

> Truth is stranger than fiction — to some people, but I am measurably familiar with it.
>
> Mark Twain

Guidance presented in this chapter is generally applicable to the preparation of both EAs and EISs. This guidance presumes that the analyst has the appropriate technical knowledge and skills for performing such analyses.

Section 7.1 provides direction on performing a cumulative impact analysis that meets analytical and regulatory requirements; guidance on addressing cumulative impacts to minority and low-income populations is also presented in Section 7.3. Section 7.2 provides an overview of guidance for performing an accident analysis in both EAs and EISs, while Section 7.3 describes the relatively new issue of environmental justice.

7.1 Cumulative Effects Analysis in EAs

The combined incremental effects of human activity pose a serious threat to environmental quality. While such impacts may be nonsignificant by themselves, over time they can compound, from one or more sources, to the point where they significantly degrade environmental resources. The analysis of cumulative effects provides a powerful tool for taking such effects into account as federal officials plan future actions.

In recent years, the issue of cumulative impacts has become the subject of increasing interest and litigation. This section has been prepared to assist practitioners in preparing a cumulative impact analysis that meets analytical and regulatory requirements. For a more in-depth discussion of the analytical and regulatory requirements, the reader is referred to the companion texts, *The NEPA Planning Process,* and *Environmental Impact Statements.*[1]

A "cumulative impact" is defined as "...the impact on the environment that results from the incremental impact of the action when added to other past, present, and reasonably foreseeable future actions, regardless of which agency (federal or non-federal) or person undertakes such other actions. Cumulative impacts can result from individually minor but collectively significant actions taking place over a period of time"

(§1508.7).

Definition of a Cumulative Impact

A cumulative impact assessment

- *Involves consideration of the incremental impact of the action when added to other (1) past, (2) present, and (3) reasonably foreseeable future actions.*

- *Involves consideration of both federal or non-federal actions.*

- *Can result from individually minor but collectively significant actions taking place over a period of time.*

7.1.1 Purpose of the Analysis

By mandating an analysis of cumulative impacts, the regulations ensure that the range of actions considered in National Environmental Policy Act (NEPA) documents takes into account not only the project proposal but also other actions that could cumulatively harm environmental quality. The results of the cumulative impact analysis should be incorporated into the agency's overall environmental planning. The conclusions can also be incorporated into the plans of state and private entities. To this end, federal agencies are to use results obtained from the cumulative impact analysis as a tool for evaluating the implications of a proposal in even **project-specific Environmental Assessments** (EAs).[2]

Focusing on What's Important

Actions affecting resources that have sustained cumulatively insignificant impacts do not add value to the cumulative impact analysis.

Emphasis must be placed on identifying activities occurring outside the jurisdiction of their respective agencies. However, the analysis should be limited to considering those past, present, and future actions that incrementally contribute to the cumulative effects on resources affected by the proposed action; actions affecting resources that have sustained cumulatively insignificant impacts generally do not add to the value of the analysis.[3]

7.1.2 Performing the Analysis

This section briefly summarizes the process used in performing a cumulative impact analysis.

Components of an adequate cumulative impact analysis — A cumulative effects study must identify:[4]

- The specific area in which effects of the proposed project would be felt
- Impacts that are expected in that area from the proposal
- Other past, proposed, and reasonably foreseeable future actions that have had or could be expected to impact the same area
- Expected impacts from these other actions
- Overall expected impact if the individual impacts were allowed to accumulate

In one case, a court deferred judgment in favor of the U.S. Forest Service, finding that it had considered effects of a timber sale in the context of past and reasonably foreseeable logging; the agency had constructed mathematical models and had performed extensive field investigations to calibrate and verify its models. It had also actively sought public comment.[5]

Beginning the cumulative impact analysis — To ensure the inclusion of resources that are most susceptible to degradation, cumulative impacts can be anticipated by considering where cumulative effects are likely to occur and what actions would most likely produce cumulative effects. In initiating a cumulative impact analysis, practitioners should:[6]

- Determine the area that would be affected.
- Make a list of the resources within that zone that could be affected by the proposed action.
- Determine the geographic areas occupied by those resources outside of the project impact zone.

Disregarding Future Actions

Future actions can generally be disregarded if[7]

- They lie outside the geographic boundaries or time frame established for the cumulative effects analysis.
- They will not affect resources that are the subject of the cumulative effects analysis.
- Their inclusion in the analysis is considered to be arbitrary (i.e., lacks a logical basis for inclusion).

The courts have struggled with the problem of determining when future actions can be disregarded as being "remote or speculative" versus those that must be analyzed. In one case, a court concluded that an EA prepared for mining operations did not need to consider the cumulative impact of other planned mines. This decision was premised on the fact that there was no practical commitment to future mining operations. The court concluded that a NEPA analysis must generally consider impacts of other proposals "... only if the projects are so interdependent that it would be unwise or irrational to complete one without the others."[8]

Evaluating cause-and-effect relationships — Determining how a particular resource responds to an environmental disturbance is essential in determining the cumulative effect of multiple actions. Analysts must therefore gather information about cause-and-effect relationships. Once all the important cause-and-effect pathways are identified, the analyst determines how the environmental resource responds to a potential disturbance. Cause-and-effect relationships for each resource are used in computing the cumulative effect resulting from all actions that will be evaluated.

> **Process for Computing Cumulative Impacts**
>
> *Analysts typically determine the separate effects of the proposed action, and past, present and reasonably foreseeable future actions. The cumulative effect can then be calculated once each group of effects is determined.*

Typically, analysts will determine the separate effects of the proposed action, and other past, present, and reasonably foreseeable future actions. The cumulative effect can then be summated once each group of effects is determined.

Performing a qualitative analysis — A cumulative impact analysis is sometimes limited to a qualitative evaluation because the cause- and-effect relationships are poorly understood. In still other cases, there may not be sufficient site-specific data available to permit a quantitative analysis. Faced with such constraints, analysts might want to consider performing a qualitative analysis in a manner similar to that shown in the following example (Table 7.1). If no numbers are available, the analyst may categorize the magnitude of effects using qualitative descriptors such as "high," "medium," or "low."

7.1.3   Considering Related Actions

> **Related Actions**
>
> *Once the cumulative impact of a number of related actions crosses the threshold of significance, a discussion of those cumulative effects in individual EAs no longer complies with NEPA.*

In one case, an agency was sued for preparing individual EAs on separate mining claims that involved a cumulatively significant impact. The court concluded that an Environmental Impact Statement (EIS) was necessary when a number of related actions cumulatively have a significant environmental impact, even if the separate

TABLE 7.1

Example of a Cumulative Effects Analysis Using Quantitative Descriptions

Environmental resources	Past actions	Present actions	Proposed actions	Future actions	Cumulative impacts
Vegetation	30% of presettlement vegetation lost	1% of vegetation lost this year	3% of existing vegetation would be lost	1% of vegetation lost yearly for next 15 years	49% of presettlement vegetation lost over next 15 years
Wetlands	30% of presettlement vegetation lost	1% of wetlands lost this year	9% of existing wetlands would be lost	3% of wetlands lost yearly for next 15 years	85% of presettlement wetlands lost over next 15 years
Turtles	20% of presettlement turtles lost	2% of turtles lost this year	6% of existing turtles would be lost	3% of turtles lost annually for next 15 years	73% of presettlement turtle population lost over next 15 years

actions, by themselves, would not. In the words of the court, "...once the cumulative impact of a number of mining claims crosses the threshold of [a] significant effect on the environment, a discussion of those cumulative effects in individual EAs no longer complies with NEPA."[9]

Actions on private lands — Case law indicates that unrelated actions on private lands must still be considered. In one case, the Eighth Circuit Court ruled that an EA must consider the impacts of activities reasonably expected to occur on private lands.[10]

> **Actions on Private Lands**
>
> *An EA must consider the impacts of activities reasonably expected to occur on private lands.*

7.1.4 Considering Connected Actions

In 1988, the U.S. Forest Service was challenged for preparing nine separate EAs on connected actions. In reviewing the case, the court found that the plaintiffs had raised serious questions as to whether these timber sales would result in a cumulatively significant impact. The court found that the agency's Findings of No Significant Impact (FONSI) were inappropriate because the EAs did not adequately address connected actions and the cumulative effects of proposed and contemplated actions. The court

> **Connected Actions**
>
> *Connected actions, in addition to other reasonably foreseeable future actions, can result in a cumulatively significant impact requiring preparation of an EIS.*

concluded that the scope of these connected actions were broad enough so as to require preparation of an EIS.[11]

In the same year, another proposed action was challenged on similar grounds. The EA failed to evaluate the cumulative effects of connected actions involving reconstruction of a 17-mile segment of a 70-mile road, as well as other segments of the road reconstruction project, related timber sales that justified the entire project, and other reasonably foreseeable future actions. The court found that the connected actions, in addition to other reasonably foreseeable future actions, could result in a cumulatively significant impact. This was because there was an inextricable nexus between the logging operations and the road construction. The court concluded that the EA failed to evaluate the ongoing and future timber harvest and the road reconstruction.[12]

One of the most important cumulative impact cases involved the U.S. Fish and Wildlife Service. The agency had prepared other independent documents indicating that related and cumulative impacts might be leading to aquatic habitat degradation. Such degradation was unaccounted for in the individual EAs that it had prepared. The court found that lack of an overall effort to evaluate cumulative impacts could result in detrimental effects on the recovery of the wolf population. This was sufficient to raise serious questions regarding whether the road and the timber sales would result in a significant cumulative impact. The agency was ordered to prepare an EIS to analyze such effects.[13]

7.1.5   Differences in Cumulative Impact Analyses Between EAs and EISs

An agency bears the burden of proof in demonstrating that no significant impact would result from a proposed action. In determining whether an EIS needs to be prepared, the analysis of cumulative effects in an EA may sometimes require more rigor than is generally called for in an EIS.

> **Analysis in an EA vs. that in an EIS**
>
> *Using EA to determine whether there are significant impacts that might sometimes necessitate a broader analysis of cumulative impacts than is normally neeeded in an EIS.*

The Fifth Circuit Court decision — The Fifth Circuit Court concluded that the question of determining when an EIS is required may necessitate a broader analysis of cumulative impacts than is generally necessary in an EIS. According to the court, an EA "should consider (1) past and present actions without regard to whether they themselves triggered NEPA responsibilities, and (2) future actions that are reasonably foreseeable, even if they are not yet proposals and may never trigger NEPA-review requirements."[14]

Specifically, the Fifth Circuit found that

> "...although cumulative impacts may sometimes demand the preparation of a comprehensive EIS, only the impacts of proposed, as distinguished from contemplated, actions need be considered in scoping an EIS. In a case like this one, on the other hand, where an EA constitutes the only environmental review undertaken thus far, the cumulative impacts analysis plays a different role..."

> "[The NEPA Regulations] require an analysis, when making the NEPA-threshold decision, [to determine if] it is reasonable to anticipate cumulatively significant impacts from the specific impacts of the proposed project...[W]hen deciding the potential significance of a single proposed action ... a broader analysis of cumulative impacts is required. The regulations clearly mandate consideration of the impacts from actions that are not yet proposals and from actions – past, present, or future – that are not themselves subject to the requirements of NEPA."

The court cautioned that it did not mean to imply that "...consideration of cumulative impacts at the threshold stage will necessarily involve extensive study or analysis of the impacts of other actions." Instead, the court emphasized that the EA analysis should be limited to determining whether "...the specific proposal under consideration may have a significant impact."

> **Cumulative Impact Analysis in an EA**
>
> *Consideration of cumulative impacts in an EA does not necessarily involve an in-depth and extensive analysis of the impacts of other actions. Instead, this analysis should focus on determining whether the specific proposal under consideration might have a significant cumulative impact.*

The court went on to state that, at a minimum, an EA must demonstrate that the agency considered impacts from "past, present, and reasonably foreseeable future actions regardless of what agency, (federal or non-federal), or person undertakes such other actions."

> **Extent of the Analysis**
>
> *The extent of the cumulative impact analysis depends on the scope of the affected area and the extent of other past, present, and future activities.*

According to the court, the extent of the analysis depends on the scope of the affected area and the extent of other past, present, and future activities.

7.2 Accident Analyses

A NEPA analysis might need to address the effects of a potential accident so as to inform the decisionmaker and public about reasonably foreseeable adverse consequences associated with the proposal. The term "reasonably foreseeable" extends to events that might have catastrophic consequences, even if their probability of occurrence is low, provided that the analysis of the impacts is supported by credible scientific evidence, is not based on pure conjecture, and is within the rule of reason (§1502.22).

With the exception of 40 Code of Federal Regulations §1502.22, the Council on Environmental Quality (CEQ) has not issued detailed guidance for addressing accident analyses. This section is intended to fill this gap, as it provides guidance for performing an accident analysis that is generally applicable to the preparation of both EAs and EISs.[15]

Definition of an Accident

For the purposes of NEPA, an accident can be viewed as an unplanned event or sequence of events that results in undesirable consequences. Accidents can be caused by equipment malfunction, human error, or natural phenomena.

For the purposes of NEPA, an accident can be viewed as an unplanned event or sequence of events that results in undesirable consequences. Accidents can be caused by equipment malfunction, human error, or natural phenomena.

7.2.1 Sliding Scale

Sliding-Scale Approach

Use a sliding-scale approach in determining:

1. *Whether an accident analysis is appropriate*

2. *The degree of effort that should be expended in performing such an analysis*

Consistent with the principle that impacts be discussed in proportion to their significance (§1502.2[b]), analysts should use a sliding-scale approach in determining whether an accident analysis is appropriate, as well as the degree of effort that should be expended in performing such an analysis. Practitioners must apply professional judgment in determining the appropriate scope and analytical requirements. For example, practitioners need to determine the appropriate range and number of accident scenarios to consider, the level of analytical detail, and degree of conservatism that should be applied. A sliding-scale approach is particularly useful in making these determinations (Table 7.2).

TABLE 7.2

Key Factors to Consider in Applying a Sliding-Scale Approach to an Accident Analysis

---

- Severity of the potential accident impacts in terms of the estimated consequences

- Probability of occurrence and overall risk

- Context of the proposal

- Degree of uncertainty regarding the analyses

- Level of technical controversy regarding the potential impacts of the proposal

---

The term "risk" used in the second bullet in Table 7.2 can be used to express the general concept that an adverse effect could occur. However, in quantitative assessments, it is more commonly understood to refer to the numeric product of the probability and consequences. "Risk" is used in the latter way in this text.

> **Definition of Risk**
>
> *Risk is most commonly understood to refer to the numeric product of the probability and consequences.*

7.2.2 Overview

An accident is an event or sequence of events that is not intended to happen, and indeed might not happen during the course of operations. The probability that a given accident will occur within a given time frame, however, can be estimated. The probability of occurrence is expressed by a number between 0 (no chance of occurring) and 1 (virtually certain to occur). Alternatively, instead of a probability of occurrence, one can specify the frequency of occurrence (e.g., once in 200 years, which also can be expressed as 0.005 times per year).

An accident scenario is the sequence of events, starting with an initiator, that makes up the accident. It is important to distinguish the probability (or frequency) of the accident initiator from that of the entire scenario; the latter quantity is of primary interest in NEPA accident analyses as it expresses the chance (or rate) that the environmental consequences will occur.

As used in this chapter, the environmental consequences of an accident are the effects on human health and the environment. In discussing an accident's effects on human health, it is both conventional and adequately informative to consider three categories of people: involved workers, noninvolved workers, and the general public. For each of these categories, effects should be evaluated for the maximally exposed individuals in these categories and the collective harm to each population; for example, this might involve identifying and quantifying, as appropriate, potential health effects (e.g., number of latent cancer fatalities).

> **Evaluating Populations and Resources**
>
> • *Evaluate impacts for involved workers, noninvolved workers, and the general public.*
>
> • *Evaluate the effects for the maximally exposed individuals in these categories and the collective harm to each population.*
>
> • *Evaluate the environmental impacts on biota and environmental media, such as land and water.*

In the context of analyzing accidents, the environment includes biota and environmental media, such as land and water, which can become contaminated as the result of an accident. The following guidance refers to effects on biota as ecological effects.

7.2.3　Accident Scenarios and Probabilities

This section provides guidance on addressing accident scenarios and probabilities.

Low-Probability High-Consequence Accidents

As appropriate, accident scenarios should represent the range, or "spectrum," of reasonably foreseeable accidents, which can include both low-probability/ high-consequence accidents and higher-probability/ (usually) lower-consequence accidents.

Range of accident scenarios — Development of realistic accident scenarios that address a reasonable range of event probabilities and consequences is the key to an informative accident analysis. The set of accident scenarios considered should serve to inform the decisionmaker and public of the overall accident risks associated with a proposal. As appropriate, accident scenarios should represent the range, or "spectrum," of reasonably foreseeable accidents, which can include both low-probability/high-consequence accidents and higher-probability/(usually) lower-consequence accidents.

Remember that the purpose for preparing an accident analysis in an EA is different from that of an EIS. In an EA, the purpose of the accident analysis is to determine whether a significant impact could result, requiring preparation of an EIS. Thus, the EA analysis normally focuses on the accident that could result in the maximum reasonable consequences. In contrast, an EIS seeks to explore a range of different accidents and consequences that will assist the decisionmaker in choosing among various alternatives.

Accident Analyses in an EA

Where there is a potential for significant consequences, an EA normally analyzes the maximum reasonably foreseeable accident(s) that represent potential scenarios at the high-consequence end of the spectrum. A maximum reasonably foreseeable accident is usually an accident with the most severe consequences that can reasonably be expected to occur for a given proposal. Such accidents usually have very low probabilities of occurrence.

Thus, where there is a potential for significant consequences, an EA normally focuses on the maximum reasonably foreseeable accident(s) that represent potential scenarios at the high-consequence end of the spectrum. A maximum reasonably foreseeable accident is usually an accident with the most severe consequences that can reasonably be expected to occur for a given proposal. Such accidents usually have very low probabilities of occurrence. Note that a maximum reasonably foreseeable accident is not the same as a "worst-case" accident, which almost always include scenarios so remote or speculative that they are not reasonably foreseeable. Analysis of worst-case accidents is not required under NEPA.

An accident analysis does not necessarily end here. Accidents in the middle of the spectrum might also need to be evaluated, as they often contribute to, or even dominate, the overall accident risks. An exception to this guidance might involve circumstances where the consequences of the maximum reasonably foreseeable accident are small. In that case, analyzing only

> **Range of Accident Scenarios**
>
> • *As appropriate, analyze the consequences of maximum reasonably foreseeable accidents.*
>
> • *As appropriate, analyze accidents in the middle of the spectrum, in most cases.*
>
> • *"Bounding" approaches may tend to mask differences among alternatives.*

the maximum reasonably foreseeable accident would provide sufficient information regarding the overall accident risks of the proposal.

Equally important, a "bounding" approach that considers only the maximum reasonably foreseeable accident might not adequately represent the overall accident risks associated with the proposal. Further, bounding analyses might not enable a reasoned choice among alternatives and appropriate consideration of mitigation, because they tend to mask real differences among the alternatives.

Scenario probabilities — Accident scenarios can involve a series of events for which an initiating event is postulated. The initiating event would be followed by a sequence of other events or circumstances that result in adverse consequences. If these secondary events always occur when the initiator occurs (i.e., the secondary events have a probability of 1 given that the initiator occurs), then the probability (or frequency) of the entire accident scenario is that of the initiator. Otherwise, the scenario probability would be the product of the conditional probabilities of the individual events.

Risk — It is generally insufficient to simply present the reader with the "risk" of an accident (calculated by multiplying the probability of occurrence times the consequence). Pre-

> **Presenting Risk**
>
> *Present the estimated accident consequences and probabilities separately.*

senting only the product of these two factors masks their individual magnitudes. Accordingly, risk should augment and not substitute for the presentations of both the probability of occurrence and the consequence of the accident.

Conservatisms — Practitioners must exercise professional judgment in determining the appropriate degree of conservatism to apply. Preparers should consider the fundamental purposes of the

> **Accident Analyses Should be Realistic**
>
> *Accident analyses should be realistic enough to be informative and technically defensible.*

analysis (e.g., purpose of an EA versus that of an EIS), the degree of uncertainty regarding the proposal and its potential impacts (see further discussion of uncertainty below), and the degree of technical controversy. In short, accident analyses should be realistic enough to be informative and technically defensible.

Using Realistic Approaches
• *Scenarios based on pure conjecture are to be avoided.*
• *Avoid being overly conservative, in compounding conservatisms; evaluating a scenario by using multiple conservative values of parameters will yield unrealistic results.*

Scenarios based on pure conjecture are to be avoided (§1502.22). Exercise good judgment in compounding conservatisms - evaluating a scenario by using multiple conservative values of parameters can yield unrealistic results. For example, in air dispersion modeling, it is nearly always unrealistic to assume only extremely unfavorable meteorologic conditions (i.e., stable or 95% most unfavorable conditions). In many cases, it would be appropriate to estimate and present accident consequences based on both neutral (50%, such as Pasquill Stability Class D) and unusually stable (95%, such as Pasquill Stability Class F) conditions. It is generally inappropriate, however, to assume only the most severe conditions for an otherwise appropriate and credible accident scenario and then fail to analyze the scenario because, by taking into account the lower probability associated with the stable atmospheric conditions, the overall probability is judged to be not reasonably foreseeable. Similarly, using estimates of plume centerline concentrations might be appropriate for evaluating impacts to maximally exposed individuals, whereas average plume concentrations would yield more realistic results for population impacts.

7.2.4   Accident Consequences

Guidance for addressing the consequences of an accident is provided in the following sections.

Involved and non-involved workers — Non-involved workers are those who would be within the vicinity of the proposed action, but not involved in the action. Any potential impacts to non-involved workers should always be considered as part of an accident analysis.

Impacts to any involved workers should also be evaluated as part of an accident analysis. For example, fatal or serious non-fatal injuries might be expected because of a worker's close proximity to the accident. In some cases, credibly estimating exposures for involved workers can require more details about an accident than could reasonably be foreseen or meaningfully modeled. As a substitute, the effects can be described semi-quantitatively or qualitatively, based on the likely number of people who would be involved and the general character of the accident scenario. For example, a qualitative analysis might indicate

> "seven workers would normally be stationed in the room where the accident could occur. While a few such workers might escape the room in time to avoid being seriously harmed, several would likely die within hours from exposure to toxic substances, and the exposed survivors might have permanent debilitating injuries, such as persistent shortness of breath."

A more detailed, semi-quantitative discussion might be appropriate for analyzing proposals with substantially greater risks.

Uncertainty — A decisionmaker needs to understand the nature and extent of uncertainty in choosing among alternatives and considering potential mitigation measures. Thus, where uncertainties preclude quantitative analysis, the unavailability of relevant information should be explicitly acknowledged. The NEPA document should describe the analysis that is used, and the effect that the incomplete or unavailable information has on the ability to estimate the probabilities or consequences of reasonably foreseeable accidents (§1502.22).

Uncertainty
• *Explain the nature and relevance of uncertainties in presenting the results of accident analyses.*
• *Describe the effect of incomplete or unavailable information on the results of accident analyses.*

Based on the prevailing circumstances, practitioners can compensate for analytical uncertainty by using conservative or "bounding" approaches that tend to overestimate potential impacts (*see* the earlier section that discusses some of the pitfalls of such approaches). In other circumstances, such as where substantial uncertainty exists regarding the validity of estimates, a qualitative description may suffice. In all cases, however, the NEPA document should explain the nature and relevance of the uncertainty.

Regardless of whether a qualitative or quantitative analysis is performed, references supporting scenario probabilities, and other data and assumptions used in the accident analysis, should be provided.

Key Data and Assumptions
Provide references for key data and assumptions used in accident analyses.

7.2.5 Intentionally Destructive Acts

A NEPA document might need to address potential environmental impacts that could result from intentionally destructive acts (i.e., acts of sabotage or terrorism). Intentionally destructive acts are not accidents per se.

Analysis of such acts (fire, explosion, missile, or other impact force) poses a challenge because the potential number of scenarios is limitless and the likelihood of attack is unknowable. Nevertheless, the physical effects of such destructive acts are generally similar to, or "bounded" by, the effects of accidents. For this reason, where intentionally destructive acts are reasonably foreseeable, a qualitative or semi-quantitative discussion of the potential consequences of intentionally destructive acts should be included in the accident analysis.

Intentionally Destructive Acts
• *Intentionally destructive acts are generally the same as, or "bounded" by the effects of, accidents.* • *Where intentionally destructive acts are reasonably foreseeable, a qualitative or semi-quantitative discussion might be all that is necessary for addressing such acts.*

The following is an example of a qualitative discussion that might be appropriate for a hypothetical proposal involving a terrorist act against a truck transporting a chemical agent:

"Explosion of a bomb beneath the transportation truck or an attack by an armor-piercing weapon is possible. However, analysis shows that the consequences of such acts would be less than or equal to those associated with a maximum reasonably foreseeable transportation accident."

7.3   Incorporating Environmental Justice

Environmental Justice (EJ) has become a topic of special interest in recent years. Accordingly, this section provides practitioners with guidance on incorporating EJ considerations into both the preparation of EAs and EISs. This direction is based principally on a Presidential Executive Order, and guidance developed by the President's Council of Environmental Quality, the U.S. Department of Energy, and the U.S. Environmental Protection Agency (EPA).[16]
    The EPA defines EJ as:[17]

"The **fair treatment** and meaningful involvement of all people regardless of race, color, national origin, or income with respect to the development, implementation, and enforcement of environmental laws, regulations, and policies. Fair treatment means that no group of people, including racial, ethnic, or socioeconomic group, should bear a **disproportionate share of the negative environmental consequences** resulting from industrial, municipal, and commercial operations or the execution of federal, state, local, and tribal programs and policies" (emphasis added).

Implemented without prudence, EJ considerations can easily get out of hand, significantly adding to what might already be a costly and resource-intensive environmental process. Emphasis is therefore placed on providing the reader with a practical and balanced approach for addressing EJ.

To this end, a sliding-scale approach should be applied in determining the level of effort most appropriate for addressing EJ considerations; that is to say, tasks such as identifying populations, assessing impacts, and enhancing participation should be performed commensurate with the potential for sustaining disproportionately significant impacts.

> **Applying a Sliding-Scale Approach**
>
> *A sliding-scale approach should be applied in determining the level of effort most appropriate for addressing EJ considerations.*
>
> *Tasks such as identifying populations, assessing impacts, and enhancing participation should be performed commensurate with the potential for sustaining disproportionately significant impacts.*

Application of the sliding-scale approach

- Helps ensure that impacts on minority and low-income populations are not overlooked.
- May be qualitative or quantitative.
- May be more or less detailed than analyses for impacts on the general population, depending upon the significance of impacts on minority or low-income populations.

7.3.1 Background

With respect to NEPA, the issue of EJ began surfacing around 1994. Some of the principal direction and guidance documents that have since been issued are outlined below.

Executive order — In February 1994, President Clinton signed Executive Order 12898, Federal Actions to Address Environmental Justice in Minority Populations and Low-Income Populations.[18] This Executive Order requires that each federal agency:

> "make achieving environmental justice part of its mission by identifying and addressing, as appropriate, disproportionately high and adverse human health or environmental effects of its programs, policies, and activities on minority and low-income populations."

> **Federal Direction for Incorporating EJ into NEPA**
>
> *Executive Order 12898, Federal Actions to Address Environmental Justice in Minority Populations and Low-Income Populations; Presidential memorandum accompanying the Executive Order.*
>
> *Environmental Justice Guidance under the National Environmental Policy Act, CEQ.*
>
> *Guidance for Incorporating Environmental Justice Concerns in EPA's NEPA Compliance Analyses, EPA.*
>
> *EPA Guidance for Consideration of Environmental Justice in Clean Air Act Section 309 Reviews, EPA.*

A Presidential Memorandum accompanying this Executive Order directs federal agencies to "...analyze the environmental effects...of federal actions, including effects on minority communities and low-income communities, when such analysis is required by the National Environmental Policy Act."

Federal agencies are also instructed to "...provide opportunities for community input in the NEPA process, including identifying potential effects and mitigation measures in consultation with affected communities and improving the accessibility of meetings, crucial documents, and notices."

CEQ guidance — In 1997, the CEQ issued *Environmental Justice Guidance under the National Environmental Policy Act.* This guidance document provides direction on incorporating EJ into the NEPA process.[19] This guidance states that the Presidential Executive Order signed in 1994 does not change prevailing legal thresholds and statutory interpretations under NEPA and existing case law. However, it emphasizes that agency consideration of impacts on minority or low-income populations can identify disproportionately high and adverse impacts that are significant and that might otherwise be overlooked.

The guidance goes on to point out that environmental justice issues encompass a broad range of impacts covered by NEPA, including impacts on the natural or physical environment and related social, cultural, and economic impacts. This guidance acknowledges that environmental justice issues can arise at any step of the NEPA process, and agencies should consider these issues at each and every step of the process, as appropriate.

This guidance states that environmental impacts to minority and low-income populations do not have a different threshold for "significance" from other impacts, but specific consideration of impacts on minority and low-income populations can identify "disproportionately high and adverse human health or environmental effects that are significant and that otherwise would be overlooked."

Environmental Protection Agency guidance — In 1998, EPA issued Guidance for Incorporating Environmental Justice Concerns in EPA's NEPA Compliance Analyses.[20] This guidance applies to NEPA reviews conducted by EPA.

Soon after, EPA issued EPA Guidance for Consideration of Environmental Justice in Clean Air Act Section 309 Reviews.[21] This guidance applies to EPA reviews (under Section 309 of the Clean Air Act) of EISs prepared by other federal agencies.

Department of Energy guidance — As of this writing, the U.S. Department of Energy has issued draft guidance on incorporating EJ into its NEPA process. The document concentrates on how to assesses impacts and how to enhance public participation among minority and low-income groups.[22]

7.3.2 Determining the Appropriate Level of NEPA Review

Environmental impacts to minority and low-income populations can affect the level of required NEPA review. Specifically, they can affect whether an action can be implemented under a categorical exclusion or EA.

Categorical exclusions — Before an agency can categorically exclude a proposed action from NEPA review, it must determine, among other things, that there are no extraordinary circumstances related to the proposal that might affect the significance of its environmental effects. Extraordinary circumstances represent unique situations presented by specific proposals, such as scientific controversy about the proposal's environmental effects, uncertain effects, or unresolved conflicts concerning alternative uses of available resources.

An extraordinary circumstance exists if a proposal that would normally be categorically excluded would have a disproportionately high and adverse affect on a minority or low-income population. In such cases, preparation of at least an EA would normally be required.

> **Extraordinary Circumstance**
>
> *An extraordinary circumstance exists if a proposal that would normally be categorically excluded would have a disproportionately high and adverse affect on a minority or low-income population. In such cases, preparation of at least an EA would normally be required.*

Environmental assessments — To issue a FONSI for a proposed action, the EA needs to demonstrate that there would be no disproportionately high and adverse impacts to minority or low-income populations. Alternatively, if such an impact should occur, the FONSI could commit to specific measures that would ensure that such an impact would be mitigated to the point of nonsignificance.

> **The FONSI and EJ**
>
> *In issuing a FONSI, the EA might need to demonstrate that there would be no disproportionately high and adverse impacts to minority or low-income populations.*
>
> *If such an impact would occur, the FONSI could commit to specific measures that would ensure that such an impact would be mitigated to the point of nonsignificance.*

7.3.3 Analyzing Environmental Impacts

Agencies are instructed to evaluate proposals for their potential to produce disproportionately high and adverse human health or environmental effects.

Practitioners should note that there might be cultural differences among stakeholders regarding what constitutes an impact or the severity of an impact. For example, an Indian tribe might regard providing general public access to a particular mountain as desecration of its sacred site. Agency officials should also recognize that risk perceptions can vary widely, and commenters might disagree with the agency's underlying assumptions concerning risk factors.

Factors used in determining whether an impact is disproportionately high and adverse — The CEQ Guidance document, *Environmental Justice Guidance under the National Environmental Policy Act*, presents factors to consider when judging the importance of **disproportionately high and adverse** health and environmental impacts on minority and low- income populations (see Table 7.3).

TABLE 7.3

Factors Useful in Judging Whether Impacts Are Disproportionately High and Adverse

Disproportionately high and adverse human health effects:

When determining whether human health effects are disproportionately high and adverse, agencies are to consider the following three factors, to the extent practicable:

1. Whether the health effects, which might be measured in risks and rates, are significant (as employed by NEPA), or above generally accepted norms. Adverse health effects may include bodily impairment, infirmity, illness, or death.

2. Whether the risk or rate of hazard exposure by a minority population, low-income population, or Indian tribe to an environmental hazard is significant (with respect to NEPA) and appreciably exceeds or is likely to appreciably exceed the risk or rate to the general population or other appropriate comparison group.

3. Whether health effects occur in a minority population, low-income population, or Indian tribe affected by cumulative or multiple adverse exposures from environmental hazards.

Disproportionately high and adverse environmental effects:

When determining whether environmental effects are disproportionately high and adverse, agencies are to consider the following three factors, to the extent practicable:

1. Whether there is or will be an impact on the natural or physical environment that significantly (with respect to NEPA) and adversely affects a minority population, low-income population, or Indian tribe. Such effects may include ecological, cultural, human health, economic, or social impacts on minority communities, low-income communities, or Indian tribes when those impacts are interrelated to impacts on the natural or physical environment.

2. Whether environmental effects are significant (with respect to NEPA) and are or may be having an adverse impact on minority populations, low-income populations, or Indian tribes that appreciably exceeds or is likely to appreciably exceed that on the general population or other appropriate comparison group.

3. Whether the environmental effects occur or would occur in a minority population, low-income population, or Indian tribe affected by cumulative or multiple adverse exposures from environmental hazards.

It is important to note that economic or social effects are **not by themselves considered significant** (i.e., requiring preparation of an EIS). However, when an environmental impact statement is prepared, and economic or social, and natural or physical environmental effects are interrelated, then the EIS will discuss all of these effects on the human environment (§1508.14). Based on

> ### Addressing Social and Economic Impacts in EAs
>
> *Economic or social effects are not by themselves considered significant (i.e., requiring preparation of an EIS). However, when an EIS is prepared, and economic or social, and natural or physical environmental effects are interrelated, then the EIS will discuss all of these effects on the human environment. Thus, this provision can be interpreted to mean that an EA does not need to consider EJ where the impacts are simply social and economic in nature.*

this provision, it is reasonable to conclude that an EA does not need to consider EJ where the impacts are simply social and economic in nature.

Evaluating high and adverse impacts — On completing the analysis of impacts on the general population, analysts should determine, consistent with the CEQ guidance, whether any impacts on a minority or low-income population has the potential to be "disproportionately high and adverse."[23] A two-step approach is warranted. Specifically, practitioners should judge whether:

1. The impacts on a minority or low-income population would be potentially significant, within the meaning of NEPA (i.e., high and adverse).

2. Any potentially significant (i.e., high and adverse) impacts would disproportionately affect a minority or low-income population relative to the general population.

Impacts to a minority or low-income population that are considered to have a potential to be disproportionately high and adverse are analyzed. As with the analysis of impacts to the general population,

> ### Applying a Sliding-Scale Approach
>
> *The sliding-scale approach should be employed in determining the level of assessment necessary to make judgments regarding the significance of impacts on minority and low-income populations.*

a sliding-scale approach should be utilized in determining the level of assessment necessary to make judgments regarding the significance of impacts on minority and low-income populations. That is, one should perform a less rigorous analysis of proposals with a clearly small potential for impact, while devoting correspondingly more attention to actions where the potential for significance is greater.

> ### Focusing on Impacts that Would be Different
>
> *The analysis should focus on identifying and evaluating impacts to minority and low-income populations that would be different from the impact on the general population. A qualitative assessment can often be sufficient to provide the decisionmaker with information on which to base informed decisions.*

Attention should focus on identifying and evaluating impacts to minority and low-income populations that would be different from the impact on the general population. A qualitative assessment can often be sufficient to provide the decisionmaker with information on which to base informed decisions. Where such differences are trivial, include only enough discussion to show why more study is not warranted.

Special Mechanisms
Any special mechanisms by which an impact could affect a minority or low-income population might also need to be described.

Approach any investigation of impacts to minority or low-income populations as a subset of impacts on the general population. If appropriate, this should be done on a resource-by-resource basis (e.g., air quality, water quality) or impact area (e.g., health impacts, facility accidents, cumulative impacts). Any special mechanisms by which an impact could affect a minority or low-income population might also need to be described. The size of the population and its geographic area should be indicated.

Consider, as appropriate, whether the proposal would

- Affect or deny access to any natural resource on which the minority or low-income population (but not the general population) depends for cultural, religious, or economic reasons (e.g., a plant from which art is made, and perhaps sold for profit).

- Affect a minority or low-income population's food source, by reducing its abundance (e.g., development that would eliminate land habitat where food animals forage, or that would increase silt in a stream that is fished).

The Investigation
Analysis should investigate effects based on considerations such as special pathways, exposures, and cultural practices.

Unique Pathways, Exposures, and Cultural Practices   In assessing environmental impacts on minority and low-income populations, one should investigate the effects, based on considerations such as special pathways, exposures, and cultural practices. Table 7.4 presents factors useful in identifying unique pathways, exposures, or cultural practices that might need to be considered. Table 7.5 provides a definition of the term, "subsistence consumption," which is used in Table 7.4.

Considering Cumulative Impacts — As appropriate, agencies should consider the potential for *multiple* or *cumulative exposures*.[24] The terms multiple or cumulative exposure are defined in Table 7.6.

In assessing cumulative impacts, practitioners need to recognize that minority and low-income populations might be affected by past, present, or reasonably foreseeable future actions in a manner different from that experienced by the general population.

TABLE 7.4

Factors Useful in Identifying Unique Pathways, Exposures, or Cultural Practices that Might Need to be Considered

- Are exposure pathways or rates of exposure for minority and low-income populations different from exposure pathways or rates for the general population? Different pathways or rates could result from variations in:
  — physical location of a population's residences, workplaces, or schools.
  — dietary practices, such as:
    • Consumption of wild plants, or subsistence hunting, fishing, or farming
    • Differential selection of foods that might have high concentrations of contaminants (for example, bottom-feeding fish or fish that feed on bottom-feeding organisms can bioconcentrate fat-soluble contaminants from sediments, and organ meats, such as elk liver, might have bioaccumulated such contaminants)
  — Water supplies, such as use of surface or well water for drinking or irrigation
- are there any known health, social, or economic conditions of a minority or low-income population that would result in a greater impact? For example, would there be a greater frequency of dose or greater impact from a dose, over a pathway shared with the general population? Such conditions could involve:
  — Different access to public services such as paved roads (unpaved roads increase exposure to contaminated fugitive dust).
  — Different access to health care (e.g., poor control of asthma can increase susceptibility to particulate matter in air).
- Does the minority or low-income population (but not the general population) use a natural resource or area for cultural, religious, or economic reasons? Such uses could include:
  — Plants for ceremonial or medicinal purposes, or from which art is made and perhaps sold for a profit.
  — Plant-gathering or clay-procurement areas.
  — Natural features mentioned in legends.
  — Ceremonial sites.

TABLE 7.5

Definitions of the Term Subsistence Consumption

In its 1997 Environmental Justice Guidance under the National Environmental Policy Act, the CEQ issued guidance on key terms related to subsistence consumption as inserts in a reprinting of Executive Order 12898. The following guidance was developed by an Interagency Working Group on Environmental Justice, established by the Executive Order and chaired by the EPA.

Subsistence consumption of fish and wildlife — "Dependence by a minority population, low- income population, Indian tribe or subgroup of such populations on indigenous fish, vegetation, and/or wildlife, as the principal portion of their diet."

Differential patterns of subsistence consumption — "Differences in rates and/or patterns of subsistence consumption by minority populations, low-income populations, and Indian tribes as compared to rates and patterns of consumption of the general population."

TABLE 7.6

Definitions of Terms Multiple and Cumulative Environmental Exposure

In its 1997 Environmental Justice Guidance under the National Environmental Policy Act, the CEQ presented definitions related to multiple and cumulative environmental exposure as inserts in a reprinting of Executive Order 12898. The following proposed definitions were developed by the Interagency Working Group on Environmental Justice, established by the Executive Order and chaired by the EPA.

Multiple environmental exposure — "Exposure to any combination of two or more chemical, biological, physical, or radiological agents (or two or more agents from two or more of these categories) from single or multiple sources that have the potential for deleterious effects to the environment and/or human health."

Cumulative environmental exposure — "Exposure to one or more chemical, biological, physical, or radiological agents across environmental media (e.g., air, water, soil) from single or multiple sources, over time in one or more locations, that have the potential for deleterious effects to the environment and/or human health."

Assess Significance
On completing the analysis, determine whether potential impacts to minority and low-income populations are high and adverse, using the criteria specified for assessing significance in the CEQ NEPA Regulations. (§1508.27)

Assessing significance and mitigation measures — On completing the analysis, the agency needs to determine whether potential impacts to minority and low-income populations are high and adverse, using the criteria specified for assessing significance in the CEQ NEPA Regulations (§1508.27).

Mitigating Disproportionate Impacts
The goal in mitigating disproportionately high and adverse effects is not to distribute the impacts proportionally or divert them to a non-minority or higher-income population. Instead, measures or alternatives should be developed to mitigate effects on both the general population and minority or low-income populations.

The Presidential Memorandum accompanying Executive Order 12898 states, "Mitigation measures outlined or analyzed in an environmental assessment, environmental impact statement, or record of decision, whenever feasible, should address significant and adverse environmental effects of proposed federal actions on minority communities and low-income communities."[25] Mitigation measures include steps to avoid, minimize, rectify, reduce, or eliminate the adverse impact (§1508.20).

The goal in mitigating disproportionately high and adverse effects is not to distribute the impacts proportionally or divert them to a non-minority or higher-income population. Instead, measures or alternatives should be developed to mitigate effects on both the general population and minority or low-income populations. In other words, the goal of mitigation is not to move the impacts around, but to identify practicable means to meet the purpose and need for taking action while avoiding or reducing undesirable environmental effects.[26]

Documenting the analysis — Agencies are expected to integrate analyses of environmental justice (EJ) concerns in an appropriate manner so as to be clear and concise within the general format of the EA/EIS outlined in the Regulations (§1502.10, §1508.9[b]).[27]

The analysis should clearly indicate, along with the basis for the conclusion, whether there are any significant impacts to a minority or low-income popula- tion. The analysis

> **Describing EJ Impacts**
>
> *The analysis should clearly indicate whether there are any significant impacts to a minority or low-income population. The analysis should also indicate whether high and adverse impacts on a minority or low-income population would appreciably exceed the same type of impacts to the general population.*

should also indicate whether high and adverse impacts on a minority or low-income population would appreciably exceed the same type of impacts to the general population.

Determining Whether Impacts Are Disproportionate: Compare any high and adverse impacts on minority and low-income populations to:

- The same type of impacts on the general population (e.g., air quality to air qual- ity), even if the impacts on the general population are nonsignificant

> **Comparing Impacts**
>
> *Compare any high and adverse effects on minority and low-income populations to the impacts on the general population, not to another subset of the general population.*

- Impacts on the general pop- ulation, not to another subset of the general population (e.g., not to a non-minority or high-income population)

Bear in mind that a potentially beneficial impact on the general population may present a disproportionately high and adverse impact to a minority and low-income population. For example, a highway might benefit people as a whole, yet so disrupt a minority or low-income population as to constitute a disproportionately high and adverse impact.

7.3.4 Suggestions for Enhancing Public Participation

Presidential Executive Order 12898 discusses the importance of public partic- ipation in addressing environmental justice issues for proposed federal actions.[28]

Notifications — Both notices of EA preparation and notices of intent (NOIs) to prepare an EIS provide useful mechanisms for generating early participation in the NEPA process by minority and low-income populations. To be effective, however, the affected populations must receive the notice and understand their role in the NEPA process.

For EAs, it can be useful to notify not only the host and any potentially affected Indian tribes and states of the agency's determination to prepare an EA, but also to notify potentially interested minority and low-income populations (and other stakeholders). When practical, explain the role of stakeholder participation in the EA process.

For EISs, consider disseminating NOIs not only through the Federal Register and major media outlets, but also through local distribution outlets.

Public meetings — To help ensure that meeting places and times are appropriate for the minority and low-income populations, it is advisable to check on meeting times and places of local community groups to minimize scheduling conflicts with any NEPA-related public meetings.

> **Meeting Places**
>
> *Churches or other places of worship, schools, community centers, public meeting halls, or local restaurants or hotels can be used as meeting places.*

Agencies should consider scheduling public meetings to accommodate people who work night or weekend shifts. Churches or other places of worship, schools, community centers, public meeting halls, or local restaurants or hotels can be used as meeting places.

Agency officials might want to consider giving special consideration to segments of potentially affected populations that live in remote locations. For example, it might be beneficial to conduct a "road-show" style of public scoping meetings, traveling to several towns of various sizes in a short period to broaden the opportunities for participation among minority and low-income populations in small towns and rural areas.

Seeking comments — As appropriate, involve minority and low-income populations in identifying alternatives (including the environmentally preferable alternative) and issues for analysis that would address their special concerns. Minority and low-income populations may be the sole authoritative source of information on their cultural characteristics.

Where it is clear that an interested or potentially affected minority or low-income population holds opposing views, it may be useful to invite a bona fide representative of the population group to present its perspective in a document that can either be included as a distinct section (or appendix) or referenced in the NEPA document.

References

1. Eccleston C. H., The NEPA Planning Process: A Comprehensive Guide with Emphasis on Efficiency, Section 12.4 and pp281–282, John Wiley & Sons Inc, New York N.Y., 1999; Eccleston C.H., Environmental Impact Statements: A Comprehensive Guide to Project and Strategic Planning, John Wiley & Sons Inc, New York N.Y., 2000.
2. CEQ, Considering Cumulative Effects Under the National Environmental Policy Act, p4, January 1997.
3. EPA, Consideration of Cumulative Impacts in EPA Review of NEPA Documents, pp 10–11, (EPA 315-R-99-002), May 1999.
4. *Fritiofson v. Alexander*, 772 F.2d 1225, 1243, 1245–6 (5th Cir 1985).
5. *Inland Empire Public Lands Council v. Schultz*, 992 F.2d 977, 982 (9th Cir 1993).
6. CEQ, Considering Cumulative Effects Under the National Environmental Policy Act, p15, January 1997.
7. CEQ, Considering Cumulative Effects Under the National Environmental Policy Act, p19, January 1997.
8. *Webb v. Gorsuch*, 699 F.2d 157, 161 (4th Cir 1983).
9. *Northern Alaska Environmental Center v. Lujan*, 15 ELR 21048 (D. Alaska, 1985).
10. *Sierra Club v. Forest Service*, 46 F.3d 835, 839 (8th Cir. 1995).
11. *Sierra Club v. U.S. Forest Service* (9th Cir. June 24, 1988).
12. *Save the Yaak Committee v. Block*, 840 F.2d 714 (9th Cir. 1988).
13. *Thomas v. Peterson*, 753 F.2d 754 (9th Cir. 1985).
14. *Fritiofson v. Alexander*, 772 F.2d 1225, 1243, 1245–1246, (5th Cir. 1985).
15. Department of Energy, Analyzing Accidents Under NEPA, draft guidance prepared by the Department of Energy's Office of NEPA Policy and Assistance, April 21, 2000.
16. Presidential Executive Order 12898, Federal Actions to Address Environmental Justice in Minority Populations and Low-Income Populations, signed by President Clinton in February 1994; CEQ Guidance, Environmental Justice Guidance under the National Environmental Policy Act, December 1997; US. Environmental Protection Agency, Guidance for Incorporating Environmental Justice Concerns in EPA's NEPA Compliance Analyses, 1998; EPA Guidance for Consideration of Environmental Justice in Clean Air Act Section 309 Reviews, 1999; U.S. Department of Energy, Draft Guidance on Incorporating Environmental Justice Considerations into the Department of Energy's National Environmental Policy Act Process, April 2000.
17. US. Environmental Protection Agency, Guidance for Incorporating Environmental Justice Concerns in EPA's NEPA Compliance Analyses, 1998; EPA Guidance for Consideration of Environmental Justice in Clean Air Act Section 309 Reviews, 1999.
18 Presidential Executive Order 12898, Federal Actions to Address Environmental Justice in Minority Populations and Low-Income Populations, February 1994.
19. CEQ, Environmental Justice Guidance under the National Environmental Policy Act, December 1997.
20. U.S. Environmental Protection Agency, Guidance for Incorporating Environmental Justice Concerns in EPA's NEPA Compliance Analyses, 1998.

21. U.S. Environmental Protection Agency, EPA Guidance for Consideration of Environmental Justice in Clean Air Act Section 309 Reviews, 1999.
22. US. Department of Energy, Draft Guidance on Incorporating Environmental Justice Considerations into the Department of Energy's National Environmental Policy Act Process, April 2000.
23. CEQ, Environmental Justice Guidance under the National Environmental Policy Act, December 1997.
24. CEQ, Environmental Justice Guidance under the National Environmental Policy Act, December 1997.
25. Presidential Executive Order 12898, Federal Actions to Address Environmental Justice in Minority Populations and Low-Income Populations, February 1994.
26. EPA, Guidance for Incorporating Environmental Justice Concerns in EPA's NEPA Compliance Analysis, 1998.
27. CEQ, Environmental Justice Guidance under the National Environmental Policy Act, December 1997.
28. Presidential Executive Order 12898, Federal Actions to Address Environmental Justice in Minority Populations and Low-Income Populations, signed by President Clinton in February 1994.

8

Assessing Significance

A Systematic Procedure

> At 50, a man can be an ass without being an optimist but not an optimist
> without being an ass.
>
> Mark Twain

Determining the significance of a potential environmental impact is both a
complex problem and central to the National Environmental Policy Act's
(NEPA) goal of protecting environmental quality. A systematic and defensible procedure for determining the threshold of significance is also crucial as
it dictates when an Environmental Impact Statement (EIS) is required, as well
as the depth of study of each issue within an EIS.

This chapter examines the concept of significance and provides the reader
with a systematic framework for making such determinations. This article is
reprinted with permission from an article by Frederic March (Sandia
National Laboratories) that appeared in the conference proceedings of the
National Association of Environmental Professionals (NAEP).[1]

8.1   Definitions and Use of the Term Significance

The concept of significance determines when:

- An action is considered "major."
- An action can be categorically excluded.
- An Environmental Assessment (EA) should be prepared for a
  proposal.
- A Finding of No Significant Impact (FONSI) can be issued.
- An Environmental Impact Statement (EIS) needs to be prepared.
- The degree of treatment (range and depth) that should be devoted
  to a particular significant issue or impact addressed in an EIS (i.e.
  sliding-scale approach).

8.1.1 Significance Determines the Depth of Analysis

The following selected citations explicitly indicate the pivotal importance and central role that significance plays:

- "All agencies of the federal government shall ...include in every recommendation or report on proposals for legislation and other major federal actions *significantly* affecting the quality of the human environment a detailed statement ..." (NEPA, 102[2][A], emphasis added).
- "Most important, NEPA documents must concentrate on the issues that are truly *significant* to the action in question rather than amassing needless detail" (§1500.1[b], emphasis added).
- "Using a *finding of no significant impact* when an action not otherwise excluded will not have a *significant effect* on the human environment and is therefore exempt from requirements to prepare an environmental impact statement" (§1500.4[q] & §1500.5[l], emphasis added).

The terms "significant effect," "significant issue," and "significant impact" are used in reinforcing the intent of the above citations (§1501.1(d), §1501.7, §1501.7(a)(2), §1501.7(a)(3), §1502.1, §1502.2(b), §1505.1(b), §1508.4).

8.1.2 The Need for a Systematic Approach

> **Definition of Significance**
>
> *Factors for assessing significance are defined in 40 CFR. 40CRF§1508.27*

With respect to the problem of determining significance, NEPA practice can fall short in two ways. Specifically, the practitioner may fail to:

1. Provide adequate evidence that impacts are nonsignificant, resulting in challenges to FONSIs or in challenges to the adequacy of EISs.
2. Identify what is truly significant and what constitutes needless detail resulting in NEPA documents of excessive cost and delay in implementation of federal programs.

The following section advances a systematic approach for determining significance, based on strict adherence to the following factors:

1. "Significantly" is understood in the strict sense defined by §1508.27.
2. Application of rigorous and systematic tests for assessing significance, based on both letter of the law and spirit of the NEPA act.

8.2   General Procedure for Determining Significance

The concept of significance is defined in the NEPA Regulations (Regulations) as having two separate but closely related aspects: (1) context and (2) intensity.

A close study of the intensity factors yields seven specific tests, depicted in Table 8.1, that can be systematically applied in deter-

<blockquote>
Seven Tests

A close study of the factors cited in 40 CFR 1508.27 yields seven specific tests that can be used to assess the significance of an impact.
</blockquote>

mining the significance of any given action. Specific metrics (criteria and threshold values) can and should be established for assessing these intensity factors to determine whether a significant impact would be incurred. Additional information on significance and its interpretation can be found in Chapter 8 of the author's companion text, *The NEPA Planning Process*.[2]

TABLE 8.1

Seven Tests for Assessing Significance

1. **Receptor Test:** Identifying the presence or absence of certain environmental receptors, and then considering the impacts only on those receptors that are potentially affected (§1508.27[b][2,3,8,9]).

2. **Activity Test:** Identifying the presence or absence of certain activities and then considering the impacts only of those activities that are likely to affect the above receptors.

3. **Regulatory Compliance Test:** Determining whether the impacts include a threatened violation of any or local law, regulation, or requirement respecting the environment. In most cases, these laws, regulations, and requirements provide threshold tests for compliance. This test covers a wide range of effects including health and safety, as well as other regulations protecting air, water, land, and ecological resources (§1508.27[b][10]).

4. **Risk/Uncertainty Test:** Determining the degree to which the possible effects on the environment are highly uncertain or involve unique or unknown risks. This requirement includes risks resulting from natural hazards and from accidents (§1508.27[b][5]).

5. **Cumulative Test:** Determining whether the action is related to other actions with individually nonsignificant but cumulative significant impacts (§1508.27[b][7]).

6. **Precedence Test:** Determining the degree to which the action might establish a precedent for future actions with significant effects or represents a decision in principle about a future consideration (§1508.27[b][6]).

7. **Controversy Test:** Determining the degree to which the effects on the quality of the human environment are likely to be highly controversial (§1508.27[b][4]).

8.2.1   Description and Basis of the Tests

To qualify for a FONSI, a proposed federal action must pass all of the tests exhibited in Table 8.1 (i.e., passing each test means qualifying as nonsignificant in relation to that test.)

First two tests — The first two tests shown in Table 8.1 merely establish certain receptors and activities that are considered in the subsequent tests. While Regulations do not identify activities per se, they instruct agencies to develop procedures, including lists of activities that can be categorically excluded, or that would normally require preparation of an EA or an EIS.

Regulatory Compliance Test — This test (see Table 8.1, Test #3) is an excellent first screen. As a practical matter, many small actions that pass this test will often pass the subsequent tests as well. However, a common error has been to assume that meeting this factor alone is sufficient to establish nonsignificance.

Risk/Uncertainty Test — This test (see Table 8.1, Test #4) is applied in cases where the action could create unavoidable risks. Such risks might involve those associated with a nuclear reactor, a treatment plant for hazardous or radioactive wastes, or projects in areas subject to natural hazards such as earthquakes, tornadoes, and hurricanes. As a practical matter, socially acceptable risks are often specified in regulations, design codes, and standards of practice, so if the action passes the Regulatory Compliance Test, it will frequently pass the risk test. The risk/uncertainty issue can also be triggered by considerations related to the Cumulative Test, the Precedence Test, and the Controversy Test.

Cumulative and Precedence Tests — The Cumulative and Precedence Tests (see Table 8.1, Test #5 and #6) least lend themselves to objective criteria as the methodological basis for predicting future effects in a complex setting is often subjective. As a result, when these criteria are at issue, they will likely be associated with the risk/uncertainty and controversy factors.

Controversy Test — The Controversy Test (see Table 8.1, Test #7) might reopen the Risk/Uncertainty Test, as many controversies center around the risk issue. Controversy is less likely to be associated with regulatory compliance (although it is possible) and more likely to relate to the action's ability to pass the Cumulative and Precedence Tests.

8.2.2 Assessing the Seven Tests of Significance

Receptor Test

The Receptor Test is based on the fact that certain environmental receptors can be used in testing a given action for significance (§1508.27[b][2,3,8,9]).

The balance of this chapter provides a specific methodology for assessing each of the seven tests described in Table 8.1.

Receptor Test — The Receptor Test (see Table 8.1) is based on the fact that

TABLE 8.2

The Receptor Test can be Assessed in Terms of Certain Significance Factors
Prescribed in §1508.27(b)(2),(3),(8), and (9)

2. The degree to which the proposed action affects **public health** or **safety**.

3. Unique characteristics of the geographic area such as proximity to **historic** or **cultural resources, park lands, prime farmlands, wetlands, wild** and **scenic rivers,** or **ecologically critical areas.**

8. The degree to which the action may adversely affect **districts, sites, highways, structures** or objects listed in the **National Register of Historic Places** or may cause loss or destruction of significant **scientific, cultural** or **historical resources.**

9. The degree to which the action may adversely affect an **endangered species** or its **habitat** that has been determined to be critical under the *Endangered Species Act of 1973*.

certain environmental receptors (shown in bold, Table 8.2) that can be used in testing a given action for significance are specifically prescribed by §1508.27(b)(2),(3),(8), and (9).

Definition of Environmental Receptor: The factors cited in Table 8.2 form the core of a checklist together with a number of other receptors subject to potentially significant effects. These include receptors cited elsewhere in CEQ's regulation, receptors associated with Executive Orders (EOs), and receptors associated with the specific context in which an action is proposed. The term "environmental resource" is commonly used as a synonym for the term "environmental receptors." As used herein, the term environmental receptor means

1. any feature of the environment, natural or manmade;

2. any renewable or depletable resource;

3. any measure or statistic describing a feature of the environment, such as population density or monitored air quality, that might be impacted by a proposed action.

Checklist of Receptors: Developing and applying a checklist of receptors is a preliminary step that enables the practitioner to document the comprehensive consideration given to the assessment of potential effects. For example, the U.S. Air Force has used a checklist known as Form 813 - Preliminary Environmental Impact Analysis. Such checklists can be helpful in determining whether the action qualifies for a categorical exclusion, is likely to quality for a FONSI if an EA is written, or is likely to require an EIS.

Checklists should contain at least all of the items in the Regulations, Executive Orders, and other directives requiring that the receptor somehow be considered. A good classification scheme is logical and hierarchical so that, in principle, it can be expanded to whatever level of detail is needed.

Table 8.3 provides a receptor checklist that systematically encompasses every receptor mentioned in the Regulations and certain other environmental

TABLE 8.3

Receptor-Related Requirements

Receptor	CEQ Citation	Other Federal Laws, Reg., or Reqs
AIR		
Air Resources	§1508.8(b)	CAA, Pollution Prevention Act
Non-Attainment Area		CAA Conformity (40 CFR 6,51& 93)
Class I Control Region		CAA Conformity (40 CFR 6,51& 93)
Radioactive Exposures		CAA-NESHAPS (40 CFR 61)
Hazardous Exposures		CAA- ESHAPS (40 CFR 61)
Acid Rain		CAA
Ozone Layer		CAA
Odor		Nuisance Case Law
Visibility		CAA
Navigable Air Space		14 CFR 77.13a
WATER		
Water Resources	§1505.8(b)	CWA, SDWA, Pollution Prevention Act
Wild and Scenic Rivers	1508.27(b)(3), 1506.8	CEQ Memoranda, FR Vol. 45, No. 175, 9/8/80
Special Sources (Ground)		Safe Drinking Water Act, 40 CFR 149
Navigable Waters		Clean Water Act, Oil Pollution Act
LAND		
Land Use	§1502.16(c), §1508.8(b)	
Park Lands	§1508.27(b)(3)	
Prime Farmlands	§1508.27(b)(3)	CEQ Memoranda, FR Vol. 45, No. 175, 9/8/80
Wetlands	§1508.27(b)(3)	Executive Order 11990
Floodplains		Executive Order 11988
Coastal Zone		Coastal Zone Management Act, 15 CFR 930
Soils/Geologic		RCRA/CERCLA
National Priority Sites		CERCLA
HUMAN HEALTH/SAFETY		
Public Health & Safety	§1508.27(b)(2)	NEPA - Sec. 2, 101(b)(2,3), 101(c)
Radioactive Exposures		CAA- NESHAPS (40 CFR 61),
		Atomic Energy Act,
		Nuclear Reg. Comm., 10 CFR 20 - Rad. Stds.
Hazardous Exposures		CAA-NESHAPS (40 CFR 61)
		OSHA, CWA, SDWA, TSCA, RCRA, CERCLA, EPCRA,
		Pollution Prevention Act,
		Hazardous Materials Transportation Act

TABLE 8.3

Receptor-Related Requirements (continued)

Receptor	CEQ Citation	Other Federal Laws, Reg., or Reqs
SPECIES/HABITAT		
Endangered Spec/Hab	1508.27(b)(9)	NEPA - Sec. 2, 102(H)
Ecologically Critical Areas	1508.27(b)(3),1508.25	Endangered Species Act
Ecosystems	1508.8(b),1508.25	Fish and Wildlife Coordination Act
Ecological Resources	1508.8(b)),1507.2(e)	Marine Mammal Protection Act
		Marine Protection, Research & Sanctuaries Act,
Biosphere		NEPA - Sec. 2
Wilderness	1506.8(b)(2)(ii)	Wilderness Act
		Also see: Environmental Protection Agency: Habitat Evaluation, Guidance for the Review of Environmental Impact Assessment Documents, January 1993.
SOCIAL/ECONOMIC		
Social	1508.8(b)	NEPA 101(a), 101(b)(4)
Economic/Standard of Living	1508.8(b)	NEPA 101(a), 101(b)(5)
Aesthetic	1508.8(b)	NEPA 101(b)(2)
Growth	1508.8(b)	NEPA 101(a)
Urban Quality	1502.16(g)	NEPA 101(a)
Built Environment	1502.16(g)	
Population Density	1508.8(b)	NEPA 101(a)
American Indians	1502.16(c)	American Indian Religious Freedom Act
Minority/Low-Income		Executive Order 12898 - Environmental Justice
Life's Amenities		NEPA 101(b)(5)
HISTORICAL/CULTURAL/SCIENTIFIC/NATIONAL HERITAGE		
Historical/Cultural	1508.27(b)(3,8), 1502.25, 1508.8(b), 1502.16(g)	NEPA 101(a), 101(b)(2,4), Nat. Hist. Preservation Act of 1966, Historic Sites Act of 1935, Antiquities Act of 1906, Natl. Regist. of Hist. Places
Archeological		Archeological Recovery Act of 1960, Archeol. Resources Protection Act, Archeol. and Historic Preservation Act of 1974
Scientific	1508.27(b)(8)	
Diversity/Variety of Choice		NEPA 101(b)(4)

(continues)

TABLE 8.3

Receptor-Related Requirements (continued)

Receptor	CEQ Citation	Other Federal Laws, Reg., or Reqs
RESOURCE COMMITMENTS		
Energy Requirements	1502.16(e)	NEPA - 102(2)(C)(v)
Natural/Depletable Res.	1502.16(f)	NEPA - 102(2)(C)(v), 101(b)(5,6)

laws, regulations, or requirements. While only some of these receptors are explicitly stated in the factors for significance (§1508.27), all of the items listed in Table 8.3 should be considered because a significant impact on any one receptor can render the entire action significant.

Procedure for Applying Receptor Test   Recommended practice applying the Receptor Test is indicated in Table 8.4.

Table 8.4

Applying the Receptor Test

Step 1: Construct (or expand an agency prescribed) receptor checklist using Table 8.3 as a guide in response to the context of the action

Step 2: Indicate which receptors are not affected because they are outside the zone of influence of the proposed action, or would otherwise not be affected.

Step 3: Indicate which receptors would be affected at a trivial and clearly nonsignificant level and provide appropriate evidence to this effect.

Step 4: For all other receptors, potentially affected at a non-trivial or potentially significant level, list the issues and the evidence needed to demonstrate non-significance or significance.

Activity Test

The presence or absence of certain activities can be used in assessing significance of an impact.

Activity Test — Table 8.5 lists a number of activities suitable for inclusion on a checklist. The table indicates the federal law, regulation, or requirement that corresponds to each activity. A thorough checklist exercise screens receptors and activities together and takes note of appropriate cause and effects relationships.

The recommended practice for applying the Activity Test is indicated in Table 8.6.

Regulatory Compliance Test — The following significance factor lends itself to establishing a number of "threshold" tests of whether an impact is significant (§1508.27[b][10]):

Whether the action threatens a violation of federal, state, or local law or requirement imposed for the protection of the environment.

TABLE 8.5

Activity-Related Requirements

Activity	Other Federal Laws, Reg., or Reqs
Generation of Criteria Air Pollutants	Clean Air Act
Generation of Radioactive Air Pollutants	Clean Air Act - NESHAP
Generation of Hazardous/Toxic Air Pollutants	Clean Air Act - NESHAP
Release of Wastewater to Sewerage Systems	Clean Water Act
Release of Wastewater to Surface Waters	Clean Water Act
Underground Injection of Waste	40 CFR 146.5
Dredged Materials Discharged to Waters of the U.S.	Sec. 404 Clean Water Act
Underground Storage Tank Installation or Removal	RCRA
Generation/Management of Conventional Waste	RCRA
Generation/Management of Hazardous Waste	RCRA/CERCLA
Generation/Management of Radioactive &/or Mixed Waste	10 CFR 20
Management and Use of Toxic Substances	40 CFR 700-799
Pesticide Use	FIFRA
PCB Use	40 CFR 761
Asbestos Removal	40 CFR 61 and 763
Disturbance of Pre-Existing Contamination	RCRA, CERCLA
Decontamination/Decommissioning	RCRA Closure Reqmnts.
Noise Propagation	Nuisance Case Law
Generation of Light/Lasers	
Generation of Ionizing Radiation	
Generation of non-Ionizing Radiation/EMF	
Alteration of Landscape	Zoning and Land Use Laws
Alteration of Water Course	Various Federal Laws

TABLE 8.6

Applying the Activity Test

Step 1: Refer to that portion of the agency's NEPA implementation procedures that specify which activities are categorically excluded, which normally require EAs, and which normally require EISs.

Step 2: Construct an activities checklist using Table 8.5 as a guide in conformity with agency practice (such as in Step 1, on prescribed checklists and additional relevant activities meriting consideration).

Step 3: Indicate which activities have no effects because they do not occur in the zone of influence of the proposed action, or other reasons.

Step 4: Indicate which activities occur at a trivial and clearly nonsignificant level, and provide some evidence to this effect.

Step 5: For all other activities that occur at a non-trivial or potentially significant level, correlate with affected receptors and state the evidence needed to demonstrate non-significance or significance.

TABLE 8.7

Applying the Regulatory Compliance Test

Step 1: Determine regulatory compliance requirements using Tables 8.3 and 8.5 as an initial guide for applicable receptors and activities per the previous two tests. Expand the list of applicable compliance requirements to include any appropriate federal, state and local regulations.

Step 2: Determine whether existing compliance programs (such as permits, licenses, monitoring, and control (programs) would also cover the proposal, and cite evidence (such as permit numbers, conditions etc.).

Step 3: For all receptors not currently covered by compliance programs (i.e., those not cited in steps 1 or 2), determine whether the proposal threatens a violation in principle. This step might require quantifying effects to demonstrate absence of "threat."

Regulatory Compliance Test
The Regulatory Compliance Test is used in determining whether an action threatens to violate any federal, state, or local law, regulation, or requirement respecting the environment.
(1508.27[b][10])

Tables 8.3 and 8.5 provide a partial list of federal laws, regulations, and requirements associated with respective receptors and activities. A comprehensive guide would cite all applicable regulations and highlight those sections that prescribe standards or other tests of compliance. In addition to the federal list, the practitioner should be aware of applicable state and local laws, regulations, and ordinances.

Table 8.7 shows the recommended practice for applying the Regulatory Compliance Test.

As indicated earlier, a common error in NEPA practice is to assume that an action simply meeting the Regulatory Compliance Test is not significant. Therefore, it must be stressed that regulatory compliance is only one of seven distinct tests that must be applied in determining significance. For example, the following circumstances could result in significant impacts in spite of passing the regulatory compliance:

- The applicable federal, state and local laws, regulations, and requirements are arguably deficient and do not prevent significant impacts per the other tests.

For example, there are many toxic and hazardous emissions for which EPA has provided no specific exposure or dose limitation in its National Emission Standards for Hazardous Air Pollutants (40 CFR 61). In this case, the agency needs to go beyond the regulation in assessing significance, and consider the literature on recommended dose and exposure limits such as those published by the Environmental Protection Agency and the American Council of Governmental Industrial Hygienists for the workplace (Threshold Limit Values for Chemical Substances and Physical Agents and Biological Exposure Indices, periodically updated);

- the agency may arguably be unable or unwilling to take the necessary steps to be in compliance with all applicable laws, regulations, and requirements.

For example, the agency might have inadequate resources and a poor track record in enforcing compliance.

Risk/Uncertainty Test — The following CEQ significance factor lends itself to objective analysis within certain limitations (§1508.27[b][5]):

> The degree to which the possible effects on the human environment are highly uncertain or involve unique or unknown risks.

This significance factor typically involves two types of cases:

Case 1: This significance factor often involves projects such as nuclear power plants, and hazardous or radioactive treatment facilities in which various accidents can produce consequences ranging up to the catastrophic. Probabilistic risk analysis constitutes a body of available methodology to characterize the risk with some degree of scientific integrity. A procedure based on probabilistic risk analysis is recommended for assessing potential significance in these cases. The reader is referred to Section 7.2 in Chapter 7, which discusses accident analyses;

Case 2: A somewhat less tractable analysis involves cumulative actions whose effects must be considered together with other actions over which the agency has no direct control. In addition, when an action becomes a precedent for other actions, the eventual results can be difficult to predict and defend. Accordingly, both situations can be characterized as highly uncertain. (A discussion of "cumulative effects" and "precedent" is presented shortly.)

Tables 8.8, 8.9 and 8.10 apply to Case 1. Table 8.8 is a modified version of a recommended severity-probability ranking that was used by the U.S. Department of Defense.[3] Table 8.9 is a hazard-frequency scale that has been used by the Department of Energy (DOE).[4] Table 8.10 combines the results of ranking magnitude and frequency (Tables 8.8 and 8.9) into a recommended significance determination. Table 8.10 has been modified slightly by the author.

> **Risk/Uncertainty Test**
>
> *The Risk/Uncertainty Test is used in determining the degree to which the possible effects on the environment are highly uncertain or involve unique or unknown risks.*
>
> **(§1508.27[b][5])**

Recommended practice (for Case 1) for applying the Risk/Uncertainty Test is shown in Table 8.11.

In applying this test, it is important to base a determination of significance on whether risk is increased or decreased compared to an initial or baseline condition. For example, a new hazardous waste storage facility may of itself

TABLE 8.8

Severity Scale for Risk Analysis

Severity	Scale	Consequences (Human/Environmental)
Catastrophic	IV	Human: Loss of ten or more lives and/or large-scale and severe human injury or illness. Environmental: Large-scale damage involving destruction of ecosystems, infrastructure, or property with long-term effects, and/or major loss of human life.
Critical	III	Human: Loss of fewer than ten lives and/or small-scale severe human injury or illness. Environmental: Moderate (medium-scale and/or long-term duration) damage to ecosystems, infrastructure, and property.
Subcritical	II	Human: Minor human injury or illness. Environmental: Minor (small-scale and short-term) damage to ecosystems, infrastructure, or property.
Negligible	I	Human: No reportable human injury or illness. Environmental: Negligible or no damage to ecosystems, infrastructure, or property.

TABLE 8.9

Frequency Scale for Risk Analysis

Category	Level	Description	Frequency (f)
Frequent	A	At least once per year	$f > 10^0$
Likely	B	Once in 1 to 10 years	$10^0 > f > 10^{-1}$
Occasional	C	Once in 10 to 100 years	$10^{-1} > f > 10^{-2}$
Unlikely	D	Once in 100 to 1,000 years	$10^{-2} > f > 10^{-3}$
Remote	E	Once in 1,000 to 1,000,000 years	$10^{-3} > f > 10^{-6}$
Incredible	F	Less than once in 1,000,000 years	$10^{-6} > f$

constitute a severe risk to the environment associated with possible explosions or fires. However, if the proposed action is to move such waste from storage conditions of higher risk to lower risk, the action might arguably be considered nonsignificant even if Table 8.10 indicates otherwise.

The issue of appropriate evidence, including database, methodology, precision, and quality control for characterizing the expected magnitude and frequency of events will vary with the environmental context. As a first-order approach, use of the tables can be based on professional judgment and rule of reason. However, there are circumstances in which regulations or an agency might require a formal probability and risk assessment such as DOE's Safety Analysis Reports. Recommended practice is to base the significance determination on such formal analysis when it is available.

When a formal risk analysis is not required by the agency or regulations, the NEPA practitioner must determine whether it is needed in the analysis, and what methodology is to be used. The potential for controversy should be a factor in making these decisions.

TABLE 8.10

Recommended Significance Assignments for Severity-Frequency Determinations

Severity	Frequency					
	A (Frequent)	B (Likely)	C (Occasionally)	D (Unlikely)	E (Remote)	F (Incredible)
IV (Catastrophic)	S	S	S	S	M	M
III (Critical)	S	S	S	M	N	N
II (Subcritical)	S	S	M	N	N	N
I (Negligible)	N	N	N	N	N	N

Legend:
S = Usually significant.
M = Marginally significant or nonsignificant, depending on context.
N = Usually nonsignificant.

TABLE 8.11

Applying the Risk/Uncertainty Test

Step 1: Determine the expected severity of the impacts that could occur for each risk-prone situation associated with an action (i.e. accidents and natural hazards) by referring to Table 8.8.

Step 2: Determine the expected frequency of the impacts that could occur for each risk-prone situation associated with an action (i.e., accidents and natural hazards) by referring to Table 8.9.

Step 3: Determine potential significance as a function of severity and frequency together by referring to Table 8.10.

Cumulative Test
The Cumulative Test is used in determining whether the action is related to other actions with individually nonsignificant but cumulative significant impacts.   (§1508.27[b][7])

The Cumulative Test — The issue of cumulative impacts is one of the most difficult in NEPA practice. Accordingly, application of this test is more complex than the others and merits somewhat more consideration.

Regulatory Guidance   The Regulations specify consideration of the following cumulative factor (§1508.27[b][7]):

> Whether the action is related to other actions with individually insignificant but cumulatively significant impacts. Significance exists if it is reasonable to anticipate a cumulatively significant impact on the environment. Significance cannot be avoided by terming an action temporary or by breaking it down into small component parts.

The Regulations define a cumulative impact as (§1508.7, emphasis added):

> "Cumulative impact" is the impact on the environment that results from the **incremental impact of the action** when added to other **past, present,** and **reasonably foreseeable future actions,** regardless of which agency (federal or non-federal) or person undertakes such other actions. Cumulative impacts can result from individually minor but collectively significant actions taking place over a period of time.

The language of the regulation thus defines a three-step logic in determining significance of cumulative impacts:

- Create a baseline defined by the environmental impacts of past, present and reasonably foreseeable future actions independent of the proposed action.
- Determine the incremental impact of the proposed action and superimpose (i.e., "add") it to the baseline.
- Apply the same other six significance tests described in this chapter to the evaluation of cumulative impacts. As described earlier, these are applied in a similar fashion to the way they would be used in assessing the significance of any individual impact.

Tools for assessing cumulative impacts — The common features of analytic tools available for the cumulative test are

- They characterize a current baseline situation representing the effects of past and present actions.
- They provide a theory for predicting reasonably foreseeable future actions that perturb the current baseline condition.

- Some of the principle tools for cumulative impact analysis include:
  - Econometric forecasting models for predicting levels of economic activities that would cause environmental impacts.
  - Geographic analysis methods that predict land use and related effects.
  - Ecosystem models that predict the future state of interdependent living systems.

These tools are often not used to their full potential because

- Many NEPA practitioners are not familiar with the tools, or are untrained in the associated disciplines of econometrics, economic geography, or ecology.
- Use of these tools may be too data intensive, time consuming and expensive to apply under many circumstances, particularly for EAs.
- The tools require skilled, experienced specialists who may not be locally available, which further adds to the expense.
- Application of the tools can become controversial in that they are based on assumptions that can be challenged, and their predictive capability is often difficult to demonstrate.

Nevertheless, such tools have a place in environmental planning, and can be valuable if used appropriately.

Applying the Cumulative Test — If a finding of nonsignificance with respect to the cumulative test is challenged upon publication of a FONSI, the issues of risk/uncertainty, precedence, and controversy might also be raised.

In applying the Cumulative Test, the recommended practice is outlined in Table 8.12. In applying this test:

- Carefully characterize the environmental issues important for cumulative analysis. Use professional judgement and common sense (i.e. rule of reason) to qualitatively characterize the cumulative effects, or to approximate them using a quantitative method such as one of the tools described above.
- Obtain assistance from an appropriate expert on the problem area (e.g. economics, ecology, geography, etc.) in selecting and applying an analytic method

If certain cumulative effects are sufficient to cause a significant impact, and if an EIS is subsequently prepared, expert assistance may continue to be

TABLE 8.12

Applying the Cumulative Test

Step 1: Create a baseline defined by environmental impacts of past and present actions. This baseline is used for comparing the effects of the reasonably foreseeable future actions (including the proposed action).

    a. Define the current environmental baseline. The represents the effects of past and present actions.

    b. Identify all the reasonably foreseeable future actions. Define the impact of reasonably foreseeable future actions, including the proposed action.

    c. Superimpose (i.e., add) the impact of reasonably foreseeable future actions (including the proposed action) on the environmental baseline defined in 1a.

Step 2: Determine whether the impact of the reasonably foreseeable future action, including the proposed action, exceeds the threshold of significance. Apply the other six tests of significance as they would be used in assessing individual impact.

Step 3: Determine whether the incremental impact of the proposed action exceeds the threshold of significance for a given incremental action. The incremental threshold is a measure of the proposed action's contribution relative to the total cumulative effect (past, present, proposed, and reasonably foreseeable actions). By measuring the incremental impact relative to the total cumulative impact, decisionmakers gain important insight that can enhance project planning and development of mitigation measures. Apply the other six significance tests in a fashion similar to the way they would be used in assessing an individual impact. Note that Step 3 may not be necessary if the significance tests in Step 2 reveal that the total cumulative effect is not significant.

---

needed in formulating an analytic methodology as part of scoping, and in implementing it during the analysis.

Precedence Test—The CEQ regulation specifies the following significance factor (§1508.27[b][6]): "The degree to which the action may establish a precedent for future actions with significant effects or represents a decision in principle about a future consideration."

> **Precedence Test**
>
> *The Precedence Test is used in determining the degree to which an action might establish a precedent for future actions with significant effects, or represents a decision in principle about a future consideration.*
> (§1508.27[b][6])

CEQ has provided no further guidance on how to interpret and implement this significance factor. The first part of this factor is similar to the Cumulative Test in that it requires that the proposal be considered together with (reasonably foreseeable) future actions. Accordingly, appropriate practice for this part of the significance factor is similar to that described for cumulative actions. However, consideration can overlap with the risk/uncertainty factor, given the problems inherent in predicting the future.

The second part of this factor (i.e., decision in principle) is somewhat unique. Appropriate practice can be inferred from court precedent. Fogelman reports:[5]

"This type of effect may occur when construction of a facility — such as a
port — ensures that an area will continue to be developed in lieu of other
areas. Once the plans are initiated and begun, it is probable that decisionmakers will order the project continued."

For the practitioner, the "rule of reason" can be used in assessing the "decision in principle." If a finding of insignificance with respect to the Precedence
Test is challenged upon publication of the FONSI, the issues of risk/uncer-
tainty, cumulative effects, and controversy might also be raised.

Recommended practice for applying the Cumulative Test is outlined in
Table 8.13.

TABLE 8.13

Applying the Cumulative Test

Step 1: With respect to precedent for future actions factor, follow the same practice rec-ommended for cumulative impacts.
Step 2: With respect to a decision in principle about a future consideration, use common sense based on an understanding of the action's context.

The Controversy Test — The regulations specify the following factor be
considered in assessing significance (§1508.2 [b][4]):

The degree to which the effects on the quality of the human environment
are likely to be highly controversial.

No further guidance is offered by the CEQ in regulation or other published
guidance. However, some clear guidance on this issue is provided in the case
of *Hanly v Kleindienst* (II):[6]

"The term 'controversial' apparently refers to cases where a substantial
dispute exists as to the size, nature, or effect of the major federal action
rather than to the existence of opposition to a use, the effect of which is
relatively undisputed...The suggestion that 'controversial' must be equat-
ed with neighborhood opposition has also been rejected by others [i.e.,
other cases]."

Thus, opposition to the proposed action per se does not constitute evidence of "controversy" in the NEPA sense of the term, an extremely important interpretation.

Table 8.14 depicts the recommended practice for applying the Controversy Test.

> **Controversy Test**
>
> *The Controversy Test is used in determining the degree to which the effects on the quality of the human environment are likely to be highly controversial.*
>
> (§1508.27[b][4])

TABLE 8.14

Applying the Controversy Test

Step 1: Determine which issues have the potential to generate controversy based on issues considered in the previous tests.

Step 2: Assess the potential for controversy by knowing the affected stakeholders and potential for project opposition. This assessment should also be based on available public response to notices or news about the proposed action and any history of past opposition on similar issues.

Step 3: As necessary, obtain the assistance of legal counsel to determine whether potential opposition meets controversy criteria per *Hanly v Kleindienst*.

8.2.3 Additional Significance Considerations in Practice

> **Additional Factors**
>
> *In carrying out the aforementioned tests, the focus was on the language in the regulatory definition of significance per §1508.27. Other factors (including the context) might also need to be considered.*

In carrying out the above tests, the focus was on the language in the regulatory definition of significance per §1508.27[a]. Other factors might also need to be considered. One of these factors, the context, is described in the following section.

> **Context**
>
> *The Regulations requires that context and intensity must be considered together. It provides specific factors for assessing intensity but little specific guidance on how to consider context.*

Context — The Regulations require that context and intensity be considered together. It provides intensity but little specific guidance on how to consider context. The language of the Regulations does not imply a test of context per se, but merely requires that context be considered in applying the various tests defined under intensity.

Context is in part established by the environment in which the action is taking place and can therefore be represented by an environmental receptor checklist as in Table 8.3. It is also established in part by certain features of the action that can also be identified in checklists such as Table 8.5. Thus, use of checklists provides evidence that context is considered in part. How much further should the practitioner go in explicitly documenting consideration given to context? This becomes a matter of professional judgment in concert with the "rule of reason," and taking into account any special agency or public concerns likely to be raised during the NEPA process.

8.2.4 Evidence of Significance

Evidence of significance is specifically required in an EA to support a FONSI (§1508.9[a][1], §1500.2[b]). Table 8.15 depicts forms of evidence that have

TABLE 8.15

Forms of Evidence That Have Been Used in Support of FONSIs

- Evidence that the document preparer reviewed items on a checklist
- Accessible databases and literature pertaining to the affected environment and to individual environmental receptors.
- Official certification or representation by federal, state or local officials as to whether a given receptor would likely be affected by the action.
- Regulatory, scientific, and professional practice literature pertaining to the effects of certain disturbances (e.g. radiological, chemical) on human health, and on other environmental receptors.
- Computation and modeling to test whether a given impact exceeds a threshold (defined by law, regulation, requirement, or recommended practice) or to characterize intensity of impact.
- Original data collected in the field or laboratory as direct evidence in support of significance or as input to computation and modeling.

---

been used in support of FONSIs (listed in approximate order of rigor, time, and expense).

The particular choice of evidence for each potential impact considered is at the discretion of the agency, as CEQ has not provided any further guidance other than §1500.1(b), in which the agency is admonished to concentrate on the issues that are truly significant to the action in question rather than amassing needless detail.

References

1. March F, *National Association of Environmental Professionals, 21st Annual Conference Proceedings,* "Determining the Significance of Proposed Actions," pp 421-436, 1996.p
2. Eccleston C. H., *The NEPA Planning Process: A Comprehensive Guide with Emphasis on Efficiency,* John Wiley & Sons Inc, New York, N.Y., 1999.
3. U.S. Department of Defense, DOD's MIL-STD-882B.
4. U.S. Department of Energy, Order 5481.1B.
5. Fogelman V., *Guide to the National Policy Act: Interpretations, Applications and Compliance,* Quorum, 1990.
6. *Hanly v Kleindienst* (II), 471 F 2nd 823 (2nd Cir. 1972), cert. denied, 412 U.S. 908 (1973)

9

The Finding of No Significant Impact

It takes your enemy and your friend, working together, to hurt you to the
heart; the one to slander you and the other to get the news to you.

Mark Twain

Once the Environmental Assessment (EA) has been completed and read by
the decisionmaker, the agency is in a position to reach a final determination
regarding the significance or nonsignificance of the potential impacts. If the
decisionmaker concludes that the action would not result in a significant
impact, this determination is recorded in the Finding of No Significant
Impact (FONSI). This chapter presents the reader with the regulatory
requirement for preparing and issuing a FONSI.

Once the FONSI has been issued, circumstances might arise in which a
change might need to be made to the proposed action described in the EA; a
decisionmaking tool (i.e., Smithsonian Solution) for determining when a change
to a proposed action requires additional analysis is presented in Chapter 4 of the
author's companion text, *Environmental Impact Statements*,[1] where a discussion
of post-monitoring and a methodology for integrating implementation of the
agency's action with an International Organization for Standardization (ISO)-
14000 consistent environmental management system is also provided.

9.1 Reaching a Determination of Nonsignificance

The purpose of a FONSI is to
publicly document a decisionmaker's determination that,
based on review of the EA, the
proposal will not result in a significant impact, and, therefore,
preparation of an Environmental Impact Statement (EIS) is
not required (§1501.4[e]). The decisionmaker must assess the environmental
facts impartially, avoiding even the appearance of being biased or prejudicial.

> **Definition of a FONSI**
>
> *"...a document by a federal agency briefly presenting
> the reasons why an action, not otherwise excluded
> (§1508.4), will not have a significant effect on the
> human environment and for which an environmental
> impact statement therefore will not be prepared..."*
> **(§1508.13)**

127

The definition of significance and the factors used in reaching such a determination are described in Chapter 4 of this text. Chapter 8 of this text provides the reader with a systematic framework and methodology for assessing the significance of an impact.

9.1.1   Principles Governing Sound Decisionmaking

A FONSI is a finding of fact. As such, the FONSI should be a citation of specific facts that the decisionmaker finds to be true and leads to the conclusion that the proposed action would not result in a significant environmental impact. Principles governing the decisionmaker's conduct in reaching a finding of no significant impact are presented in Table 9.1.

TABLE 9.1

Principles Governing the Decisionmaker's Assessment

The agency bears the *burden of proof* in demonstrating that no significant impacts would result.
The decisionmaker's role is **not** to be a *committee of compassion* for the project proponent.
The final decision **must** be based on the *assessment of facts*.
Information (particularly with respect to an applicant) is not the same as facts.
Opinions or best professional judgment without some factual basis are *without merit*.

9.1.2   Criteria That Must be Met in Reaching
a Decision to not Prepare an EIS

As indicated in Table 9.2, the courts appear to have established four criteria that must be met in the process of reaching a decision to not prepare an EIS:[2]

The decisionmaker must carefully weigh the evidence presented in the EA before concluding that a FONSI is appropriate. The FONSI should be written to clearly demonstrate to the reader that the responsible decisionmaker signing the FONSI thoroughly understands (1) The scope of the action, (2) Implications of the proposal, and (3) that no significant impacts would result.

TABLE 9.2

Four Criteria Established by the Courts That Agencies Must Meet in Reaching a Decision not to Prepare an EIS

1. Whether the agency took a "hard look" at the problem.
2. Whether the agency identified the relevant areas of environmental concern.
3. Whether the agency made a convincing case that the impact was nonsignificant.
4. If there is an impact of true significance, has the agency convincingly shown that changes in the proposed action would sufficiently reduce the effect to the point of nonsignificance?

An example of a poor decisionmaking process — A recent court case illustrates both the importance of preparing an EA that is readable and the need for the decisionmaker to thoroughly understand the analysis before reaching a final decision. This case involved a state-sponsored action that required federal authorization. The action was subsequently challenged in court. In a deposition, the decisionmaker was asked to read a highly technical section from the document involving the analysis of an environmental hazard. The decisionmaker acknowledged that he did not completely understand this analysis. When asked about the significance of certain impacts that had been analyzed, he also acknowledged that he was not sure how many people could be harmed in the event of an accident. The decisionmaker did not appear to thoroughly understand the analysis for which he had made a decision. Sensing that it was in a weak position, the agency eventually agreed to settle the case out of court.

9.1.3 The Agency's Administrative Record

The agency's administrative record should fully, clearly, and accurately document how the decisionmaker progressed through the facts to arrive at a final determination. At a minimum, the administrative record's written documentation should:

- Include all factual material that is necessary to support the agency's finding (e.g., field studies, analyses, reports, memorandums)
- List any environmental requirements or standards that were considered
- Record the process used in weighing the evidence
- Present the final determination with any conditions or mitigation measures

The text *Environmental Impact Statements* details the requirements for preparing and maintaining a defensible administrative record. Table 9.3 provides a generalized checklist for that can used by agency officals in reaching a determination of nonsignificance.

9.2 Preparing the FONSI

A poorly prepared FONSI is vulnerable to legal attack. As the agency is faced with the burden of proving that no significant impacts would occur, all conclusions presented in the FONSI must be directly tied to the analysis presented in the EA.

TABLE 9.3

Generalized Checklist for Reaching a Determination of Nonsignificance

Has the agency taken a "hard look" at the problem?

Have all relevant areas of environmental concern been adequately identified?

Has the decisionmaker impartially assessed the facts with an open mind?

Did the decisionmaker carefully weigh the evidence and facts presented to him before concluding that a FONSI was appropriate?

Does the decisionmaker thoroughly understand the scope of the action, its implications on environmental quality, and the fact that no significant impact would occur?

Has the agency made a convincing case that all impacts would be nonsignificant?

If there is a significant impact, has the agency convincingly shown that changes in the proposal or other mitigation actions would sufficiently reduce the effect below the threshold of significance?

Was the final determination made based on an assessment of the facts?

Does the agency's administrative record fully, clearly, and accurately document how the decisionmaker progressed through the facts to arrive at a final determination?

9.2.1 Documentation Requirements

At a minimum, the FONSI must contain the items indicated in Table 9.4 (§1508.13). With respect to appending a copy of the EA, the FONSI may either (1) include, (2) summarize, or (3) incorporate it by reference.[3]

TABLE 9.4

Regulatory Requirements for the FONSI

1. A brief explanation of the reasons why the action will not have a significant effect on the human environment. If the EA is included, this discussion need not repeat discussion in the assessment but can incorporate it by reference.
2. Include the EA or a summary of it.
3. Note any other environmental documents related to the scope of the proposed action.

Recommendations — In addition to the regulatory requirements described above, every FONSI should

- Explicitly state that the proposed action *would not result in a significant impact*.
- Specifically, cite or summarize evidence indicating that no significant impacts would occur. This evidence should discuss both the (1) intensity and (2) context of the impacts that were considered.

- Indicate which factors were weighted most heavily in making the determination.

> **Where, Who, When, and Why**
>
> *The FONSI should include information clearly indicating the scope of the action and where it is to take place. It should also indicate who has proposed the action, why the action was proposed, and when it is scheduled to be carried out.*

9.2.2   A Checklist for Preparing the FONSI

A generalized checklist for assisting an agency in preparing the FONSI is presented in Table 9.5.

TABLE 9.5

Generalized Checklist for Preparing the FONSI

Has the FONSI been specifically prepared and tailored to the action described in the EA?

Does it clearly demonstrate that the responsible decisionmaker thoroughly understands the scope and comprehends the implications of the action, and that no significant impacts would result?

Does it explain the scope of the action (e.g., who, what, when, where, why, and how)?

Does it include the EA or a summary of it?

Does it explicitly state that no significant impacts will result from pursuing the action?

Does it conclusively demonstrate and explain why the action will not result in any significant environmental impacts (direct, indirect, cumulative)?

Does it demonstrate that both the intensity and context were taken into account in reaching a decision of nonsignificance?

Does it indicate which factors were weighed most heavily in reaching the determination of nonsignificance?

Are all conclusions directly tied to the analysis presented in the EA?

Does it note any other environmental documents related to the scope of the action?

Does it describe any mitigation measures that will be adopted? Were such measures designed and customized to address the specific impacts? If applicable, is a specific monitoring and implementation plan described to ensure that any mitigation measures are successfully adopted?

Has the EA adequately evaluated the effectiveness of any mitigation measures committed to in the FONSI? Would these measures mitigate any significant impacts to the point of nonsignificance? Are these mitigation measures free from scientific controversy?

Do funding and technical means exist for implementing any mitigation commitments made in the FONSI?

9.3   Mitigated FONSIs

Mitigation Versus Impact Reduction
Mitigation measures are additional steps, not part of the proposed action, that can be taken to reduce impacts. The term "impact reduction" refers to measures used for reducing adverse impacts regardless of whether such impacts are significant or not.

Mitigation measures are additional steps, not part of the proposed action, that can be taken to eliminate or reduce potential impacts (§1508.25[b][3]). Actions considered to be standard engineering practice or required under law or regulation are not normally considered mitigation measures.[4] The term "impact reduction" is frequently used in referring to measures used for reducing adverse impacts regardless of whether such impacts are significant or not.

Agencies are encouraged to include impact reduction measures in an EA, even if the effects are not considered significant.[5] Such measures may allow an agency to further reduce impacts even though they might already lie below the threshold of significance. Additional information on mitigation can be found in the author's companion text, *The NEPA Planning Process*.[6]

The decisionmaker should carefully consider any mitigation measures that are to be adopted, as such commitments are legally binding. The agency is responsible for ensuring that funds and technical means exist for implementing such measures.

9.3.1   Mitigation and the Courts

At one time, CEQ discouraged the use of the mitigated EA/FONSI.[7] During the 1980s and 1990s, numerous courts upheld an agency's right to prepare mitigated FONSIs.[8] Today, a majority of appellate courts have accepted the use of mitigated FONSIs.

Mitigation may be deemed insufficient if the agency has no control over how it will be implemented, due to reliance on a third party to implement the measures. When mitigation measures will be implemented by a third party, the commitment, while it need not be contractual, must be more than vague statements of good intentions.[9] Mitigation is generally deemed legally sufficient if it reduces the impact to the point of nonsignificance and is offered by an agency with control to legally enforce the measures.[10]

Examples where mitigation has been upheld — The courts have upheld mitigation in cases such as airport noise suppression, incorporation of water release procedures for maintaining lake levels during spawning season, and replacing lost wetlands.[11]

In 1993, an EA prepared by the Army Corp of Engineers for a hydroelectric project was challenged. The court found that the proposed mitigation was sufficient to compensate for the adverse impacts, even though these measures would not completely compensate for the adverse effects.[12]

In another case, a mitigated EA was prepared for transportation of hazardous materials by rail. When challenged, the court concluded that the mitigation measures were sufficient to support a FONSI.[13] The mitigation included such measures as:

1. Notifying surrounding communities of the train schedule
2. Providing surrounding communities with copies of the emergency response plans
3. Providing a toll-free telephone number to local emergency response groups
4. Developing the proposed action in accordance with the Department of Transportation's regulations for transport of hazardous materials

Examples where mitigation has been rejected — In contrast, the courts are unlikely to accept mitigation measures that (1) have not been adequately investigated or (2) can not conclusively demonstrate that the impacts would be reduced to the point of nonsignificance. For example, certain types of mitigation for reducing traffic and wetlands impacts have been rejected because they were found to be inadequate. Similarly, preservation of a canyon from a river diversion project was found to be inadequate because it did not mitigate effects on species located outside the preserved area.[14]

In a recent case, an EA was prepared for construction of a parking lot. The proposed action involved removal of vegetation and 500-year-old cedar trees that the agency's administrative record characterized as significant. The EA included certain mitigation measures. The court addressed the issue of mitigation by noting that an agency "may reach a FONSI if mitigation measures are proposed that **directly address the impacts** identified in the Environmental Assessment." In this case, however, the agency's mitigation — removal of a nearby picnic area and its regeneration as forest (that would take more than 500 years to recover) — lacked adequate scientific analysis to constitute sufficient mitigation to support a FONSI.[15]

9.3.2 Criteria for Adopting Mitigation Measures

The courts appear to be imposing six criteria (Table 9.6) in the analysis of mitigated EAs.[16] Before adopting any mitigation measures, it is recommended that the decisionmaker ensure they are consistent with these six criteria.

TABLE 9.6

Six Basic Criteria that the Analysis of Mitigation Measures Must Address

1. Proposed mitigation measures must be demonstrably effective. Cursory statements regarding the effectiveness of mitigation measures are generally insufficient; an EA must present sufficient evidence to support mitigation claims.

2. The effectiveness of mitigation measures should be free from scientific controversy. Disagreement or controversy among experts can substantially weaken an agency's ability to prove that no significant impacts will occur.

3. Mitigation measures should address specific environmental issues and concerns, including cumulative impacts. Mitigation measures that are vague or general in nature are normally inadequate. More to the point, mitigation measures must be designed to address specific impacts and issues.

4. Mitigation measures should be fully identified and defined prior to filing the FONSI. An agency cannot rely on future or to-be-determined mitigation measures because there is no way to adequately assess whether such methods would *effectively* mitigate the environmental impacts.

5. Mitigation measures must effectively reduce impacts to the point of non-significance.

6. A specific monitoring or implementation plan should be included to ensure that mitigation measures are effectively carried out.

References

1. *Environmental Impact Statements: A Comprehensive Guide to Project and Strategic Planning*, John Wiley & Sons Inc., New York, N. Y., 2000.

2. *Cabinet Mountains Wilderness/Scotchman's Peak Grizzly Bears v Peterson*, 685 F.2d 678 (DC Cir. 1982).

3. CEQ, Council on Environmental Quality - Forty Most Asked Questions Concerning CEQ's National Environmental Policy Act Regulations (40 CFR 1500–1508), Federal Register Vol. 46, No. 55, 18026–18038, March, 23, 1981, Question number 37a.

4. CEQ, Public memorandum titled, Talking Points on CEQ's Oversight of Agency Compliance with the NEPA Regulations, 1980.

5. CEQ, Council on Environmental Quality - Forty Most Asked Questions Concerning CEQ's National Environmental Policy Act Regulations (40 CFR 1500–1508), Federal Register Vol. 46, No. 55, 18026–18038, March, 23, 1981, Question number 39.

6. Eccleston C. H., *The NEPA Planning Process: A Comprehensive Guide with Emphasis on Efficiency*, John Wiley & Sons Inc, New York, N. Y., 1999.

7. CEQ, Council on Environmental Quality, Forty Most Asked Questions Concerning CEQ'S National Environmental Policy Act Regulations, 40 CFR 1500–1508, Federal Register Vol. 46, No.55, 18026–18038, March 23, 1981.

8. *Hawksbill Sea Turtle v Fema*, 126 F.3d 461 (3rd Cir. 1997); *City of Waltham v. US Postal Service*, 11 F.3d 235 (1st Cir. 1993); *Roanoke River Basin Ass'n v. Hudson*, 940 F.2d 58 (4th Cir. 1991); *Abenaki Nation of Mississquoi v. Hughes*, 805 F.Supp 234 (D.Vt 1992).

9. *Preservation Coalition v Pierce*, 667 F.2d 851 (9th Cir. 1982).

10. *Louisiana v Lee*, 758 F.2d 1081 (5th Cir. 1985).

11. *CARE Now, Inc. v. Federal Aviation Administration*, 844 F.2d 1569 (11th Cir. 1988); *Roanoke River Basin Association v. Hudson*, 940 F.2d 58 (4th Cir. 1991); *Abenaki Indian Nation v. Hughes*, 805 F. Supp. 234 (D. Vt. 1992).

12. *Friends of Pavette v Horseshoe Bend Hydroelectric*, 988 F.2d 989 (9th Cir. 1993).

13. *City of Auburn v US Government*, 154 F.3d 1025 (9th Cir. 1998).

14. *United States v. 27.09 Acres of Land (II)*, 760 F. Supp. 345 (S.D.N.Y 1991); *Morgan v. Walter*, 728 F. Supp. 1483 (D. Idaho 1989).

15. *Coalition for Canyon Preservation and Wildlands Center for Preventing Roads v Department of Transportation*, No. CV 98-84-M-DWM, 1999, U.S. District., LEXIS 835 (D. Mont. January 19, 1999).

16. Daniels S. E. and Kelly C. M., Deciding between an EA and an EIS may be a question of mitigation, *Western Journal of Applied Forestry*, Vol 5, No 4, March 1991.

10

Summary

Clothes make the man. Naked people have little or no influence in society.

Mark Twain

Given such a disproportionate number of Environmental Assessments (EA) to Environmental Impact Statements (EIS), agencies are clearly going to great lengths to avoid preparing EISs. The increased reliance on EAs can also be partly attributed to the fact that agencies are learning to integrate the National Environmental Policy Act's (NEPA) objectives (e.g., proposing lower impact alternatives, integrating mitigation measures into project designs) into early planning, thus reducing potential impacts before actions mature to the permitting stage.

When challenged, an EA is normally more difficult to defend than an EIS, because the burden of proof is on the agency to demonstrate that significant impacts would not result or that any such impacts can be mitigated. An EIS does not carry this burden of proof. For this reason, adversaries have increasingly focused efforts on challenging EAs, which are considered more vulnerable targets; agencies can defend against this strategy by devoting particular care to the preparation of their EAs and in exercising prudence in reaching a conclusion regarding significance. Alternatively, adversaries may increase their odds of success by directing efforts at flawed or poorly prepared EAs and Findings of No Significant Impact (FONSI).

Conclusions regarding significance are reserved for the decisionmaker, based on the information presented in the EA. The decisionmaker is responsible for ensuring that a FONSI is never issued for an action where an EIS is warranted.

EAs that make cursory, sweeping, or unsubstantiated conclusions are at particular risk of a successful challenge. The Council on Environmental Quality (CEQ) has also cautioned that repeated failure to involve the public can constitute a violation of the NEPA regulations, providing adversaries with sufficient grounds for a successful legal challenge.

An EA must provide a rigorous, scientifically accurate, and unbiased basis for assessing potential significance. The analysis should be prepared so as to clearly assist the reader in drawing conclusions regarding significance. All investigated impacts should be clearly documented and based on professionally accepted scientific methodologies.

   With respect to the regulatory vacuum that exists in providing detailed direction on preparing EAs, it is the author's hope that this text has helped bridge this void; applying the "rule of reason," this text identifies relevant EIS regulatory requirements that can also be logically interpreted to apply to preparation of EAs. The text has also attempted to capture the professional experience and best professional practices from seasoned practitioners who have spent years preparing EAs. Lacking detailed regulatory direction, the focus of this book has been on providing the reader with a reasonable, definitive, consistent, and comprehensive methodology for managing, analyzing, and writing EAs.

Appendices

Appendix A: The NEPA Act

THE NATIONAL ENVIRONMENTAL POLICY ACT OF 1969, as amended (Pub. L. 91-190, 42 U.S.C. 4321-4347, January 1, 1970, as amended by Pub. L. 94-52, July 3, 1975, Pub. L. 94-83, August 9, 1975, and Pub. L. 97-258, § 4(b), Sept. 13, 1982)

An Act to establish a national policy for the environment, to provide for the establishment of a Council on Environmental Quality, and for other purposes.

Be it enacted by the Senate and House of Representatives of the United States of America in Congress assembled, That this Act may be cited as the "National Environmental Policy Act of 1969."

PURPOSE

Sec. 2 [42 USC § 4321]. The purposes of this Act are: To declare a national policy which will encourage productive and enjoyable harmony between man and his environment; to promote efforts which will prevent or eliminate damage to the environment and biosphere and stimulate the health and welfare of man; to enrich the understanding of the ecological systems and natural resources important to the Nation; and to establish a Council on Environmental Quality.

TITLE I

Congressional Declaration of National Environmental Policy

Sec. 101 [42 USC § 4331].

(a) The Congress, recognizing the profound impact of man's activity on the interrelations of all components of the natural environment, particularly the profound influences of population growth, high-density urbanization, industrial expansion, resource exploitation, and new and expanding technological advances and recognizing further the critical importance of restoring and maintaining environmental quality to the overall welfare and development of man, declares that it is the continuing policy of the Federal Government, in cooperation with State and local governments, and other concerned public and private organizations, to use all practicable means and measures, including financial and technical assistance, in a manner calculated to foster and promote the general welfare, to create and maintain conditions under which man and nature can exist in productive harmony, and fulfill the social, economic, and other requirements of present and future generations of Americans.

(b) In order to carry out the policy set forth in this Act, it is the continuing responsibility of the Federal Government to use all practicable means, consist with other essential considerations of national policy, to improve and coordinate Federal plans, functions, programs, and resources to the end that the Nation may —

    (1) fulfill the responsibilities of each generation as trustee of the environment for succeeding generations;

    (2) assure for all Americans safe, healthful, productive, and aesthetically and culturally pleasing surroundings;

141

(3) attain the widest range of beneficial uses of the environment without degradation, risk to health or safety, or other undesirable and unintended consequences;

(4) preserve important historic, cultural and natural aspects of our national heritage, and maintain, wherever possible, an environment which supports diversity, and variety of individual choice;

(5) achieve a balance between population and resource use which will permit high standards of living and a wide sharing of life's amenities; and

(6) enhance the quality of renewable resources and approach the maximum attainable recycling of depletable resources.

(c) The Congress recognizes that each person should enjoy a healthful environment and that each person has a responsibility to contribute to the preservation and enhancement of the environment.

Sec. 102 [42 USC § 4332]. The Congress authorizes and directs that, to the fullest extent possible: (1) the policies, regulations, and public laws of the United States shall be interpreted and administered in accordance with the policies set forth in this Act, and (2) all agencies of the Federal Government shall —

(A) utilize a systematic, interdisciplinary approach which will insure the integrated use of the natural and social sciences and the environmental design arts in planning and in decisionmaking which may have an impact on man's environment;

(B) identify and develop methods and procedures, in consultation with the Council on Environmental Quality established by title II of this Act, which will insure that presently unquantified environmental amenities and values may be given appropriate consideration in decisionmaking along with economic and technical considerations;

(C) include in every recommendation or report on proposals for legislation and other major Federal actions significantly affecting the quality of the human environment, a detailed statement by the responsible official on —

(i) the environmental impact of the proposed action,

(ii) any adverse environmental effects which cannot be avoided should the proposal be implemented,

(iii) alternatives to the proposed action,

(iv) the relationship between local short-term uses of man's environment and the maintenance and enhancement of long-term productivity, and

(v) any irreversible and irretrievable commitments of resources which would be involved in the proposed action should it be implemented.

Prior to making any detailed statement, the responsible Federal official shall consult with and obtain the comments of any Federal agency which has jurisdiction by law or special expertise with respect to any environmental impact involved. Copies of such statement and the comments and views of the appropriate Federal, State, and local agencies, which are authorized to develop and enforce environmental standards, shall be made available to the President, the Council on Environmental Quality and to the public as provided by section 552 of title 5, United States Code, and shall accompany the proposal through the existing agency review processes;

(D) Any detailed statement required under subparagraph (C) after January 1, 1970, for any major Federal action funded under a program of grants to States shall not be deemed to be legally insufficient solely by reason of having been prepared by a State agency or official, if:

(i) the State agency or official has statewide jurisdiction and has the responsibility for such action,

(ii) the responsible Federal official furnishes guidance and participates in such preparation,

(iii) the responsible Federal official independently evaluates such statement prior to its approval and adoption, and

(iv) after January 1, 1976, the responsible Federal official provides early notification to, and solicits the views of, any other State or any Federal land management entity of any action or any alternative thereto which may have significant impacts upon such State or affected Federal land management entity and, if there is any disagreement on such impacts, prepares a written assessment of such impacts and views for incorporation into such detailed statement.

The procedures in this subparagraph shall not relieve the Federal official of his responsibilities for the scope, objectivity, and content of the entire statement or of any other responsibility under this Act; and further, this subparagraph does not affect the legal sufficiency of statements prepared by State agencies with less than statewide jurisdiction.

(E) study, develop, and describe appropriate alternatives to recommended courses of action in any proposal which involves unresolved conflicts concerning alternative uses of available resources;

(F) recognize the worldwide and long-range character of environmental problems and, where consistent with the foreign policy of the United States, lend appropriate support to initiatives, resolutions, and programs designed to maximize international cooperation in anticipating and preventing a decline in the quality of mankind's world environment;

(G) make available to States, counties, municipalities, institutions, and individuals, advice and information useful in restoring, maintaining, and enhancing the quality of the environment;

(H) initiate and utilize ecological information in the planning and development of resource-oriented projects; and

(I) assist the Council on Environmental Quality established by title II of this Act.

Sec. 103 [42 USC § 4333]. All agencies of the Federal Government shall review their present statutory authority, administrative regulations, and current policies and procedures for the purpose of determining whether there are any deficiencies or inconsistencies therein which prohibit full compliance with the purposes and provisions of this Act and shall propose to the President not later than July 1, 1971, such measures as may be necessary to bring their authority and policies into conformity with the intent, purposes, and procedures set forth in this Act.

Sec. 104 [42 USC § 4334]. Nothing in section 102 [42 USC § 4332] or 103 [42 USC § 4333] shall in any way affect the specific statutory obligations of any Federal agency (1) to comply with criteria or standards of environmental quality, (2) to coordinate or consult with any other Federal or State agency, or (3) to act, or refrain from acting contingent upon the recommendations or certification of any other Federal or State agency.

Sec. 105 [42 USC § 4335]. The policies and goals set forth in this Act are supplementary to those set forth in existing authorizations of Federal agencies.

TITLE II

Council on Environmental Quality

Sec. 201 [42 USC § 4341]. The President shall transmit to the Congress annually beginning July 1, 1970, an Environmental Quality Report (hereinafter referred to as the "report") which shall set forth (1) the status and condition of the major natural, manmade, or altered environmental classes of the Nation, including, but not limited to, the air, the aquatic, including marine, estuarine, and fresh water, and the terrestrial environment, including, but not limited to, the forest, dryland, wetland, range, urban, suburban and rural environment; (2) current and foreseeable trends in the quality, management and utilization of such environments and the effects of those trends on the social, economic, and other requirements of the Nation; (3) the adequacy of available natural resources for fulfilling human and economic requirements of the Nation in the light of expected population pressures; (4) a review of the programs and activities (including regulatory activities) of the Federal Government, the State and local governments, and nongovernmental entities or individuals with particular reference to their effect on the environment and on the conservation, development and utilization of natural resources; and (5) a program for remedying the deficiencies of existing programs and activities, together with recommendations for legislation.

Sec. 202 [42 USC § 4342]. There is created in the Executive Office of the President a Council on Environmental Quality (hereinafter referred to as the "Council"). The Council shall be composed of three members who shall be appointed by the President to serve at his pleasure,by and with the advice and consent of the Senate. The President shall designate one of the members of the Council to serve as Chairman. Each member shall be a person who, as a result of his training, experience, and attainments, is exceptionally well qualified to analyze and interpret environmental trends and information of all kinds; to appraise programs and activities of the Federal Government in the light of the policy set forth in title I of this Act; to be conscious of and responsive to the scientific, economic, social, aesthetic, and cultural needs and interests of the Nation; and to formulate and recommend national policies to promote the improvement of the quality of the environment.

Sec. 203 [42 USC § 4343].

(a) The Council may employ such officers and employees as may be necessary to carry out its functions under this Act. In addition, the Council may employ and fix the compensation of such experts and consultants as may be necessary for the carrying out of its functions under this Act, in accordance with section 3109 of title 5, United States Code (but without regard to the last sentence thereof).

(b) Notwithstanding section 1342 of Title 31, the Council may accept and employ voluntary and uncompensated services in furtherance of the purposes of the Council:

Sec. 204 [42 USC § 4344]. It shall be the duty and function of the Council.

(1) to assist and advise the President in the preparation of the Environmental Quality Report required by section 201 [42 USC § 4341] of this title;

(2) to gather timely and authoritative information concerning the conditions and trends in the quality of the environment both current and prospective, to analyze and interpret such information for the purpose of determining whether such conditions and trends are interfering, or are likely to interfere, with the achievement of the policy set forth in title I of this Act, and to compile and submit to the President studies relating to such conditions and trends;

(3) to review and appraise the various programs and activities of the Federal Government in light of the policy set forth in title I of this Act for the purpose of determining the extent to which such programs and activities are contributing to the achievement of such policy, and to make recommendations to the President with respect thereto;

(4) to develop and recommend to the President national policies to foster and promote the improvement of environmental quality to meet the conservation, social, economic, health, and other requirements and goals of the Nation;

(5) to conduct investigations, studies, surveys, research, and analyses relating to environmental quality;

(6) to document and define changes in the natural environment, including the plant and animal systems, and to accumulate necessary data and other information for a continuing analysis of these changes or trends and an interpretation of their underlying causes;

(7) to report at least once each year to the President on the state and condition of the environment; and

(8) to make and furnish such studies, reports thereon, and recommendations with respect to matters of policy and legislation as the President may request.

Sec. 205 [42 USC § 4345]. In exercising its powers, functions, and duties under this Act, the Council shall —

(1) consult with the Citizens' Advisory Committee on Environmental Quality established by Executive Order No. 11472, dated May 29, 1969, and with such representatives of science, industry, agriculture, labor, conservation organizations, State and local governments and other groups, as it deems advisable; and

(2) utilize, to the fullest extent possible, the services, facilities and information (including statistical information) of public and private agencies and organizations, and individuals, in order that duplication of effort and expense may be avoided, thus assuring that the Council's activities will not unnecessarily overlap or conflict with similar activities authorized by law and performed by established agencies.

Sec. 206 [42 USC § 4346]. Members of the Council shall serve full time and the Chairman of the Council shall be compensated at the rate provided for Level II of the Executive Schedule Pay Rates [5 USC § 5313]. The other members of the Council shall be compensated at the rate provided for Level IV of the Executive Schedule Pay Rates [5 USC § 5315].

Sec. 207 [42 USC § 4346a]. The Council may accept reimbursements from any private nonprofit organization or from any department, agency, or instrumentality of the Federal Government, any State, or local government, for the reasonable travel expenses incurred by an officer or employee of the Council in connection with his attendance at any conference, seminar, or similar meeting conducted for the benefit of the Council.

Sec. 208 [42 USC § 4346b]. The Council may make expenditures in support of its international activities, including expenditures for: (1) international travel; (2) activities in implementation of international agreements; and (3) the support of international exchange programs in the United States and in foreign countries.

Sec. 209 [42 USC § 4347]. There are authorized to be appropriated to carry out the provisions of this chapter not to exceed $300,000 for fiscal year 1970, $700,000 for fiscal year 1971, and $1,000,000 for each fiscal year thereafter.

Appendix B: The CEQ NEPA Regulations

TABLE OF CONTENTS

PART 1500—PURPOSE, POLICY, AND MANDATE

Sec.
1500.1 Purpose.
1500.2 Policy.
1500.3 Mandate.
1500.4 Reducing paperwork.
1500.5 Reducing delay.
1500.6 Agency authority.

AUTHORITY: NEPA, the Environmental Quality Improvement Act of 1970, as amended (42 U.S.C. 4371 *et seq.*), sec. 309 of the Clean Air Act, as amended (42 U.S.C. 7609) and E.O. 11514, Mar. 5, 1970, as amended by E.O. 11991, May 24, 1977).

SOURCE: 43 FR 55990, Nov. 28, 1978, unless otherwise noted.

§ 1500.1 Purpose.

(a) The National Environmental Policy Act (NEPA) is our basic national charter for protection of the environment. It establishes policy, sets goals (section 101), and provides means (section 102) for carrying out the policy. Section 102(2) contains "action-forcing" provisions to make sure that federal agencies act according to the letter and spirit of the Act. The regulations that follow implement section 102(2). Their purpose is to tell federal agencies what they must do to comply with the procedures and achieve the goals of the Act. The President, the federal agencies, and the courts share responsibility for enforcing the Act so as to achieve the substantive requirements of section 101.

(b) NEPA procedures must insure that environmental information is available to public officials and citizens before decisions are made and before actions are taken. The information must be of high quality. Accurate scientific analysis, expert agency comments, and public scrutiny are essential to implementing NEPA. Most important, NEPA documents must concentrate on the issues that are truly significant to the action in question, rather than amassing needless detail.

(c) Ultimately, of course, it is not better documents but better decisions that count. NEPA's purpose is not to generate paperwork—even excellent paperwork—but to foster excellent action. The NEPA process is intended to help public officials make decisions that are based on understanding of environmental consequences, and take actions that protect, restore, and enhance the environment. These regulations provide the direction to achieve this purpose.

§ 1500.2 Policy.

Federal agencies shall to the fullest extent possible:

(a) Interpret and administer the policies, regulations, and public laws of the United States in accordance with the policies set forth in the Act and in these regulations.

(b) Implement procedures to make the NEPA process more useful to decisionmakers and the public; to reduce paperwork and the accumulation of extraneous background data; and to emphasize real environmental issues and alternatives. Environmental impact statements shall be concise, clear, and to the point, and shall be supported by evidence that agencies have made the necessary environmental analyses.

(c) Integrate the requirements of NEPA with other planning and environmental review procedures required by law or by agency practice so that all such procedures run concurrently rather than consecutively.

(d) Encourage and facilitate public involvement in decisions which affect the quality of the human environment.

(e) Use the NEPA process to identify and assess the reasonable alternatives to proposed actions that will avoid or minimize adverse effects of these actions upon the quality of the human environment.

(f) Use all practicable means, consistent with the requirements of the Act and other essential considerations of national policy, to restore and enhance the quality of the human environment and avoid or minimize any possible adverse effects of their actions upon the quality of the human environment.

§ 1500.3 Mandate.

Parts 1500 through 1508 of this title provide regulations applicable to and binding on all Federal agencies for implementing the procedural provisions of the National Environmental Policy Act of 1969, as amended (Pub. L. 91–190, 42 U.S.C. 4321 et seq.) (NEPA or the Act) except where compliance would be inconsistent with other statutory requirements. These regulations are issued pursuant to NEPA, the Environmental Quality Improvement Act of 1970, as amended (42 U.S.C. 4371 et seq.) section 309 of the Clean Air Act, as amended (42 U.S.C. 7609) and Executive Order 11514, Protection and Enhancement of Environmental Quality (March 5, 1970, as amended by Executive Order 11991, May 24, 1977). These regulations, unlike the predecessor guidelines, are not confined to sec. 102(2)(C) (environmental impact statements). The regulations apply to the whole of section 102(2). The provisions of the Act and of these regulations must be read together as a whole in order to comply with the spirit and letter of the law. It is the Council's intention that judicial review of agency compliance with these regulations not occur before an agency has filed the final environmental impact statement, or has made a final finding of no significant impact (when such a finding will result in action affecting the environment), or takes action that will result in irreparable injury. Furthermore, it is the Council's intention that any trivial violation of these regulations not give rise to any independent cause of action.

§ 1500.4   Reducing paperwork.

Agencies shall reduce excessive paperwork by:

(a) Reducing the length of environmental impact statements (§ 1502.2(c)), by means such as setting appropriate page limits (§§ 1501.7(b)(1) and 1502.7).

(b) Preparing analytic rather than encyclopedic environmental impact statements (§ 1502.2(a)).

(c) Discussing only briefly issues other than significant ones (§ 1502.2(b)).

(d) Writing environmental impact statements in plain language (§ 1502.8).

(e) Following a clear format for environmental impact statements (§ 1502.10).

(f) Emphasizing the portions of the environmental impact statement that are useful to decisionmakers and the public (§§ 1502.14 and 1502.15) and reducing emphasis on background material (§ 1502.16).

(g) Using the scoping process, not only to identify significant environmental issues deserving of study, but also to deemphasize insignificant issues, narrowing the scope of the environmental impact statement process accordingly (§ 1501.7).

(h) Summarizing the environmental impact statement (§ 1502.12) and circulating the summary instead of the entire environmental impact statement if the latter is unusually long (§ 1502.19).

(i) Using program, policy, or plan environmental impact statements and tiering from statements of broad scope to those of narrower scope, to eliminate repetitive discussions of the same issues (§§ 1502.4 and 1502.20).

(j) Incorporating by reference (§ 1502.21).

(k) Integrating NEPA requirements with other environmental review and consultation requirements (§ 1502.25).

(l) Requiring comments to be as specific as possible (§ 1503.3).

(m) Attaching and circulating only changes to the draft environmental impact statement, rather than rewriting and circulating the entire statement when changes are minor (§ 1503.4(c)).

(n) Eliminating duplication with State and local procedures, by providing for joint preparation (§ 1506.2), and with other Federal procedures, by providing that an agency may adopt appropriate environmental documents prepared by another agency (§ 1506.3).

(o) Combining environmental documents with other documents (§ 1506.4).

(p) Using categorical exclusions to define categories of actions which do not individually or cumulatively have a significant effect on the human environment and which are therefore exempt from requirements to prepare an environmental impact statement (§ 1508.4).

(q) Using a finding of no significant impact when an action not otherwise excluded will not have a significant effect on the human environment and is therefore exempt from requirements to prepare an environmental impact statement (§ 1508.13).

[43 FR 55990, Nov. 29, 1978; 44 FR 873, Jan. 3, 1979]

§ 1500.5 Reducing delay.

Agencies shall reduce delay by:

(a) Integrating the NEPA process into early planning (§ 1501.2).

(b) Emphasizing interagency cooperation before the environmental impact statement is prepared, rather than submission of adversary comments on a completed document (§ 1501.6).

(c) Insuring the swift and fair resolution of lead agency disputes (§ 1501.5).

(d) Using the scoping process for an early identification of what are and what are not the real issues (§ 1501.7).

(e) Establishing appropriate time limits for the environmental impact statement process (§§ 1501.7(b)(2) and 1501.8).

(f) Preparing environmental impact statements early in the process (§ 1502.5).

(g) Integrating NEPA requirements with other environmental review and consultation requirements (§ 1502.25).

(h) Eliminating duplication with State and local procedures by providing for joint preparation (§ 1506.2) and with other Federal procedures by providing that an agency may adopt appropriate environmental documents prepared by another agency (§ 1506.3).

(i) Combining environmental documents with other documents (§ 1506.4).

(j) Using accelerated procedures for proposals for legislation (§ 1506.8).

(k) Using categorical exclusions to define categories of actions which do not individually or cumulatively have a significant effect on the human environment (§ 1508.4) and which are therefore exempt from requirements to prepare an environmental impact statement.

(l) Using a finding of no significant impact when an action not otherwise excluded will not have a significant effect on the human environment (§ 1508.13) and is therefore exempt from requirements to prepare an environmental impact statement.

§ 1500.6 Agency authority.

Each agency shall interpret the provisions of the Act as a supplement to its existing authority and as a mandate to view traditional policies and missions in the light of the Act's national environmental objectives. Agencies shall review their policies, procedures, and regulations accordingly and revise them as necessary to insure full compliance with the purposes and provisions of the Act. The phrase "to the fullest extent possible" in section 102 means that each agency of the Federal Government shall comply with that section unless existing law applicable to the agency's operations expressly prohibits or makes compliance impossible.

PART 1501—NEPA AND AGENCY PLANNING

Sec.
1501.1 Purpose.
1501.2 Apply NEPA early in the process.
1501.3 When to prepare an environmental assessment.
1501.4 Whether to prepare an environmental impact statement.
1501.5 Lead agencies.
1501.6 Cooperating agencies.
1501.7 Scoping.
1501.8 Time limits.

AUTHORITY: NEPA, the Environmental Quality Improvement Act of 1970, as amended (42 U.S.C. 4371 *et seq.*), sec. 309 of the Clean Air Act, as amended (42 U.S.C. 7609, and E.O. 11514 (Mar. 5, 1970, as amended by E.O. 11991, May 24, 1977).

SOURCE: 43 FR 55992, Nov. 29, 1978, unless otherwise noted.

§ 1501.1 Purpose.

The purposes of this part include:

(a) Integrating the NEPA process into early planning to insure appropriate consideration of NEPA's policies and to eliminate delay.

(b) Emphasizing cooperative consultation among agencies before the environmental impact statement is prepared rather than submission of adversary comments on a completed document.

(c) Providing for the swift and fair resolution of lead agency disputes.

(d) Identifying at an early stage the significant environmental issues deserving of study and deemphasizing insignificant issues, narrowing the scope of the environmental impact statement accordingly.

(e) Providing a mechanism for putting appropriate time limits on the environmental impact statement process.

§ 1501.2 Apply NEPA early in the process.

Agencies shall integrate the NEPA process with other planning at the earliest possible time to insure that planning and decisions reflect environmental values, to avoid delays later in the process, and to head off potential conflicts. Each agency shall:

(a) Comply with the mandate of section 102(2)(A) to "utilize a systematic, interdisciplinary approach which will insure the integrated use of the natural and social sciences and the environmental design arts in planning and in decisionmaking which may have an impact on man's environment," as specified by § 1507.2.

(b) Identify environmental effects and values in adequate detail so they can be compared to economic and technical analyses. Environmental documents and appropriate analyses shall be circulated and reviewed at the same time as other planning documents.

(c) Study, develop, and describe appropriate alternatives to recommended courses of action in any proposal which involves unresolved conflicts concerning alternative uses of available resources as provided by section 102(2)(E) of the Act.

(d) Provide for cases where actions are planned by private applicants or other non-Federal entities before Federal involvement so that:

(1) Policies or designated staff are available to advise potential applicants of studies or other information foreseeably required for later Federal action.

(2) The Federal agency consults early with appropriate State and local agencies and Indian tribes and with interested private persons and organizations when its own involvement is reasonably foreseeable.

(3) The Federal agency commences its NEPA process at the earliest possible time.

§ 1501.3 When to prepare an environmental assessment.

(a) Agencies shall prepare an environmental assessment (§ 1508.9) when necessary under the procedures adopted by individual agencies to supplement these regulations as described in § 1507.3. An assessment is not necessary if the agency has decided to prepare an environmental impact statement.

(b) Agencies may prepare an environmental assessment on any action at any time in order to assist agency planning and decisionmaking.

§ 1501.4 Whether to prepare an environmental impact statement.

In determining whether to prepare an environmental impact statement the Federal agency shall:

(a) Determine under its procedures supplementing these regulations (described in § 1507.3) whether the proposal is one which:

(1) Normally requires an environmental impact statement, or

(2) Normally does not require either an environmental impact statement or an environmental assessment (categorical exclusion).

(b) If the proposed action is not covered by paragraph (a) of this section, prepare an environmental assessment (§ 1508.9). The agency shall involve environmental agencies, applicants, and the public, to the extent practicable, in preparing assessments required by § 1508.9(a)(1).

(c) Based on the environmental assessment make its determination whether to prepare an environmental impact statement.

(d) Commence the scoping process (§ 1501.7), if the agency will prepare an environmental impact statement.

(e) Prepare a finding of no significant impact (§ 1508.13), if the agency determines on the basis of the environmental assessment not to prepare a statement.

(1) The agency shall make the finding of no significant impact available

to the affected public as specified in § 1506.6.

(2) In certain limited circumstances, which the agency may cover in its procedures under § 1507.3, the agency shall make the finding of no significant impact available for public review (including State and areawide clearinghouses) for 30 days before the agency makes its final determination whether to prepare an environmental impact statement and before the action may begin. The circumstances are:

(i) The proposed action is, or is closely similar to, one which normally requires the preparation of an environmental impact statement under the procedures adopted by the agency pursuant to § 1507.3, or

(ii) The nature of the proposed action is one without precedent.

§ 1501.5 Lead agencies.

(a) A lead agency shall supervise the preparation of an environmental impact statement if more than one Federal agency either:

(1) Proposes or is involved in the same action; or

(2) Is involved in a group of actions directly related to each other because of their functional interdependence or geographical proximity.

(b) Federal, State, or local agencies, including at least one Federal agency, may act as joint lead agencies to prepare an environmental impact statement (§ 1506.2).

(c) If an action falls within the provisions of paragraph (a) of this section the potential lead agencies shall determine by letter or memorandum which agency shall be the lead agency and which shall be cooperating agencies. The agencies shall resolve the lead agency question so as not to cause delay. If there is disagreement among the agencies, the following factors (which are listed in order of descending importance) shall determine lead agency designation:

(1) Magnitude of agency's involvement.

(2) Project approval/disapproval authority.

(3) Expertise concerning the action's environmental effects.

(4) Duration of agency's involvement.

(5) Sequence of agency's involvement.

(d) Any Federal agency, or any State or local agency or private person substantially affected by the absence of lead agency designation, may make a written request to the potential lead agencies that a lead agency be designated.

(e) If Federal agencies are unable to agree on which agency will be the lead agency or if the procedure described in paragraph (c) of this section has not resulted within 45 days in a lead agency designation, any of the agencies or persons concerned may file a request with the Council asking it to determine which Federal agency shall be the lead agency.

A copy of the request shall be transmitted to each potential lead agency. The request shall consist of:

(1) A precise description of the nature and extent of the proposed action.

(2) A detailed statement of why each potential lead agency should or should not be the lead agency under the criteria specified in paragraph (c) of this section.

(f) A response may be filed by any potential lead agency concerned within 20 days after a request is filed with the Council. The Council shall determine as soon as possible but not later than 20 days after receiving the request and all responses to it which Federal agency shall be the lead agency and which other Federal agencies shall be cooperating agencies.

[43 FR 55992, Nov. 29, 1978; 44 FR 873, Jan. 3, 1979]

§ 1501.6 Cooperating agencies.

The purpose of this section is to emphasize agency cooperation early in the NEPA process. Upon request of the lead agency, any other Federal agency which has jurisdiction by law shall be a cooperating agency. In addition any other Federal agency which has special expertise with respect to any environmental issue, which should be addressed in the statement may be a cooperating agency upon request of the lead agency. An agency may request the lead agency to designate it a cooperating agency.

(a) The lead agency shall:

(1) Request the participation of each cooperating agency in the NEPA process at the earliest possible time.

(2) Use the environmental analysis and proposals of cooperating agencies with jurisdiction by law or special expertise, to the maximum extent possible consistent with its responsibility as lead agency.

(3) Meet with a cooperating agency at the latter's request.

(b) Each cooperating agency shall:

(1) Participate in the NEPA process at the earliest possible time.

(2) Participate in the scoping process (described below in § 1501.7).

(3) Assume on request of the lead agency responsibility for developing information and preparing environmental analyses including portions of the environmental impact statement concerning which the cooperating agency has special expertise.

(4) Make available staff support at the lead agency's request to enhance the latter's interdisciplinary capability.

(5) Normally use its own funds. The lead agency shall, to the extent available funds permit, fund those major activities or analyses it requests from cooperating agencies. Potential lead agencies shall include such funding requirements in their budget requests.

(c) A cooperating agency may in response to a lead agency's request for assistance in preparing the environmental impact statement (described in paragraph (b) (3), (4), or (5) of this section) reply that other program commitments preclude any involvement or the degree of involvement requested in the action that is the subject of the environmental impact statement. A copy of this reply shall be submitted to the Council.

§ 1501.7 Scoping.

There shall be an early and open process for determining the scope of issues to be addressed and for identifying the significant issues related to a proposed action. This process shall be termed scoping. As soon as practicable after its decision to prepare an environmental impact statement and

before the scoping process the lead agency shall publish a notice of intent (§ 1508.22) in the Federal Register except as provided in § 1507.3(e).

(a) As part of the scoping process the lead agency shall:

(1) Invite the participation of affected Federal, State, and local agencies, any affected Indian tribe, the proponent of the action, and other interested persons (including those who might not be in accord with the action on environmental grounds), unless there is a limited exception under § 1507.3(c). An agency may give notice in accordance with § 1506.6.

(2) Determine the scope (§ 1508.25) and the significant issues to be analyzed in depth in the environmental impact statement.

(3) Identify and eliminate from detailed study the issues which are not significant or which have been covered by prior environmental review (§ 1506.3), narrowing the discussion of these issues in the statement to a brief presentation of why they will not have a significant effect on the human environment or providing a reference to their coverage elsewhere.

(4) Allocate assignments for preparation of the environmental impact statement among the lead and cooperating agencies, with the lead agency retaining responsibility for the statement.

(5) Indicate any public environmental assessments and other environmental impact statements which are being or will be prepared that are related to but are not part of the scope of the impact statement under consideration.

(6) Identify other environmental review and consultation requirements so the lead and cooperating agencies may prepare other required analyses and studies concurrently with, and integrated with, the environmental impact statement as provided in § 1502.25.

(7) Indicate the relationship between the timing of the preparation of environmental analyses and the agency's tentative planning and decisionmaking schedule.

(b) As part of the scoping process the lead agency may:

(1) Set page limits on environmental documents (§ 1502.7).

§ 1501.8

(2) Set time limits (§ 1501.8).

(3) Adopt procedures under § 1507.3 to combine its environmental assessment process with its scoping process.

(4) Hold an early scoping meeting or meetings which may be integrated with any other early planning meeting the agency has. Such a scoping meeting will often be appropriate when the impacts of a particular action are confined to specific sites.

(c) An agency shall revise the determinations made under paragraphs (a) and (b) of this section if substantial changes are made later in the proposed action, or if significant new circumstances or information arise which bear on the proposal or its impacts.

§ 1501.8 Time limits.

Although the Council has decided that prescribed universal time limits for the entire NEPA process are too inflexible, Federal agencies are encouraged to set time limits appropriate to individual actions (consistent with the time intervals required by § 1506.10). When multiple agencies are involved the reference to agency below means lead agency.

(a) The agency shall set time limits if an applicant for the proposed action requests them: *Provided,* That the limits are consistent with the purposes of NEPA and other essential considerations of national policy.

(b) The agency may:

(1) Consider the following factors in determining time limits:

(i) Potential for environmental harm.

(ii) Size of the proposed action.

(iii) State of the art of analytic techniques.

(iv) Degree of public need for the proposed action, including the consequences of delay.

(v) Number of persons and agencies affected.

(vi) Degree to which relevant information is known and if not known the time required for obtaining it.

(vii) Degree to which the action is controversial.

(viii) Other time limits imposed on the agency by law, regulations, or executive order.

(2) Set overall time limits or limits for each constituent part of the NEPA process, which may include:

(i) Decision on whether to prepare an environmental impact statement (if not already decided).

(ii) Determination of the scope of the environmental impact statement.

(iii) Preparation of the draft environmental impact statement.

(iv) Review of any comments on the draft environmental impact statement from the public and agencies.

(v) Preparation of the final environmental impact statement.

(vi) Review of any comments on the final environmental impact statement.

(vii) Decision on the action based in part on the environmental impact statement.

(3) Designate a person (such as the project manager or a person in the agency's office with NEPA responsibilities) to expedite the NEPA process.

(c) State or local agencies or members of the public may request a Federal Agency to set time limits.

PART 1502—ENVIRONMENTAL IMPACT STATEMENT

Sec.
1502.23 Cost-benefit analysis.
1502.24 Methodology and scientific accuracy.
1502.25 Environmental review and consultation requirements.

AUTHORITY: NEPA, the Environmental Quality Improvement Act of 1970, as amended (42 U.S.C. 4371 *et seq.*), sec. 309 of the Clean Air Act, as amended (42 U.S.C. 7609), and E.O. 11514 (Mar. 5, 1970, as amended by E.O. 11991, May 24, 1977).

SOURCE: 43 FR 55994, Nov. 29, 1978, unless otherwise noted.

§ 1502.1 Purpose.

The primary purpose of an environmental impact statement is to serve as an action-forcing device to insure that the policies and goals defined in the Act are infused into the ongoing programs and actions of the Federal Government. It shall provide full and fair discussion of significant environmental impacts and shall inform decisionmakers and the public of the reasonable alternatives which would avoid or minimize adverse impacts or enhance the quality of the human environment. Agencies shall focus on significant environmental issues and alternatives and shall reduce paperwork and the accumulation of extraneous background data. Statements shall be concise, clear, and to the point, and shall be supported by evidence that the agency has made the necessary environmental analyses. An environmental impact statement is more than a disclosure document. It shall be used by Federal officials in conjunction with other relevant material to plan actions and make decisions.

§ 1502.2 Implementation.

To achieve the purposes set forth in § 1502.1 agencies shall prepare environmental impact statements in the following manner:

(a) Environmental impact statements shall be analytic rather than encyclopedic.

(b) Impacts shall be discussed in proportion to their significance. There shall be only brief discussion of other than significant issues. As in a finding of no significant impact, there should be only enough discussion to show why more study is not warranted.

(c) Environmental impact statements shall be kept concise and shall be no longer than absolutely necessary to comply with NEPA and with these regulations. Length should vary first with potential environmental problems and then with project size.

(d) Environmental impact statements shall state how alternatives considered in it and decisions based on it will or will not achieve the requirements of sections 101 and 102(1) of the Act and other environmental laws and policies.

(e) The range of alternatives discussed in environmental impact statements shall encompass those to be considered by the ultimate agency decisionmaker.

(f) Agencies shall not commit resources prejudicing selection of alternatives before making a final decision (§ 1506.1).

(g) Environmental impact statements shall serve as the means of assessing the environmental impact of proposed agency actions, rather than justifying decisions already made.

§ 1502.3 Statutory requirements for statements.

As required by sec. 102(2)(C) of NEPA environmental impact statements (§ 1508.11) are to be included in every recommendation or report.

On proposals (§ 1508.23).

For legislation and (§ 1508.17).

Other major Federal actions (§ 1508.18).

Significantly (§ 1508.27).

Affecting (§§ 1508.3, 1508.8).

The quality of the human environment (§ 1508.14).

§ 1502.4 Major Federal actions requiring the preparation of environmental impact statements.

(a) Agencies shall make sure the proposal which is the subject of an environmental impact statement is properly defined. Agencies shall use the criteria for scope (§ 1508.25) to determine which proposal(s) shall be the subject of a particular statement. Proposals or parts of proposals which are related to each other closely enough to be, in effect, a single course of action shall be evaluated in a single impact statement.

(b) Environmental impact statements may be prepared, and are sometimes required, for broad Federal actions such as the adoption of new agency programs or regulations (§ 1508.18). Agencies shall prepare statements on broad actions so that they are relevant to policy and are timed to coincide with meaningful points in agency planning and decisionmaking.

(c) When preparing statements on broad actions (including proposals by more than one agency), agencies may find it useful to evaluate the proposal(s) in one of the following ways:

(1) Geographically, including actions occurring in the same general location, such as body of water, region, or metropolitan area.

(2) Generically, including actions which have relevant similarities, such as common timing, impacts, alternatives, methods of implementation, media, or subject matter.

(3) By stage of technological development including federal or federally assisted research, development or demonstration programs for new technologies which, if applied, could significantly affect the quality of the human environment. Statements shall be prepared on such programs and shall be available before the program has reached a stage of investment or commitment to implementation likely to determine subsequent development or restrict later alternatives.

(d) Agencies shall as appropriate employ scoping (§ 1501.7), tiering (§ 1502.20), and other methods listed in §§ 1500.4 and 1500.5 to relate broad and narrow actions and to avoid duplication and delay.

§ 1502.5 Timing.

An agency shall commence preparation of an environmental impact statement as close as possible to the time the agency is developing or is presented with a proposal (§ 1508.23) so that preparation can be completed in time for the final statement to be included in any recommendation or report on the proposal. The statement shall be prepared early enough so that it can serve practically as an important contribution to the decisionmaking process and will not be used to rationalize or justify decisions already made (§§ 1500.2(c), 1501.2, and 1502.2). For instance:

(a) For projects directly undertaken by Federal agencies the environmental impact statement shall be prepared at the feasibility analysis (go-no go) stage and may be supplemented at a later stage if necessary.

(b) For applications to the agency appropriate environmental assessments or statements shall be commenced no later than immediately after the application is received. Federal agencies are encouraged to begin preparation of such assessments or statements earlier, preferably jointly with applicable State or local agencies.

(c) For adjudication, the final environmental impact statement shall normally precede the final staff recommendation and that portion of the public hearing related to the impact study. In appropriate circumstances the statement may follow preliminary hearings designed to gather information for use in the statements.

(d) For informal rulemaking the draft environmental impact statement shall normally accompany the proposed rule.

§ 1502.6 Interdisciplinary preparation.

Environmental impact statements shall be prepared using an inter-disciplinary approach which will insure the integrated use of the natural and social sciences and the environmental design arts (section 102(2)(A) of the Act). The disciplines of the preparers shall be appropriate to the scope and issues identified in the scoping process (§ 1501.7).

§ 1502.7 Page limits.

The text of final environmental impact statements (e.g., paragraphs (d) through (g) of § 1502.10) shall normally be less than 150 pages and for proposals of unusual scope or complexity shall normally be less than 300 pages.

§ 1502.8 Writing.

Environmental impact statements shall be written in plain language and may use appropriate graphics so that decisionmakers and the public can readily understand them. Agencies should employ writers of clear prose or editors to write, review, or edit statements, which will be based upon the analysis and supporting data from the natural and social sciences and the environmental design arts.

§ 1502.9 Draft, final, and supplemental statements.

Except for proposals for legislation as provided in § 1506.8 environmental impact statements shall be prepared in two stages and may be supplemented.

(a) Draft environmental impact statements shall be prepared in accordance with the scope decided upon in the scoping process. The lead agency shall work with the cooperating agencies and shall obtain comments as required in Part 1503 of this chapter. The draft statement must fulfill and satisfy to the fullest extent possible the requirements established for final statements in section 102(2)(C) of the Act. If a draft statement is so inadequate as to preclude meaningful analysis, the agency shall prepare and circulate a revised draft of the appropriate portion. The agency shall make every effort to disclose and discuss at appropriate points in the draft statement all major points of view on the environmental impacts of the alternatives including the proposed action.

(b) Final environmental impact statements shall respond to comments as required in Part 1503 of this chapter. The agency shall discuss at appropriate points in the final statement any responsible opposing view which was not adequately discussed in the draft statement and shall indicate the agency's response to the issues raised.

(c) Agencies:

(1) Shall prepare supplements to either draft or final environmental impact statements if:

(i) The agency makes substantial changes in the proposed action that are relevant to environmental concerns; or

(ii) There are significant new circumstances or information relevant to environmental concerns and bearing on the proposed action or its impacts.

(2) May also prepare supplements when the agency determines that the purposes of the Act will be furthered by doing so.

(3) Shall adopt procedures for introducing a supplement into its formal administrative record, if such a record exists.

(4) Shall prepare, circulate, and file a supplement to a statement in the same fashion (exclusive of scoping) as a draft and final statement unless alternative procedures are approved by the Council.

§ 1502.10 Recommended format.

Agencies shall use a format for environmental impact statements which will encourage good analysis and clear presentation of the alternatives including the proposed action. The following standard format for environmental impact statements should be followed unless the agency determines that there is a compelling reason to do otherwise:

(a) Cover sheet.

(b) Summary.

(c) Table of contents.

(d) Purpose of and need for action.

(e) Alternatives including proposed action (sections 102(2)(C)(iii) and 102(2)(E) of the Act).

(f) Affected environment.

(g) Environmental consequences (especially sections 102(2)(C) (i), (ii), (iv), and (v) of the Act).

(h) List of preparers.

(i) List of Agencies, Organizations, and persons to whom copies of the statement are sent.

(j) Index.

(k) Appendices (if any).

If a different format is used, it shall include paragraphs (a), (b), (c), (h), (i), and (j), of this section and shall include the substance of paragraphs (d), (e), (f), (g), and (k) of this section, as further described in §§ 1502.11 through 1502.18, in any appropriate format.

§ 1502.11 Cover sheet.

The cover sheet shall not exceed one page. It shall include:

(a) A list of the responsible agencies including the lead agency and any co-operating agencies.

(b) The title of the proposed action that is the subject of the statement (and if appropriate the titles of related cooperating agency actions), together with the State(s) and county(ies) (or other jurisdiction if applicable) where the action is located.

(c) The name, address, and telephone number of the person at the agency who can supply further information.

(d) A designation of the statement as a draft, final, or draft or final supplement.

(e) A one paragraph abstract of the statement.

(f) The date by which comments must be received (computed in cooperation with EPA under § 1506.10).

The information required by this section may be entered on Standard Form 424 (in items 4, 6, 7, 10, and 18).

§ 1502.12 Summary.

Each environmental impact statement shall contain a summary which adequately and accurately summarizes the statement. The summary shall stress the major conclusions, areas of controversy (including issues raised by agencies and the public), and the issues to be resolved (including the choice among alternatives). The summary will normally not exceed 15 pages.

§ 1502.13 Purpose and need.

The statement shall briefly specify the underlying purpose and need to which the agency is responding in proposing the alternatives including the proposed action.

§ 1502.14 Alternatives including the proposed action.

This section is the heart of the environmental impact statement. Based on the information and analysis presented in the sections on the Affected Environment (§ 1502.15) and the Environmental Consequences (§ 1502.16), it should present the environmental impacts of the proposal and the alternatives in comparative form, thus sharply defining the issues and providing a clear basis for choice among options by the decisionmaker and the public. In this section agencies shall:

(a) Rigorously explore and objectively evaluate all reasonable alternatives, and for alternatives which were eliminated from detailed study, briefly discuss the reasons for their having been eliminated.

(b) Devote substantial treatment to each alternative considered in detail including the proposed action so that reviewers may evaluate their comparative merits.

(c) Include reasonable alternatives not within the jurisdiction of the lead agency.

(d) Include the alternative of no action.

(e) Identify the agency's preferred alternative or alternatives, if one or more exists, in the draft statement and identify such alternative in the final statement unless another law prohibits the expression of such a preference.

(f) Include appropriate mitigation measures not already included in the proposed action or alternatives.

§ 1502.15 Affected environment.

The environmental impact statement shall succinctly describe the environment of the area(s) to be affected or created by the alternatives under consideration. The descriptions shall be no longer than is necessary to understand the effects of the alternatives. Data and analyses in a statement shall be commensurate with the importance of the impact, with less important material summarized, consolidated, or simply referenced. Agencies shall avoid useless bulk in statements and shall concentrate effort and attention on important issues. Verbose descriptions of the affected environment are themselves no measure of the adequacy of an environmental impact statement.

§ 1502.16 Environmental consequences.

This section forms the scientific and analytic basis for the comparisons under § 1502.14. It shall consolidate the discussions of those elements required by sections 102(2)(C) (i), (ii), (iv), and (v) of NEPA which are within the scope of the statement and as much of section 102(2)(C)(iii) as is necessary to support the comparisons. The discussion will include the environmental impacts of the alternatives including the proposed action, any adverse environmental effects which cannot be avoided should the proposal be implemented, the relationship between short-term uses of man's environment and the maintenance and enhancement of long-term productivity, and any irreversible or irretrievable commitments of resources which would be involved in the proposal should it be implemented. This section should not duplicate discussions in § 1502.14. It shall include discussions of:

(a) Direct effects and their significance (§ 1508.8).

(b) Indirect effects and their significance (§ 1508.8).

(c) Possible conflicts between the proposed action and the objectives of Federal, regional, State, and local (and in the case of a reservation, Indian tribe) land use plans, policies and controls for the area concerned. (See § 1506.2(d).)

(d) The environmental effects of alternatives including the proposed action. The comparisons under § 1502.14 will be based on this discussion.

(e) Energy requirements and conservation potential of various alternatives and mitigation measures.

(f) Natural or depletable resource requirements and conservation potential of various alternatives and mitigation measures.

(g) Urban quality, historic and cultural resources, and the design of the built environment, including the reuse and conservation potential of various alternatives and mitigation measures.

(h) Means to mitigate adverse environmental impacts (if not fully covered under § 1502.14(f)).

[43 FR 55994, Nov. 29, 1978; 44 FR 873, Jan. 3, 1979]

§ 1502.17   List of preparers.

The environmental impact statement shall list the names, together with their qualifications (expertise, experience, professional disciplines), of the persons who were primarily responsible for preparing the environmental impact statement or significant background papers, including basic components of the statement (§§ 1502.6 and 1502.8). Where possible the persons who are responsible for a particular analysis, including analyses in background papers, shall be identified. Normally the list will not exceed two pages.

§ 1502.18   Appendix.

If an agency prepares an appendix to an environmental impact statement the appendix shall:

(a) Consist of material prepared in connection with an environmental impact statement (as distinct from material which is not so prepared and which is incorporated by reference (§ 1502.21)).

(b) Normally consist of material which substantiates any analysis fundamental to the impact statement.

(c) Normally be analytic and relevant to the decision to be made.

(d) Be circulated with the environmental impact statement or be readily available on request.

§ 1502.19   Circulation of the environmental impact statement.

Agencies shall circulate the entire draft and final environmental impact statements except for certain appendices as provided in § 1502.18(d) and unchanged statements as provided in § 1503.4(c). However, if the statement is unusually long, the agency may circulate the summary instead, except that the entire statement shall be furnished to:

(a) Any Federal agency which has jurisdiction by law or special expertise with respect to any environmental impact involved and any appropriate Federal, State or local agency authorized to develop and enforce environmental standards.

(b) The applicant, if any.

(c) Any person, organization, or agency requesting the entire environmental impact statement.

(d) In the case of a final environmental impact statement any person, organization, or agency which submitted substantive comments on the draft.

If the agency circulates the summary and thereafter receives a timely request for the entire statement and for additional time to comment, the time for that requestor only shall be extended by at least 15 days beyond the minimum period.

§ 1502.20 Tiering.

Agencies are encouraged to tier their environmental impact statements to eliminate repetitive discussions of the same issues and to focus on the actual issues ripe for decision at each level of environmental review (§ 1508.28). Whenever a broad environmental impact statement has been prepared (such as a program or policy statement) and a subsequent statement or environmental assessment is then prepared on an action included within the entire program or policy (such as a site specific action) the subsequent statement or environmental assessment need only summarize the issues discussed in the broader statement and incorporate discussions from the broader statement by reference and shall concentrate on the issues specific to the subsequent action. The subsequent document shall state where the earlier document is available. Tiering may also be appropriate for different stages of actions. (Section 1508.28).

§ 1502.21 Incorporation by reference.

Agencies shall incorporate material into an environmental impact statement by reference when the effect will be to cut down on bulk without impeding agency and public review of the action. The incorporated material shall be cited in the statement and its content briefly described. No material may be incorporated by reference unless it is reasonably available for inspection by potentially interested persons within the time allowed for comment. Material based on proprietary data which is itself not available for review and comment shall not be incorporated by reference.

§ 1502.22 Incomplete or unavailable information.

When an agency is evaluating reasonably foreseeable significant adverse effects on the human environment in an environmental impact statement and there is incomplete or unavailable information, the agency shall always make clear that such information is lacking.

(a) If the incomplete information relevant to reasonably foreseeable significant adverse impacts is essential to a reasoned choice among alternatives and the overall costs of obtaining it are not exorbitant, the agency shall include the information in the environmental impact statement.

(b) If the information relevant to reasonably foreseeable significant adverse impacts cannot be obtained because the overall costs of obtaining it are exorbitant or the means to obtain it are not known, the agency shall include within the environmental impact statement: (1) A statement that such information is incomplete or unavailable; (2) a statement of the relevance of the incomplete or unavailable information to evaluating reasonably foreseeable significant adverse impacts on the human environment; (3) a summary of existing credible scientific evidence which is relevant to evaluating the reasonably foreseeable significant adverse impacts on the human environment, and (4) the agency's evaluation of such impacts based upon theoretical approaches or research methods generally accepted in the scientific community. For the purposes of this section, "reasonably foreseeable" includes impacts which have catastrophic consequences, even if their probability of occurrence is low, provided that the analysis of the impacts is supported by credible scientific evidence, is not based on pure conjecture, and is within the rule of reason.

(c) The amended regulation will be applicable to all environmental impact statements for which a Notice of Intent (40 CFR 1508.22) is published in the FEDERAL REGISTER on or after May 27, 1986. For environmental impact statements in progress, agencies may choose to comply with the requirements of either the original or amended regulation.

[51 FR 15625, Apr. 25, 1986]

§ 1502.23 Cost-benefit analysis.

If a cost-benefit analysis relevant to the choice among environmentally different alternatives is being considered for the proposed action, it shall be incorporated by reference or appended to the statement as an aid in evaluating the environmental consequences. To assess the adequacy of compliance with section 102(2)(B) of the Act the statement shall, when a cost-benefit analysis is prepared, discuss the relationship between that analysis and any analyses of unquantified environmental impacts, values, and amenities. For purposes of complying with the Act, the weighing of the merits and drawbacks of the various alternatives need not be displayed in a monetary cost-benefit analysis and should not be when there are important qualitative considerations. In any event, an environmental impact statement should at least indicate those considerations, including factors not related to environmental quality, which are likely to be relevant and important to a decision.

§ 1502.24 Methodology and scientific accuracy.

Agencies shall insure the professional integrity, including scientific integrity, of the discussions and analyses in environmental impact statements. They shall identify any methodologies used and shall make explicit reference by footnote to the scientific and other sources relied upon for conclusions in the statement. An agency may place discussion of methodology in an appendix.

§ 1502.25 Environmental review and consultation requirements.

(a) To the fullest extent possible, agencies shall prepare draft environmental impact statements concurrently with and integrated with environmental impact analyses and related surveys and studies required by the Fish and Wildlife Coordination Act (16 U.S.C. 661 et seq.), the National Historic Preservation Act of 1966 (16 U.S.C. 470 et seq.), the Endangered Species Act of 1973 (16 U.S.C. 1531 et seq.), and other environmental review laws and executive orders.

(b) The draft environmental impact statement shall list all Federal permits, licenses, and other entitlements which must be obtained in implementing the proposal. If it is uncertain whether a Federal permit, license, or other entitlement is necessary, the draft environmental impact statement shall so indicate.

PART 1503—COMMENTING

Sec.
1503.1 Inviting comments.
1503.2 Duty to comment.
1503.3 Specificity of comments.
1503.4 Response to comments.

AUTHORITY: NEPA, the Environmental Quality Improvement Act of 1970, as amended (42 U.S.C. 4371 *et seq.*), sec. 309 of the Clean Air Act, as amended (42 U.S.C. 7609), and E.O. 11514 (Mar. 5, 1970, as amended by E.O. 11991, May 24, 1977).

SOURCE: 43 FR 55997, Nov. 29, 1978, unless otherwise noted.

§ 1503.1 Inviting comments.

(a) After preparing a draft environmental impact statement and before preparing a final environmental impact statement the agency shall:

(1) Obtain the comments of any Federal agency which has jurisdiction by law or special expertise with respect to any environmental impact involved or which is authorized to develop and enforce environmental standards.

(2) Request the comments of:

(i) Appropriate State and local agencies which are authorized to develop and enforce environmental standards;

(ii) Indian tribes, when the effects may be on a reservation; and

(iii) Any agency which has requested that it receive statements on actions of the kind proposed.

Under Executive Order No. 12372, the Office of Management and Budget, through its system of clearinghouses, provides a means of securing the views of State and local environmental agencies. The clearinghouses may be used, by mutual agreement of the lead agency and the clearinghouse, for securing

State and local reviews of the draft environmental impact statements.

(3) Request comments from the applicant, if any.

(4) Request comments from the public, affirmatively soliciting comments from those persons or organizations who may be interested or affected.

(b) An agency may request comments on a final environmental impact statement before the decision is finally made. In any case other agencies or persons may make comments before the final decision unless a different time is provided under § 1506.10.

§ 1503.2 Duty to comment.

Federal agencies with jurisdiction by law or special expertise with respect to any environmental impact involved and agencies which are authorized to develop and enforce environmental standards shall comment on statements within their jurisdiction, expertise, or authority. Agencies shall comment within the time period specified for comment in § 1506.10. A Federal agency may reply that it has no comment. If a cooperating agency is satisfied that its views are adequately reflected in the environmental impact statement, it should reply that it has no comment.

§ 1503.3 Specificity of comments.

(a) Comments on an environmental impact statement or on a proposed action shall be as specific as possible and may address either the adequacy of the statement or the merits of the alternatives discussed or both.

(b) When a commenting agency criticizes a lead agency's predictive methodology, the commenting agency should describe the alternative methodology which it prefers and why.

(c) A cooperating agency shall specify in its comments whether it needs additional information to fulfill other applicable environmental reviews or consultation requirements and what information it needs. In particular, it shall specify any additional information it needs to comment adequately on the draft statement's analysis of significant site-specific effects associated with the granting or approving by that cooperating agency of neces-

sary Federal permits, licenses, or entitlements.

(d) When a cooperating agency with jurisdiction by law objects to or expresses reservations about the proposal on grounds of environmental impacts, the agency expressing the objection or reservation shall specify the mitigation measures it considers necessary to allow the agency to grant or approve applicable permit, license, or related requirements or concurrences.

§ 1503.4 Response to comments.

(a) An agency preparing a final environmental impact statement shall assess and consider comments both individually and collectively, and shall respond by one or more of the means listed below, stating its response in the final statement. Possible responses are to:

(1) Modify alternatives including the proposed action.

(2) Develop and evaluate alternatives not previously given serious consideration by the agency.

(3) Supplement, improve, or modify its analyses.

(4) Make factual corrections.

(5) Explain why the comments do not warrant further agency response, citing the sources, authorities, or reasons which support the agency's position and, if appropriate, indicate those circumstances which would trigger agency reappraisal or further response.

(b) All substantive comments received on the draft statement (or summaries thereof where the response has been exceptionally voluminous), should be attached to the final statement whether or not the comment is thought to merit individual discussion by the agency in the text of the statement.

(c) If changes in response to comments are minor and are confined to the responses described in paragraphs (a) (4) and (5) of this section, agencies may write them on errata sheets and attach them to the statement instead of rewriting the draft statement. In such cases only the comments, the responses, and the changes and not the final statement need be circulated (§ 1502.19). The entire document with a new cover sheet shall be filed as the final statement (§ 1506.9).

PART 1504—PREDECISION REFERRALS TO THE COUNCIL OF PROPOSED FEDERAL ACTIONS DETERMINED TO BE ENVIRONMENTALLY UNSATISFACTORY

Sec.
1504.1 Purpose.
1504.2 Criteria for referral.
1504.3 Procedure for referrals and response.

AUTHORITY: NEPA, the Environmental Quality Improvement Act of 1970, as amended (42 U.S.C. 4371 *et seq.*), sec. 309 of the Clean Air Act, as amended (42 U.S.C. 7609), and E.O. 11514 (Mar. 5, 1970, as amended by E.O. 11991, May 24, 1977).

SOURCE: 43 FR 55998, Nov. 29, 1978, unless otherwise noted.

§ 1504.1 Purpose.

(a) This part establishes procedures for referring to the Council Federal interagency disagreements concerning proposed major Federal actions that might cause unsatisfactory environmental effects. It provides means for early resolution of such disagreements.

(b) Under section 309 of the Clean Air Act (42 U.S.C. 7609), the Administrator of the Environmental Protection Agency is directed to review and comment publicly on the environmental impacts of Federal activities, including actions for which environmental impact statements are prepared. If after this review the Administrator determines that the matter is "unsatisfactory from the standpoint of public health or welfare or environmental quality," section 309 directs that the matter be referred to the Council (hereafter "environmental referrals").

(c) Under section 102(2)(C) of the Act other Federal agencies may make similar reviews of environmental impact statements, including judgments on the acceptability of anticipated environmental impacts. These reviews must be made available to the President, the Council and the public.

§ 1504.2 Criteria for referral.

Environmental referrals should be made to the Council only after concerted, timely (as early as possible in the process), but unsuccessful attempts to resolve differences with the lead agency. In determining what environmental objections to the matter are appropriate to refer to the Council, an agency should weigh potential adverse environmental impacts, considering:

(a) Possible violation of national environmental standards or policies.

(b) Severity.

(c) Geographical scope.

(d) Duration.

(e) Importance as precedents.

(f) Availability of environmentally preferable alternatives.

§ 1504.3 Procedure for referrals and response.

(a) A Federal agency making the referral to the Council shall:

(1) Advise the lead agency at the earliest possible time that it intends to refer a matter to the Council unless a satisfactory agreement is reached.

(2) Include such advice in the referring agency's comments on the draft environmental impact statement, except when the statement does not contain adequate information to permit an assessment of the matter's environmental acceptability.

(3) Identify any essential information that is lacking and request that it be made available at the earliest possible time.

(4) Send copies of such advice to the Council.

(b) The referring agency shall deliver its referral to the Council not later than twenty-five (25) days after the final environmental impact statement has been made available to the Environmental Protection Agency, commenting agencies, and the public. Except when an extension of this period has been granted by the lead agency, the Council will not accept a referral after that date.

(c) The referral shall consist of:

(1) A copy of the letter signed by the head of the referring agency and delivered to the lead agency informing the lead agency of the referral and the reasons for it, and requesting that no action be taken to implement the matter until the Council acts upon the

referral. The letter shall include a copy of the statement referred to in (c)(2) of this section.

(2) A statement supported by factual evidence leading to the conclusion that the matter is unsatisfactory from the standpoint of public health or welfare or environmental quality. The statement shall:

(i) Identify any material facts in controversy and incorporate (by reference if appropriate) agreed upon facts,

(ii) Identify any existing environmental requirements or policies which would be violated by the matter,

(iii) Present the reasons why the referring agency believes the matter is environmentally unsatisfactory,

(iv) Contain a finding by the agency whether the issue raised is of national importance because of the threat to national environmental resources or policies or for some other reason,

(v) Review the steps taken by the referring agency to bring its concerns to the attention of the lead agency at the earliest possible time, and

(vi) Give the referring agency's recommendations as to what mitigation alternative, further study, or other course of action (including abandonment of the matter) are necessary to remedy the situation.

(d) Not later than twenty-five (25) days after the referral to the Council the lead agency may deliver a response to the Council, and the referring agency. If the lead agency requests more time and gives assurance that the matter will not go forward in the interim, the Council may grant an extension. The response shall:

(1) Address fully the issues raised in the referral.

(2) Be supported by evidence.

(3) Give the lead agency's response to the referring agency's recommendations.

(e) Interested persons (including the applicant) may deliver their views in writing to the Council. Views in support of the referral should be delivered not later than the referral. Views in support of the response shall be delivered not later than the response.

(f) Not later than twenty-five (25) days after receipt of both the referral and any response or upon being informed that there will be no response

(unless the lead agency agrees to a longer time), the Council may take one or more of the following actions:

(1) Conclude that the process of referral and response has successfully resolved the problem.

(2) Initiate discussions with the agencies with the objective of mediation with referring and lead agencies.

(3) Hold public meetings or hearings to obtain additional views and information.

(4) Determine that the issue is not one of national importance and request the referring and lead agencies to pursue their decision process.

(5) Determine that the issue should be further negotiated by the referring and lead agencies and is not appropriate for Council consideration until one or more heads of agencies report to the Council that the agencies' disagreements are irreconcilable.

(6) Publish its findings and recommendations (including where appropriate a finding that the submitted evidence does not support the position of an agency).

(7) When appropriate, submit the referral and the response together with the Council's recommendation to the President for action.

(g) The Council shall take no longer than 60 days to complete the actions specified in paragraph (f) (2), (3), or (5) of this section.

(h) When the referral involves an action required by statute to be determined on the record after opportunity for agency hearing, the referral shall be conducted in a manner consistent with 5 U.S.C. 557(d) (Administrative Procedure Act).

[43 FR 55998, Nov. 29, 1978; 44 FR 873, Jan. 3, 1979]

PART 1505—NEPA AND AGENCY DECISIONMAKING

Sec.

AUTHORITY: NEPA, the Environmental Quality Improvement Act of 1970, as amended (42 U.S.C. 4371 *et seq.*), sec. 309 of the Clean Air Act, as amended (42 U.S.C.

7609), and E.O. 11514 (Mar. 5, 1970, as amended by E.O. 11991, May 24, 1977).

Source: 43 FR 55999, Nov. 29, 1978, unless otherwise noted.

§ 1505.1 Agency decisionmaking procedures.

Agencies shall adopt procedures (§ 1507.3) to ensure that decisions are made in accordance with the policies and purposes of the Act. Such procedures shall include but not be limited to:

(a) Implementing procedures under section 102(2) to achieve the requirements of sections 101 and 102(1).

(b) Designating the major decision points for the agency's principal programs likely to have a significant effect on the human environment and assuring that the NEPA process corresponds with them.

(c) Requiring that relevant environmental documents, comments, and responses be part of the record in formal rulemaking or adjudicatory proceedings.

(d) Requiring that relevant environmental documents, comments, and responses accompany the proposal through existing agency review processes so that agency officials use the statement in making decisions.

(e) Requiring that the alternatives considered by the decisionmaker are encompassed by the range of alternatives discussed in the relevant environmental documents and that the decisionmaker consider the alternatives described in the environmental impact statement. If another decision document accompanies the relevant environmental documents to the decisionmaker, agencies are encouraged to make available to the public before the decision is made any part of that document that relates to the comparison of alternatives.

§ 1505.2 Record of decision in cases requiring environmental impact statements.

At the time of its decision (§ 1506.10) or, if appropriate, its recommendation to Congress, each agency shall prepare a concise public record of decision. The record, which may be integrated into any other record prepared by the agency, shall:

(a) State what the decision was.

(b) Identify all alternatives considered by the agency in reaching its decision, specifying the alternative or alternatives which were considered to be environmentally preferable. An agency may discuss preferences among alternatives based on relevant factors including economic and technical considerations and agency statutory missions. An agency shall identify and discuss all such factors including any essential considerations of national policy which were balanced by the agency in making its decision and state how those considerations entered into its decision.

(c) State whether all practicable means to avoid or minimize environmental harm from the alternative selected have been adopted, and if not, why they were not. A monitoring and enforcement program shall be adopted and summarized where applicable for any mitigation.

§ 1505.3 Implementing the decision.

Agencies may provide for monitoring to assure that their decisions are carried out and should do so in important cases. Mitigation (§ 1505.2(c)) and other conditions established in the environmental impact statement or during its review and committed as part of the decision shall be implemented by the lead agency or other appropriate consenting agency. The lead agency shall:

(a) Include appropriate conditions in grants, permits or other approvals.

(b) Condition funding of actions on mitigation.

(c) Upon request, inform cooperating or commenting agencies on progress in carrying out mitigation measures which they have proposed and which were adopted by the agency making the decision.

(d) Upon request, make available to the public the results of relevant monitoring.

PART 1506—OTHER REQUIREMENTS OF NEPA

Sec.
1506.1 Limitations on actions during NEPA process.
1506.2 Elimination of duplication with State and local procedures.
1506.3 Adoption.
1506.4 Combining documents.
1506.5 Agency responsibility.
1506.6 Public involvement.
1506.7 Further guidance.
1506.8 Proposals for legislation.
1506.9 Filing requirements.
1506.10 Timing of agency action.
1506.11 Emergencies.
1506.12 Effective date.

AUTHORITY: NEPA, the Environmental Quality Improvement Act of 1970, as amended (42 U.S.C. 4371 *et seq.*), sec. 309 of the Clean Air Act, as amended (42 U.S.C. 7609), and E.O. 11514 (Mar. 5, 1970, as amended by E.O. 11991, May 24, 1977).

SOURCE: 43 FR 56000, Nov. 29, 1978, unless otherwise noted.

§ 1506.1 Limitations on actions during NEPA process.

(a) Until an agency issues a record of decision as provided in § 1505.2 (except as provided in' paragraph (c) of this section), no action concerning the proposal shall be taken which would:

(1) Have an adverse environmental impact; or

(2) Limit the choice of reasonable alternatives.

(b) If any agency is considering an application from a non-Federal entity, and is aware that the applicant is about to take an action within the agency's jurisdiction that would meet either of the criteria in paragraph (a) of this section, then the agency shall promptly notify the applicant that the agency will take appropriate action to insure that the objectives and procedures of NEPA are achieved.

(c) While work on a required program environmental impact statement is in progress and the action is not covered by an existing program statement, agencies shall not undertake in the interim any major Federal action covered by the program which may significantly affect the quality of the human environment unless such action:

(1) Is justified independently of the program;

(2) Is itself accompanied by an adequate environmental impact statement; and

(3) Will not prejudice the ultimate decision on the program. Interim action prejudices the ultimate decision on the program when it tends to determine subsequent development or limit alternatives.

(d) This section does not preclude development by applicants of plans or designs or performance of other work necessary to support an application for Federal, State or local permits or assistance. Nothing in this section shall preclude Rural Electrification Administration approval of minimal expenditures not affecting the environment (*e.g.* long leadtime equipment and purchase options) made by non-governmental entities seeking loan guarantees from the Administration.

§ 1506.2 Elimination of duplication with State and local procedures.

(a) Agencies authorized by law to cooperate with State agencies of statewide jurisdiction pursuant to section 102(2)(D) of the Act may do so.

(b) Agencies shall cooperate with State and local agencies to the fullest extent possible to reduce duplication between NEPA and State and local requirements, unless the agencies are specifically barred from doing so by some other law. Except for cases covered by paragraph (a) of this section, such cooperation shall to the fullest extent possible include:

(1) Joint planning processes.

(2) Joint environmental research and studies.

(3) Joint public hearings (except where otherwise provided by statute).

(4) Joint environmental assessments.

(c) Agencies shall cooperate with State and local agencies to the fullest extent possible to reduce duplication between NEPA and comparable State and local requirements, unless the agencies are specifically barred from doing so by some other law. Except for cases covered by paragraph (a) of this section, such cooperation shall to the fullest extent possible include joint environmental impact statements. In such cases one or more Federal agencies and one or more State or local agencies shall be joint lead agencies. Where State laws or local ordinances have environmental impact statement requirements in addition to but not in conflict with those in NEPA, Federal agencies shall cooperate in fulfilling these requirements as well as those of Federal laws so that one document will comply with all applicable laws.

(d) To better integrate environmental impact statements into State or local planning processes, statements shall discuss any inconsistency of a proposed action with any approved State or local plan and laws (whether or not federally sanctioned). Where an inconsistency exists, the statement should describe the extent to which the agency would reconcile its proposed action with the plan or law.

§ 1506.3 Adoption.

(a) An agency may adopt a Federal draft or final environmental impact statement or portion thereof provided that the statement or portion thereof meets the standards for an adequate statement under these regulations.

(b) If the actions covered by the original environmental impact statement and the proposed action are substantially the same, the agency adopting another agency's statement is not required to recirculate it except as a final statement. Otherwise the adopting agency shall treat the statement as a draft and recirculate it (except as provided in paragraph (c) of this section).

(c) A cooperating agency may adopt without recirculating the environmental impact statement of a lead agency when, after an independent review of the statement, the cooperating agency concludes that its comments and suggestions have been satisfied.

(d) When an agency adopts a statement which is not final within the agency that prepared it, or when the action it assesses is the subject of a referral under Part 1504, or when the statement's adequacy is the subject of a judicial action which is not final, the agency shall so specify.

§ 1506.4 Combining documents.

Any environmental document in compliance with NEPA may be combined with any other agency document to reduce duplication and paperwork.

§ 1506.5 Agency responsibility.

(a) *Information.* If an agency requires an applicant to submit environmental information for possible use by the agency in preparing an environmental impact statement, then the agency should assist the applicant by outlining the types of information required. The agency shall independently evaluate the information submitted and shall be responsible for its accuracy. If the agency chooses to use the information submitted by the applicant in the environmental impact statement, either directly or by reference, then the names of the persons responsible for the independent evaluation shall be included in the list of preparers (§ 1502.17). It is the intent of this paragraph that acceptable work not be redone, but that it be verified by the agency.

(b) *Environmental assessments.* If an agency permits an applicant to prepare an environmental assessment, the agency, besides fulfilling the requirements of paragraph (a) of this section, shall make its own evaluation of the environmental issues and take responsibility for the scope and content of the environmental assessment.

(c) *Environmental impact statements.* Except as provided in §§ 1506.2 and 1506.3 any environmental impact statement prepared pursuant to the requirements of NEPA shall be prepared directly by or by a contractor selected by the lead agency or where appropriate under § 1501.6(b), a cooperating agency. It is the intent of these regulations that the contractor be chosen solely by the lead agency, or by the lead agency in cooperation with cooperating agencies, or where appropriate by a cooperating agency to avoid any conflict of interest. Contractors shall execute a disclosure statement prepared by the lead agency, or where appropriate the cooperating agency, specifying that they have no financial or other interest in the outcome of the project. If the document is prepared by contract, the responsible Federal official shall furnish guidance and participate in the preparation and shall independently evaluate the statement prior to its approval and take responsibility for its scope and contents. Nothing in this section is intended to prohibit any agency from requesting any person to submit information to it or to prohibit any person from submitting information to any agency.

§ 1506.6 Public involvement.

Agencies shall:

(a) Make diligent efforts to involve the public in preparing and implementing their NEPA procedures.

(b) Provide public notice of NEPA-related hearings, public meetings, and the availability of environmental documents so as to inform those persons and agencies who may be interested or affected.

(1) In all cases the agency shall mail notice to those who have requested it on an individual action.

(2) In the case of an action with effects of national concern notice shall include publication in the FEDERAL REGISTER and notice by mail to national organizations reasonably expected to be interested in the matter and may include listing in the *102 Monitor*. An agency engaged in rulemaking may provide notice by mail to national organizations who have requested that notice regularly be provided. Agencies shall maintain a list of such organizations.

(3) In the case of an action with effects primarily of local concern the notice may include:

(i) Notice to State and areawide clearinghouses pursuant to EO 12372, the Intergovernmental Review Process.

(ii) Notice to Indian tribes when effects may occur on reservations.

(iii) Following the affected State's public notice procedures for comparable actions.

(iv) Publication in local newspapers (in papers of general circulation rather than legal papers).

(v) Notice through other local media.

(vi) Notice to potentially interested community organizations including small business associations.

(vii) Publication in newsletters that may be expected to reach potentially interested persons.

(viii) Direct mailing to owners and occupants of nearby or affected property.

(ix) Posting of notice on and off site in the area where the action is to be located.

(c) Hold or sponsor public hearings or public meetings whenever appropriate or in accordance with statutory requirements applicable to the agency. Criteria shall include whether there is:

(1) Substantial environmental controversy concerning the proposed action or substantial interest in holding the hearing.

(2) A request for a hearing by another agency with jurisdiction over the action supported by reasons why a hearing will be helpful. If a draft environmental impact statement is to be considered at a public hearing, the agency should make the statement available to the public at least 15 days in advance (unless the purpose of the hearing is to provide information for the draft environmental impact statement).

(d) Solicit appropriate information from the public.

(e) Explain in its procedures where interested persons can get information or status reports on environmental impact statements and other elements of the NEPA process.

(f) Make environmental impact statements, the comments received, and any underlying documents available to the public pursuant to the provisions of the Freedom of Information Act (5 U.S.C. 552), without regard to the exclusion for interagency memoranda where such memoranda transmit comments of Federal agencies on the environmental impact of the proposed action. Materials to be made available to the public shall be provided to the public without charge to the extent practicable, or at a fee which is not more than the actual costs of reproducing copies required to be sent to other Federal agencies, including the Council.

§ 1506.7 Further guidance.

The Council may provide further guidance concerning NEPA and its procedures including:

(a) A handbook which the Council may supplement from time to time, which shall in plain language provide guidance and instructions concerning the application of NEPA and these regulations.

(b) Publication of the Council's Memoranda to Heads of Agencies.

(c) In conjunction with the Environmental Protection Agency and the publication of the 102 Monitor, notice of:

(1) Research activities;

(2) Meetings and conferences related to NEPA; and

(3) Successful and innovative procedures used by agencies to implement NEPA.

§ 1506.8 Proposals for legislation.

(a) The NEPA process for proposals for legislation (§ 1508.17) significantly affecting the quality of the human environment shall be integrated with the legislative process of the Congress. A legislative environmental impact statement is the detailed statement required by law to be included in a recommendation or report on a legislative proposal to Congress. A legislative environmental impact statement shall be considered part of the formal transmittal of a legislative proposal to Congress; however, it may be transmitted to Congress up to 30 days later in order to allow time for completion of an accurate statement which can serve as the basis for public and Congressional debate. The statement must be available in time for Congressional hearings and deliberations.

(b) Preparation of a legislative environmental impact statement shall conform to the requirements of these regulations except as follows:

(1) There need not be a scoping process.

(2) The legislative statement shall be prepared in the same manner as a draft statement, but shall be considered the "detailed statement" required by statute; *Provided,* That when any of the following conditions exist both the draft and final environmental impact statement on the legislative proposal shall be prepared and circulated as provided by §§ 1503.1 and 1506.10.

(i) A Congressional Committee with jurisdiction over the proposal has a rule requiring both draft and final environmental impact statements.

(ii) The proposal results from a study process required by statute (such as those required by the Wild and Scenic Rivers Act (16 U.S.C. 1271 et seq.) and the Wilderness Act (16 U.S.C. 1131 et seq.)).

(iii) Legislative approval is sought for Federal or federally assisted construction or other projects which the agency recommends be located at specific geographic locations. For proposals requiring an environmental impact statement for the acquisition of space by the General Services Administration, a draft statement shall accompany the Prospectus or the 11(b) Report of Building Project Surveys to the Congress, and a final statement shall be completed before site acquisition.

(iv) The agency decides to prepare draft and final statements.

(c) Comments on the legislative statement shall be given to the lead agency which shall forward them along with its own responses to the Congressional committees with jurisdiction.

§ 1506.9 Filing requirements.

Environmental impact statements together with comments and responses shall be filed with the Environmental Protection Agency, attention Office of Federal Activities (A-104), 401 M Street SW., Washington, D.C. 20460. Statements shall be filed with EPA no earlier than they are also transmitted to commenting agencies and made available to the public. EPA shall deliver one copy of each statement to the Council, which shall satisfy the requirement of availability to the President. EPA may issue guidelines to agencies to implement its responsibilities under this section and § 1506.10.

§ 1506.10 Timing of agency action.

(a) The Environmental Protection Agency shall publish a notice in the FEDERAL REGISTER each week of the environmental impact statements filed during the preceding week. The minimum time periods set forth in this section shall be calculated from the date of publication of this notice.

(b) No decision on the proposed action shall be made or recorded under § 1505.2 by a Federal agency until the later of the following dates:

(1) Ninety (90) days after publication of the notice described above in paragraph (a) of this section for a draft environmental impact statement.

(2) Thirty (30) days after publication of the notice described above in paragraph (a) of this section for a final environmental impact statement.

An exception to the rules on timing may be made in the case of an agency decision which is subject to a formal internal appeal. Some agencies have a formally established appeal process which allows other agencies or the public to take appeals on a decision and make their views known, after publication of the final environmental impact statement. In such cases, where a real opportunity exists to alter the decision, the decision may be made and recorded at the same time the environmental impact statement is published. This means that the period for appeal of the decision and the 30-day period prescribed in paragraph (b)(2) of this section may run concurrently. In such cases the environmental impact statement shall explain the timing and the public's right of appeal. An agency engaged in rulemaking under the Administrative Procedure Act or other statute for the purpose of protecting the public health or safety, may waive the time period in paragraph (b)(2) of this section and publish a decision on the final rule simultaneously with publication of the notice of the availability of the final environmental impact statement as described in paragraph (a) of this section.

(c) If the final environmental impact statement is filed within ninety (90) days after a draft environmental impact statement is filed with the Environmental Protection Agency, the minimum thirty (30) day period and the minimum ninety (90) day period may run concurrently. However, subject to paragraph (d) of this section agencies shall allow not less than 45 days for comments on draft statements.

(d) The lead agency may extend prescribed periods. The Environmental Protection Agency may upon a showing by the lead agency of compelling reasons of national policy reduce the prescribed periods and may upon a showing by any other Federal agency of compelling reasons of national policy also extend prescribed periods, but only after consultation with the lead agency. (Also see § 1507.3(d).) Failure to file timely comments shall not be a sufficient reason for extending a period. If the lead agency does not concur with the extension of time, EPA may not extend it for more than 30 days. When the Environmental Protection Agency reduces or extends any period of time it shall notify the Council.

[43 FR 56000, Nov. 29, 1978; 44 FR 874, Jan. 3, 1979]

§ 1506.11 Emergencies.

Where emergency circumstances make it necessary to take an action with significant environmental impact without observing the provisions of these regulations, the Federal agency taking the action should consult with the Council about alternative arrangements. Agencies and the Council will limit such arrangements to actions necessary to control the immediate impacts of the emergency. Other actions remain subject to NEPA review.

§ 1506.12 Effective date.

The effective date of these regulations is July 30, 1979, except that for agencies that administer programs that qualify under section 102(2)(D) of the Act or under sec. 104(h) of the Housing and Community Development Act of 1974 an additional four months shall be allowed for the State or local agencies to adopt their implementing procedures.

(a) These regulations shall apply to the fullest extent practicable to ongoing activities and environmental documents begun before the effective date. These regulations do not apply to an environmental impact statement or supplement if the draft statement was filed before the effective date of these regulations. No completed environmental documents need be redone by reasons of these regulations. Until these regulations are applicable, the Council's guidelines published in the FEDERAL REGISTER of August 1, 1973, shall continue to be applicable. In cases where these regulations are applicable the guidelines are superseded. However, nothing shall prevent an agency from proceeding under these regulations at an earlier time.

(b) NEPA shall continue to be applicable to actions begun before January 1, 1970, to the fullest extent possible.

PART 1507—AGENCY COMPLIANCE

Sec.
1507.1 Compliance.
1507.2 Agency capability to comply.
1507.3 Agency procedures.

AUTHORITY: NEPA, the Environmental Quality Improvement Act of 1970, as amended (42 U.S.C. 4371 *et seq.*), sec. 309 of the Clean Air Act, as amended (42 U.S.C. 7609), and E.O. 11514 (Mar. 5, 1970, as amended by E.O. 11991, May 24, 1977).

SOURCE: 43 FR 56002, Nov. 29, 1978, unless otherwise noted.

§ 1507.1 Compliance.

All agencies of the Federal Government shall comply with these regulations. It is the intent of these regulations to allow each agency flexibility in adapting its implementing procedures authorized by § 1507.3 to the requirements of other applicable laws.

§ 1507.2 Agency capability to comply.

Each agency shall be capable (in terms of personnel and other resources) of complying with the requirements enumerated below. Such compliance may include use of other's resources, but the using agency shall itself have sufficient capability to evaluate what others do for it. Agencies shall:

(a) Fulfill the requirements of section 102(2)(A) of the Act to utilize a systematic, interdisciplinary approach which will insure the integrated use of the natural and social sciences and the environmental design arts in planning and in decisionmaking which may have an impact on the human environment. Agencies shall designate a person to be responsible for overall review of agency NEPA compliance.

(b) Identify methods and procedures required by section 102(2)(B) to insure that presently unquantified environmental amenities and values may be given appropriate consideration.

(c) Prepare adequate environmental impact statements pursuant to section 102(2)(C) and comment on statements in the areas where the agency has jurisdiction by law or special expertise or is authorized to develop and enforce environmental standards.

(d) Study, develop, and describe alternatives to recommended courses of action in any proposal which involves unresolved conflicts concerning alternative uses of available resources. This requirement of section 102(2)(E) extends to all such proposals, not just the more limited scope of section 102(2)(C)(iii) where the discussion of alternatives is confined to impact statements.

(e) Comply with the requirements of section 102(2)(H) that the agency initiate and utilize ecological information in the planning and development of resource-oriented projects.

(f) Fulfill the requirements of sections 102(2)(F), 102(2)(G), and 102(2)(I), of the Act and of Executive Order 11514, Protection and Enhancement of Environmental Quality, Sec. 2.

§ 1507.3 Agency procedures.

(a) Not later than eight months after publication of these regulations as finally adopted in the FEDERAL REGISTER, or five months after the establishment of an agency, whichever shall come later, each agency shall as necessary adopt procedures to supplement these regulations. When the agency is a department, major subunits are encouraged (with the consent of the department) to adopt their own procedures. Such procedures shall not paraphrase these regulations. They shall confine themselves to implementing procedures. Each agency shall consult with the Council while developing its procedures and before publishing them in the FEDERAL REGISTER for comment. Agencies with similar programs should consult with each other and the Council to coordinate their procedures, especially for programs requesting similar information from applicants. The procedures shall be adopted only after an opportunity for public review and after review by the Council for conformity with the Act and these regulations. The Council shall complete its review within 30 days. Once in effect they shall be filed with the Council and made readily available to the public. Agencies are encouraged to publish explanatory guidance for these regulations and their own procedures. Agencies shall continue to review their policies and procedures and in consultation with the Council to revise them as necessary to ensure full compliance with the purposes and provisions of the Act.

(b) Agency procedures shall comply with these regulations except where compliance would be inconsistent with statutory requirements and shall include:

(1) Those procedures required by §§ 1501.2(d), 1502.9(c)(3), 1505.1, 1506.6(e), and 1508.4.

(2) Specific criteria for and identification of those typical classes of action:

(i) Which normally do require environmental impact statements.

(ii) Which normally do not require either an environmental impact statement or an environmental assessment (categorical exclusions (§ 1508.4)).

(iii) Which normally require environmental assessments but not necessarily environmental impact statements.

(c) Agency procedures may include specific criteria for providing limited exceptions to the provisions of these regulations for classified proposals. They are proposed actions which are specifically authorized under criteria established by an Executive Order or statute to be kept secret in the interest of national defense or foreign policy and are in fact properly classified pursuant to such Executive Order or statute. Environmental assessments and environmental impact statements which address classified proposals may be safeguarded and restricted from public dissemination in accordance with agencies' own regulations applica-

ble to classified information. These documents may be organized so that classified portions can be included as annexes, in order that the unclassified portions can be made available to the public.

(d) Agency procedures may provide for periods of time other than those presented in § 1506.10 when necessary to comply with other specific statutory requirements.

(e) Agency procedures may provide that where there is a lengthy period between the agency's decision to prepare an environmental impact statement and the time of actual preparation, the notice of intent required by § 1501.7 may be published at a reasonable time in advance of preparation of the draft statement.

PART 1508—TERMINOLOGY AND INDEX

Sec.
1508.1   Terminology.
1508.2   Act.
1508.3   Affecting.
1508.4   Categorical exclusion.
1508.5   Cooperating agency.
1508.6   Council.
1508.7   Cumulative impact.
1508.8   Effects.
1508.9   Environmental assessment.
1508.10   Environmental document.
1508.11   Environmental impact statement.
1508.12   Federal agency.
1508.13   Finding of no significant impact.
1508.14   Human environment.
1508.15   Jurisdiction by law.
1508.16   Lead agency.
1508.17   Legislation.
1508.18   Major Federal action.
1508.19   Matter.
1508.20   Mitigation.
1508.21   NEPA process.
1508.22   Notice of intent.
1508.23   Proposal.
1508.24   Referring agency.
1508.25   Scope.
1508.26   Special expertise.
1508.27   Significantly.
1508.28   Tiering.

AUTHORITY: NEPA, the Environmental Quality Improvement Act of 1970, as amended (42 U.S.C. 4371 *et seq.*), sec. 309 of the Clean Air Act, as amended (42 U.S.C. 7609), and E.O. 11514 (Mar. 5, 1970, as amended by E.O. 11991, May 24, 1977).

SOURCE: 43 FR 56003, Nov. 29, 1978, unless otherwise noted.

§ 1508.1   Terminology.

The terminology of this part shall be uniform throughout the Federal Government.

§ 1508.2   Act.

"Act" means the National Environmental Policy Act, as amended (42 U.S.C. 4321, et seq.) which is also referred to as "NEPA."

§ 1508.3   Affecting.

"Affecting" means will or may have an effect on.

§ 1508.4   Categorical exclusion.

"Categorical exclusion" means a category of actions which do not individually or cumulatively have a significant effect on the human environment and which have been found to have no such effect in procedures adopted by a Federal agency in implementation of these regulations (§ 1507.3) and for which, therefore, neither an environmental assessment nor an environmental impact statement is required. An agency may decide in its procedures or otherwise, to prepare environmental assessments for the reasons stated in § 1508.9 even though it is not required to do so. Any procedures under this section shall provide for extraordinary circumstances in which a normally excluded action may have a significant environmental effect.

§ 1508.5   Cooperating agency.

"Cooperating agency" means any Federal agency other than a lead agency which has jurisdiction by law or special expertise with respect to any environmental impact involved in a proposal (or a reasonable alternative) for legislation or other major Federal action significantly affecting the quality of the human environment. The selection and responsibilities of a cooperating agency are described in § 1501.6. A State or local agency of similar qualifications or, when the effects are on a reservation, an Indian Tribe, may by agreement with the lead agency become a cooperating agency.

§ 1508.6   Council.

"Council" means the Council on Environmental Quality established by Title II of the Act.

§ 1508.7   Cumulative impact.

"Cumulative impact" is the impact on the environment which results from the incremental impact of the action when added to other past, present, and reasonably foreseeable future actions regardless of what agency (Federal or non-Federal) or person undertakes such other actions. Cumulative impacts can result from individually minor but collectively significant actions taking place over a period of time.

§ 1508.8   Effects.

"Effects" include:
(a) Direct effects, which are caused by the action and occur at the same time and place.
(b) Indirect effects, which are caused by the action and are later in time or farther removed in distance, but are still reasonably foreseeable. Indirect effects may include growth inducing effects and other effects related to induced changes in the pattern of land use, population density or growth rate, and related effects on air and water and other natural systems, including ecosystems.

Effects and impacts as used in these regulations are synonymous. Effects includes ecological (such as the effects on natural resources and on the components, structures, and functioning of affected ecosystems), aesthetic, historic, cultural, economic, social, or health, whether direct, indirect, or cumulative. Effects may also include those resulting from actions which may have both beneficial and detrimental effects, even if on balance the agency believes that the effect will be beneficial.

§ 1508.9   Environmental assessment.

"Environmental assessment":
(a) Means a concise public document for which a Federal agency is responsible that serves to:
(1) Briefly provide sufficient evidence and analysis for determining

whether to prepare an environmental impact statement or a finding of no significant impact.

(2) Aid an agency's compliance with the Act when no environmental impact statement is necessary.

(3) Facilitate preparation of a statement when one is necessary.

(b) Shall include brief discussions of the need for the proposal, of alternatives as required by section 102(2)(E), of the environmental impacts of the proposed action and alternatives, and a listing of agencies and persons consulted.

§ 1508.10 Environmental document.

"Environmental document" includes the documents specified in § 1508.9 (environmental assessment), § 1508.11 (environmental impact statement), § 1508.13 (finding of no significant impact), and § 1508.22 (notice of intent).

§ 1508.11 Environmental impact statement.

"Environmental impact statement" means a detailed written statement as required by section 102(2)(C) of the Act.

§ 1508.12 Federal agency.

"Federal agency" means all agencies of the Federal Government. It does not mean the Congress, the Judiciary, or the President, including the performance of staff functions for the President in his Executive Office. It also includes for purposes of these regulations States and units of general local government and Indian tribes assuming NEPA responsibilities under section 104(h) of the Housing and Community Development Act of 1974.

§ 1508.13 Finding of no significant impact.

"Finding of no significant impact" means a document by a Federal agency briefly presenting the reasons why an action, not otherwise excluded (§ 1508.4), will not have a significant effect on the human environment and for which an environmental impact statement therefore will not be prepared. It shall include the environmental assessment or a summary of it and shall note any other environmental documents related to it

(§ 1501.7(a)(5)). If the assessment is included, the finding need not repeat any of the discussion in the assessment but may incorporate it by reference.

§ 1508.14 Human environment.

"Human environment" shall be interpreted comprehensively to include the natural and physical environment and the relationship of people with that environment. (See the definition of "effects" (§ 1508.8).) This means that economic or social effects are not intended by themselves to require preparation of an environmental impact statement. When an environmental impact statement is prepared and economic or social and natural or physical environmental effects are interrelated, then the environmental impact statement will discuss all of these effects on the human environment.

§ 1508.15 Jurisdiction by law.

"Jurisdiction by law" means agency authority to approve, veto, or finance all or part of the proposal.

§ 1508.16 Lead agency.

"Lead agency" means the agency or agencies preparing or having taken primary responsibility for preparing the environmental impact statement.

§ 1508.17 Legislation.

"Legislation" includes a bill or legislative proposal to Congress developed by or with the significant cooperation and support of a Federal agency, but does not include requests for appropriations. The test for significant cooperation is whether the proposal is in fact predominantly that of the agency rather than another source. Drafting does not by itself constitute significant cooperation. Proposals for legislation include requests for ratification of treaties. Only the agency which has primary responsibility for the subject matter involved will prepare a legislative environmental impact statement.

§ 1508.18 Major Federal action.

"Major Federal action" includes actions with effects that may be major and which are potentially subject to

Federal control and responsibility. Major reinforces but does not have a meaning independent of significantly (§ 1508.27). Actions include the circumstance where the responsible officials fail to act and that failure to act is reviewable by courts or administrative tribunals under the Administrative Procedure Act or other applicable law as agency action.

(a) Actions include new and continuing activities, including projects and programs entirely or partly financed, assisted, conducted, regulated, or approved by federal agencies; new or revised agency rules, regulations, plans, policies, or procedures; and legislative proposals (§§ 1506.8, 1508.17). Actions do not include funding assistance solely in the form of general revenue sharing funds, distributed under the State and Local Fiscal Assistance Act of 1972, 31 U.S.C. 1221 et seq., with no Federal agency control over the subsequent use of such funds. Actions do not include bringing judicial or administrative civil or criminal enforcement actions.

(b) Federal actions tend to fall within one of the following categories:

(1) Adoption of official policy, such as rules, regulations, and interpretations adopted pursuant to the Administrative Procedure Act, 5 U.S.C. 551 et seq.; treaties and international conventions or agreements; formal documents establishing an agency's policies which will result in or substantially alter agency programs.

(2) Adoption of formal plans, such as official documents prepared or approved by federal agencies which guide or prescribe alternative uses of federal resources, upon which future agency actions will be based.

(3) Adoption of programs, such as a group of concerted actions to implement a specific policy or plan; systematic and connected agency decisions allocating agency resources to implement a specific statutory program or executive directive.

(4) Approval of specific projects, such as construction or management activities located in a defined geographic area. Projects include actions approved by permit or other regulatory decision as well as federal and federally assisted activities.

§ 1508.19 Matter.

"Matter" includes for purposes of Part 1504:

(a) With respect to the Environmental Protection Agency, any proposed legislation, project, action or regulation as those terms are used in section 309(a) of the Clean Air Act (42 U.S.C. 7609).

(b) With respect to all other agencies, any proposed major federal action to which section 102(2)(C) of NEPA applies.

§ 1508.20 Mitigation.

"Mitigation" includes:

(a) Avoiding the impact altogether by not taking a certain action or parts of an action.

(b) Minimizing impacts by limiting the degree or magnitude of the action and its implementation.

(c) Rectifying the impact by repairing, rehabilitating, or restoring the affected environment.

(d) Reducing or eliminating the impact over time by preservation and maintenance operations during the life of the action.

(e) Compensating for the impact by replacing or providing substitute resources or environments.

§ 1508.21 NEPA process.

"NEPA process" means all measures necessary for compliance with the requirements of section 2 and Title I of NEPA.

§ 1508.22 Notice of intent.

"Notice of intent" means a notice that an environmental impact statement will be prepared and considered. The notice shall briefly:

(a) Describe the proposed action and possible alternatives.

(b) Describe the agency's proposed scoping process including whether, when, and where any scoping meeting will be held.

(c) State the name and address of a person within the agency who can answer questions about the proposed action and the environmental impact statement.

§ 1508.23 Proposal.

"Proposal" exists at that stage in the development of an action when an agency subject to the Act has a goal and is actively preparing to make a decision on one or more alternative means of accomplishing that goal and the effects can be meaningfully evaluated. Preparation of an environmental impact statement on a proposal should be timed (§ 1502.5) so that the final statement may be completed in time for the statement to be included in any recommendation or report on the proposal. A proposal may exist in fact as well as by agency declaration that one exists.

§ 1508.24 Referring agency.

"Referring agency" means the federal agency which has referred any matter to the Council after a determination that the matter is unsatisfactory from the standpoint of public health or welfare or environmental quality.

§ 1508.25 Scope.

Scope consists of the range of actions, alternatives, and impacts to be considered in an environmental impact statement. The scope of an individual statement may depend on its relationships to other statements (§§1502.20 and 1508.28). To determine the scope of environmental impact statements, agencies shall consider 3 types of actions, 3 types of alternatives, and 3 types of impacts. They include:

(a) Actions (other than unconnected single actions) which may be:

(1) Connected actions, which means that they are closely related and therefore should be discussed in the same impact statement. Actions are connected if they:

(i) Automatically trigger other actions which may require environmental impact statements.

(ii) Cannot or will not proceed unless other actions are taken previously or simultaneously.

(iii) Are interdependent parts of a larger action and depend on the larger action for their justification.

(2) Cumulative actions, which when viewed with other proposed actions have cumulatively significant impacts and should therefore be discussed in the same impact statement.

(3) Similar actions, which when viewed with other reasonably foreseeable or proposed agency actions, have similarities that provide a basis for evaluating their environmental consequences together, such as common timing or geography. An agency may wish to analyze these actions in the same impact statement. It should do so when the best way to assess adequately the combined impacts of similar actions or reasonable alternatives to such actions is to treat them in a single impact statement.

(b) Alternatives, which include: (1) No action alternative.

(2) Other reasonable courses of actions.

(3) Mitigation measures (not in the proposed action).

(c) Impacts, which may be: (1) Direct; (2) indirect; (3) cumulative.

§ 1508.26 Special expertise.

"Special expertise" means statutory responsibility, agency mission, or related program experience.

§ 1508.27 Significantly.

"Significantly" as used in NEPA requires considerations of both context and intensity:

(a) *Context.* This means that the significance of an action must be analyzed in several contexts such as society as a whole (human, national), the affected region, the affected interests, and the locality. Significance varies with the setting of the proposed action. For instance, in the case of a site-specific action, significance would usually depend upon the effects in the locale rather than in the world as a whole. Both short- and long-term effects are relevant.

(b) *Intensity.* This refers to the severity of impact. Responsible officials must bear in mind that more than one agency may make decisions about partial aspects of a major action. The following should be considered in evaluating intensity:

(1) Impacts that may be both beneficial and adverse. A significant effect may exist even if the Federal agency

believes that on balance the effect will be beneficial.

(2) The degree to which the proposed action affects public health or safety.

(3) Unique characteristics of the geographic area such as proximity to historic or cultural resources, park lands, prime farmlands, wetlands, wild and scenic rivers, or ecologically critical areas.

(4) The degree to which the effects on the quality of the human environment are likely to be highly controversial.

(5) The degree to which the possible effects on the human environment are highly uncertain or involve unique or unknown risks.

(6) The degree to which the action may establish a precedent for future actions with significant effects or represents a decision in principle about a future consideration.

(7) Whether the action is related to other actions with individually insignificant but cumulatively significant impacts. Significance exists if it is reasonable to anticipate a cumulatively significant impact on the environment. Significance cannot be avoided by terming an action temporary or by breaking it down into small component parts.

(8) The degree to which the action may adversely affect districts, sites, highways, structures, or objects listed in or eligible for listing in the National Register of Historic Places or may cause loss or destruction of significant scientific, cultural, or historical resources.

(9) The degree to which the action may adversely affect an endangered or threatened species or its habitat that has been determined to be critical under the Endangered Species Act of 1973.

(10) Whether the action threatens a violation of Federal, State, or local law or requirements imposed for the protection of the environment.

[43 FR 56003, Nov. 29, 1978; 44 FR 874, Jan. 3, 1979]

§ 1508.28 Tiering.

"Tiering" refers to the coverage of general matters in broader environmental impact statements (such as national program or policy statements) with subsequent narrower statements or environmental analyses (such as regional or basinwide program statements or ultimately site-specific statements) incorporating by reference the general discussions and concentrating solely on the issues specific to the statement subsequently prepared. Tiering is appropriate when the sequence of statements or analyses is:

(a) From a program, plan, or policy environmental impact statement to a program, plan, or policy statement or analysis of lesser scope or to a site-specific statement or analysis.

(b) From an environmental impact statement on a specific action at an early stage (such as need and site selection) to a supplement (which is preferred) or a subsequent statement or analysis at a later stage (such as environmental mitigation). Tiering in such cases is appropriate when it helps the lead agency to focus on the issues which are ripe for decision and exclude from consideration issues already decided or not yet ripe.

Appendix C:
Environmental Assessment Checklists

A set of checklists for reviewing and preparing Environmental Assessment (EA) is provided in Tables C.1 through C.18.[1]

Not all questions presented in the following checklist will apply to all EAs. Professional judgment must be exercised in responding to each question. While these checklists lists are intended to be as comprehensive as possible, they cannot cover every conceivable regulatory requirement or issue that might need to be addressed in an EA. Prudence should be exercised because:

- The checklists should not be relied upon as the sole method for ensuring quality and compliance with regulatory requirements.
- No checklist can be prepared that is universally applicable to all circumstances;
- The checklists cannot guarantee that an EA will be adequate or be in full compliance with NEPA's regulation direction or case law;
- A price list and copies of the checklists can be obtained at a nominal fee by contacting The NEPA and Environmental Strategies Company at ecclestonc@msn.com.

---

[1] Modified from Environmental Assessment Checklist, U.S. Department of Energy, 1994.

TABLE C.1©

Summary (Optional)

	Yes	No	N/A	Page	Adequacy Evaluation and Comments
1.1 Does the summary address the entire EA?					
1.2 Is the summary consistent with information in the document?					
1.3 Does the summary highlight key differences among the alternatives (if applicable)?					
1.4 Does the summary describe the:					
Underlying need (and purpose, if applicable) for taking action?					
Proposed action?					
Each of the reasonable alternatives?					
Principal environmental issues and results?					

TABLE C.2©

Purpose and Need for Action

	Yes	No	N/A	Page	Adequacy Evaluation and Comments
2.1 Does the statement of purpose and need define the need for the action [40 CFR1508.9]?					
2.2 Does the statement of purpose and need relate to the broad requirement or desire for agency action, and not to the need for one specific proposal?					
2.3 Is the statement of purpose and need written so that it does not inappropriately narrow the range of reasonable alternatives?					
2.4 Does the statement of purpose and need identify the problem or opportunity to which the agency is responding?					

TABLE C.3©

Description of the Proposed Action and Alternatives

	Yes	No	N/A	Page	Adequacy Evaluation and Comments
3.1 Is the proposed action described in sufficient detail so that potential impacts can be identified? To the extent possible, are all phases described (e.g., construction, operation, maintenance, and decommissioning)?					
3.2 Are any discharges or releases associated with the proposed action quantified, including both the rates and durations?					
3.3 As appropriate, are mitigation measures included in the description of the proposed action?					
3.4 Is the project description written broadly enough to encompass future modifications?					
3.5 Does the proposed action exclude discussions that are more appropriately placed in the statement of purpose and need?					
3.6 Is the proposed action described in terms of the agency action to be taken (even a private action that has been federalized)?					
3.7 Does the EA address a range of reasonable alternatives that satisfy the agency's purpose and need, including reasonable alternatives outside the agency's jurisdiction?					
3.8 If there are alternatives that appear obvious or have been identified but the public but are not analyzed, does the EA explain why they were excluded?					
3.9 Does the EA include the no-action alternative [10 CFR 1021.321(c)]?					
3.10 Is the no-action alternative described in sufficient detail so that its scope is clear and potential impacts can be identified?					

(continues)

TABLE C.3©

Description of the Proposed Action and Alternatives (continued)

	Yes	No	N/A	Page	Adequacy Evaluation and Comments
3.11 Does the no action alternative include a discussion of the legal ramifications of taking no action, if appropriate?					
3.12 Does the EA take into account relationships between the proposed action and other actions to be taken by the agency in order to avoid improper segmentation?					
3.13 Does the proposed action comply with CEQ regulations for interim actions [40 CFR 1506.1]?					

| | | | | | Adequacy Evaluation and |
	Yes	No	N/A	Page	Comments

TABLE C.4©

Description Of The Affected Environment

	Yes	No	N/A	Page	Adequacy Evaluation and Comments
4.1 Does the EA identify either the presence or absence of the following within the area potentially affected by the proposed action and alternatives:					
Floodplains [EO 11988; 10 CFR 1022]?					
Wetlands [EO 11990; 10 CFR 1022; 40 CFR 1508.27(b)(3)]?					
Threatened, endangered, or candidate species and/or their critical habitat, and other special status (e.g., state-listed) species [16 USC 1531; 40 CFR 1508.27(b)(3)]?					
Prime or unique farmland [7 USC 4201; 7 CFR 658; 40 CFR 1508.27(b)(3)]?					
State or national parks, forests, conservation areas, or other areas of recreational, ecological, scenic, or aesthetic importance?					
Wild and scenic rivers [16 USC 1271; 40 CFR 1508.27(b)(3)]?					
Natural resources (e.g., timber, range, soils, minerals, wildlife, water bodies, aquifers)?					
Property of historic, archaeological, or architectural significance (including sites on or eligible for the National Register of Historic of Natural Landmarks) [16 USC 470; 36 CFR 800; 40 CFR 1508.27(b)(3)]?					
Native Americans' concerns [16 USC 470; 42 USC 1996]?					
Minority and low-income populations (including a description of their use and consumption of environmental resources) [EO 12898]?					

(continues)

TABLE C.4

Description Of The Affected Environment (continued)

	Yes	No	N/A	Page	Adequacy Evaluation and Comments
4.2 Does the description of the affected environment provide the necessary information to support the impact analysis, including cumulative impact analysis?					
4.3 Does the EA appropriately use incorporation by reference? Is/are the incorporated documents(s) up-to-date and publicly available?					
4.4 If this EA incorporates, in whole or in part, a NEPA document prepared by another federal agency, has the information been independently evaluated?					

TABLE C.5©					
Environmental Effects					
	Yes	No	N/A	Page	Adequacy Evaluation and Comments
5.1 Does the EA identify the potential effects (including cumulative effects) to the following, as identified in question 4.1:					
Floodplains [EO 11988; 10 CFR 1022]?					
Wetlands [EO 11990; 10 CFR 1022; 40 CFR 1508.27(b)(3)]?					
Threatened, endangered, or candidate species and/or their critical habitat, and other special status (e.g., state-listed) species [16 USC 1531; 40 CFR 1508.27(b)(3)]?					
Prime or unique farmland [7 USC 4201; 7 CFR 658; 40 CFR 1508.27(b)(3)]?					
State or national parks, forests, conservation areas, or other areas of recreational, ecological, scenic, aesthetic importance?					
Wild and scenic rivers [16 USC 1271; 40 CFR 1508.27(b)(3)]?					
Natural resources (e.g., timber, range, soils, minerals, fish, wildlife, water bodies, aquifers)?					
Property of historic, archaeological, or architectural significance (including sites on or eligible for the National Register of Historic Places and the National Registry of Natural Landmarks) [16 USC 470; 36 CFR 800; 40 CFR 1508.27(b)(3)]?					
Native Americans' concerns [16 USC 470; 42 USC 1996]?					
Minority and low-income populations [EO 12898]?					

(continues)

TABLE C.5

Environmental Effects (continued)

	Yes	No	N/A	Page	Adequacy Evaluation and Comments
5.2 Does the EA analyze the proposed action for					
Both short-term and long-term effects [40 CFR 1508.27(a)]?					
Both beneficial and adverse impacts [40 CFR 1508.27(b)(1)]?					
Effects on public health and safety [40 CFR 1508.27(b)(2)]?					
Disproportionately high and adverse human health or environmental effects on minority and low-income communities [EO 12898]?					
5.3 Do the discussions of environmental impacts include (as appropriate) human health effects, effects of accidents, and transportation effects?					
5.4 As appropriate, does the EA address the degree to which the possible effects on the human environment might be highly uncertain or involve unique or unknown risks [40 CFR 1508.27(b)(5)]?					
5.5 Do the discussions of environmental impacts identify possible indirect and cumulative impacts?					
5.6 Does the EA quantify environmental impacts, where practical?					
5.7 Are all potentially non-trivial impacts identified? Are impacts analyzed using a sliding-scale approach -- i.e., proportional to their potential significance?					
5.8 Does the EA identify all reasonably foreseeable impacts [40 CFR 15.08.8]?					
5.9 If information related to potential impacts is incomplete or unavailable, does the EA indicate that such information is lacking [40 CFR 1502.22]?					

TABLE C.5

Environmental Effects (continued)

	Yes	No	N/A	Page	Adequacy Evaluation and Comments
5.10 Are sufficient data and references presented to allow review of the verification of the analytical methods and results?					

TABLE C.6©

Overall Considerations

	Yes	No	N/A	Page	Adequacy Evaluation and Comments
6.1 Because conclusions of overall significance shall be reserved for the FONSI or determination to prepare an EIS, are words such as "significant" and "insignificant" absent from conclusory statements in the EA?					
6.2 Is the EA written such that the decisionmaker can reach conclusions regarding potential impacts based on the information and analyses presented in the EA?					
6.3 Does the EA avoid the implication that compliance with regulatory requirements demonstrates the absence of significant environmental?					
6.4 Are mitigation measures appropriate to the potential impacts identified in the EA [40 CFR 1500.2(f)]?					
6.5 Does the EA show that the agency has taken a "hard look" at environmental consequences [*Kleppe v. Sierra Club*, 427 US 390, 410 (1976)]?					

TABLE C.7©

Procedural Considerations

	Yes	No	N/A	Page	Adequacy Evaluation and Comments
7.1 As applicable, were host states, tribes, and the public notified of the agency's determination to prepare the EA? Does the EA address issues known to be of concern to the states, tribes, and public?					
7.2 Has the EA been made available to the agencies, states, tribes, and the public?					
7.3 Have stakeholders, including the public, been involved to the extent practicable during the preparation of the EA [CEQ (46 FR 18037); 40 CFR 1506.6; 40 CFR 1501.4(b)]? Has the agency proactively sought the involvement of minority and low-income communities in the review and preparation process [EO 12898]?					+
7.4 As applicable, have comments from host states, tribes, and the public been addressed?					
7.5 Is a Floodplain/Wetlands Assessment required and, if so, has one been completed? If required, has a Public Notice been published in the Federal Register?					
7.6 Does the EA demonstrate adequate consultation with appropriate agencies to ensure compliance with sensitive resource laws and regulations? Are letters of consultation appended [16 USC 1531; 36 CFR 800]?					
7.7 Does the EA include a listing of agencies and persons consulted [40 CFR 1508.9(b)]?					

TABLE C.8©

Format, General Document Quality, User-Friendliness

	Yes	No	N/A	Page	Adequacy Evaluation and Comments
8.1 Is the EA written precisely and concisely, using plain language, and without jargon?					
8.2 Is the agency listed as the prepare on the title page of the EA?					
8.3 As applicable, is the metric system of units used (with English units in parentheses)?					
8.4 If scientific notation is used, is an explanation provided?					
8.5 Are technical terms defined where necessary?					
8.6 Are the units consistent throughout the document?					
8.7 If regulatory terms are used, are they consistent with their regulatory definitions?					
8.8 Are visual aids used whenever possible to simplify the EA?					
8.9 Are abbreviations and acronyms defined the first time they are used?					
8.10 Is the use of abbreviations minimized to the extent practical?					
8.11 Do the appendices support the content and conclusions contained in the main body of the EA? Is information in the appendix consistent with information in the main body of the EA?					
8.12 Is information in tables and figures consistent with information in the text and appendices?					

TABLE C.9©

Key to Supplemental Topical Questions

	Yes	No	N/A	Page	Adequacy Evaluation and Comments
9.1 Does the proposed action present a potential for impacts on water resources or water quality?					
9.2 Does the proposed action present a potential for impacts related to geology or soils?					
9.3 Does the proposed action present a potential for impacts on air quality?					
9.4 Does the proposed action present a potential for impacts on wildlife or habitat?					
9.5 Does the proposed action present a potential for effects on human health?					
9.6 Does the proposed action involve transportation?					
9.7 Does the proposed action involve waste management?					
9.8 Does the proposed action present a potential for impacts on socioeconomic conditions?					
9.9 Does the proposed action present a potential for impacts to historic, archaeological, or other cultural sites or properties?					

TABLE C.10©					
Water Resources and Water Quality					
	Yes	No	N/A	Page	Adequacy Evaluation and Comments
10.1 Does the EA identify potential effects of the proposed action and alternatives on surface water quantity and quality under both normal operations and accident conditions?					
10.2 Does the EA evaluate whether the proposed action or alternatives would be subject to Water quality or effluent standards? National Interim Primary Drinking Water Regulations? National Secondary Drinking Water Regulations?					
10.3 Does the EA state whether the proposed action or alternatives would Include work in, under, over, or having an effect on navigable water of the United States? Include the discharge of dredged or fill material into waters of the United States? Include the deposit of fill material or an excavation that alters or modifies capacity of any navigable waters of the United States? Require a Rivers and Harbors Act Section 10 permit or a Clean Water Act (Section 402 or Section 404) permit?					
10.4 Does the EA identify potential effects of the proposed action and alternatives on groundwater quantity and quality (including aquifers) under both normal operations and accident conditions?					

(continues)

TABLE C.10

Water Resources and Water Quality (continued)

	Yes	No	N/A	Page	Adequacy Evaluation and Comments
10.5 Does the EA consider whether the proposed action or alternatives could affect any municipal or private drinking water supplies?					

TABLE C.11©

Geology and Soils

	Yes	No	N/A	Page	Adequacy Evaluation and Comments
11.1 Does the EA describe and quantify the land area proposed to be altered, excavated, or otherwise disturbed? Is this description consistent with other sections (e.g., land use, habitat area)?					
11.2 Are issues related to seismicity sufficiently characterized, quantified, and analyzed?					
11.3 If the action involves disturbance of surface soils, are erosion control measures addressed?					

TABLE C.12©

Air Quality

	Yes	No	N/A	Page	Adequacy Evaluation and Comments
12.1 Does the EA identify potential effects of the proposed action on ambient air quality under both normal and accident conditions?					
12.2 Are potential emissions quantified to the extent practicable (amount and rate of release)?					
12.3 Does the EA evaluate potential effects to human health and the environment from exposure to radiation and hazardous chemicals in emissions?					
12.4 Does the EA evaluate whether the proposed action and alternatives would Be in compliance with the National Ambient Air Quality Standards? Be in compliance with the State Implementation Plan? Potentially affect any area designated as Class I under the Clean Air Act? Be subject to New Source Performance Standards? Be subject to National Emissions Standards for Hazardous Air Pollutants? Be subject to emissions limitations in an Air Quality Control Region?					

TABLE C.13©

Wildlife And Habitat

	Yes	No	N/A	Page	Adequacy Evaluation and Comments
13.1 If the EA identifies potential effects of the proposed action and alternatives on threatened or endangered species and/or critical habitat, has consultation with the USFWS or NMFS been concluded? Does the EA address candidate species?					
13.2 Are state-listed species identified and, if so, are results of state consultation documented?					
13.3 Are potential effects (including cumulative effects) analyzed for fish and wildlife other than threatened and endangered species and for habitats other than critical habitat?					
13.4 Does the EA analyze the impacts of the proposed action on the biodiversity of the affected ecosystem, including genetic diversity and species diversity?					
13.5 Are habitat types identified and estimates provided by type for the amount of habitat lost or adversely affected?					

TABLE C.14©

Human Health Effects

	Yes	No	N/A	Page	Adequacy Evaluation and Comments
14.1 Have the susceptible populations been identified -- i.e., involved workers, noninvolved workers, and the public (including minority and low-income communities, as appropriate)?					
14.2 Does the EA establish the period of exposure (e.g., 30 years or 70 years) for exposed workers and the public?					

TABLE C.14

Human Health Effects (continued)

	Yes	No	N/A	Page	Adequacy Evaluation and Comments
14.3 Does the EA identify all potential routes of exposure?					
14.4 When providing quantitative estimates of impacts, does the EA use current dose-to-risk conversion factors that have been adopted by cognizant health and environmental agencies?					
14.5 When providing quantitative estimates of health effects due to radiation exposure, are collective effects expressed in estimated numbers of fatal cancers, and are maximum individual effects expressed as the estimated maximum probability of death of an individual?					
14.6 Does the EA describe assumptions used in the health effects analysis and the basis for health effects calculations?					
14.7 As appropriate, does the EA analyze radiological impacts under normal operating conditions for					
Involved workers:					
– Collective dose?					
– Maximum individual?					
– Latent cancer fatalities?					
Uninvolved workers:					
– Collective dose?					
– Maximum individual?					
– Latent cancer fatalities?					
Public:					
– Collective dose?					
– Maximum individual?					
– Latent cancer fatalities?					

(continues)

TABLE C.14

Human Health Effects (continued)

	Yes	No	N/A	Page	Adequacy Evaluation and Comments
14.8 As applicable, does the EA identify a spectrum of potential accident scenarios that could occur over the life of the proposed action?					
14.9 As appropriate, does the EA analyze radiological impacts under accident conditions for: Involved workers: – Collective dose? – Maximum individual? – Latent cancer fatalities? Uninvolved workers: – Collective dose? – Maximum individual? – Latent cancer fatalities? Public: – Collective dose? – Maximum individual? – Latent cancer fatalities?					
14.10 Are non-radiological impacts (e.g., chemical exposures) addressed for both routine and accident conditions?					

TABLE C.15©

Transportation

	Yes	No	N/A	Page	Adequacy Evaluation and Comments
15.1 If transport of hazardous or radioactive waste or materials is part of the proposed action, or if transport is a major factor, are the potential effects analyzed (including to a site, on-site, and *from* a site)?					
15.2 Does the EA analyze all reasonably foreseeable transportation links (e.g., overland transport, port transfer, marine transport, global commons) [EO 12114]?					
15.3 Does the EA avoid relying exclusively on statements that transportation will be in accordance with all applicable state and federal regulations and requirements?					
15.4 Does the EA address both routine transportation as well as reasonably foreseeable accidents?					
15.5 Are the estimation methods used for assessing radiological impacts of transportation defensible?					
15.6 Does the EA address the annual, total, and cumulative impacts of all agency and non-agency transportation on specific routes associated with the proposed action?					

TABLE C.16©

Waste Management and Waste Minimization

	Yes	No	N/A	Page	Adequacy Evaluation and Comments
16.1 Are pollution prevention and waste minimization practices incorporated into the proposed action (e.g., is pollution prevented or reduced at the source when feasible; are by-products that cannot be prevented or recycled treated in an environmentally safe manner when feasible; is disposal only used as a last resort)?					
16.2 If waste would be generated, does the EA examine the human health effects and environmental impacts of managing that waste, including waste generated during decontaminating and decommissioning?					
16.3 Are waste materials characterized by type and estimated quantity, where possible?					
16.4 Does the EA identify RCRA/CERCLA issues related to the proposed action and alternatives?					
16.5 Does the EA establish whether the proposed action and alternatives would be in compliance with federal or state laws and guidelines affecting the generation, transportation, treatment, storage, or disposal of hazardous and other waste?					

TABLE C.17©

Socioeconomic Considerations

	Yes	No	N/A	Page	Adequacy Evaluation and Comments
17.1 Does the EA consider potential effects on land use patterns, consistency with applicable land use plans, and compatibility of nearby uses?					
17.2 Does the EA consider possible changes in the local population due to the proposed action?					
17.3 Does the EA consider potential economic impacts, such as effects on jobs and housing, particularly in regard to disproportionately adverse effects on minority and low-income communities?					
17.4 Does the EA consider potential effects on public water and wastewater services, stormwater management, community services, and utilities?					
17.5 Does the EA evaluate potential noise effects of the proposed action and the application of community noise level standards?					

TABLE C.18©

Cultural Resources

	Yes	No	N/A	Page	Adequacy Evaluation and Comments
18.1 Was the SHPO consulted?					
18.2 Was an archaeological survey conducted?					
18.3 Does the EA include a provision for mitigation in the event unanticipated archaeological materials are encountered?					

Appendices D–F:
Examples and Critiques of Real EAs

The three Environmental Assessments (EAs) presented in their entirety in Appendices D - F are typical of many of the EAs produced in recent years by federal agencies. They are not outstanding with respect to content, format, scope, or controversy. Each is part of the public record and has successfully supported Findings of No Significant Impact (FONSIs). While no single EA can be identified as an ideal model for preparing other EAs, the three EAs presented here illustrate several principles inherent in a proficient analysis. They also illustrate some deficiencies common to many EAs.

Each appendix provides a critique of the subject EA, followed by a copy of the actual EA.

Appendix D.1

Example 1: Critique of the Environmental Assessment for Relocation of Station (Small) Ashtabula, Ashtabula, Ohio

Agency: U.S. Coast Guard
Preparer: Tetra Tech NUS, Gaithersburg, Maryland
Date: June 1999
Action: Relocation of a small Coast Guard Station; including a station house, garage, and two boats; to another site

Overview

The EA addresses the proposed relocation of a small Coast Guard station in Ashtabula, Ohio (on Lake Erie) from an older station house constructed in the 1930s to a newer modular station house to be constructed on a nearby vacant lot. Under the Proposed Action, the Coast Guard would lease the vacant lot from the City of Ashtabula, install the modular building, and relocate its boats used in search and rescue operations to a City-owned marina.

The EA is typical of numerous EAs prepared annually for minor construction projects. Much of the focus of the EA is on potential impacts resulting from new construction, such as soil erosion and sedimentation of surface water, loss of wetlands and other natural habitats, aesthetic changes to historic districts, disturbance of archaeological artifacts, and social and economic disruptions caused by the sudden but temporary influx of construction laborers. The impacts are all very slight; as examples, only lawns and brush would be lost to accommodate the modular structure and standard erosion control methods routinely employed at most construction sites would adequately protect adjoining surface water from sedimentation. If proposed grading activities had been expected to encroach into large areas of wetlands, mature forests, or other sensitive or regionally unique areas of natural vegetation, an Environmental Impact Statement (EIS) would have been required. Similarly, an EIS would have been required if readily implemented erosion control practices could not have been counted on to prevent sedimentation of aquatic habitats adjoining the site.

One interesting feature of this EA is that it had to address potential impacts resulting from a proposal by the City of Ashtabula to expand its city dock to offset the loss of mooring space resulting from acceptance of the Coast Guard boats. The City of Ashtabula is not a federal agency and its actions are therefore not subject to review under National Environmental Policy Act (NEPA). But the Coast Guard decided to discuss the dock expansion in the EA because it would only occur if necessitated by a Coast Guard decision. The delineation between federal and nonfederal actions is not always clear, and failure to address secondary impacts perceived by the public to be connected to a federal action can lead to a FONSI being challenged.

Like many EAs, this EA was not written directly by agency personnel. Instead it was prepared by a contractor and reviewed by environmental staff employed at the Coast Guard's Facility Design and Construction Center. The reviewing staff signed the FONSI, thereby assuming responsibility for the contents of the EA and FONSI as if they had written it. The EA properly identifies the contractor as the preparer of the EA. However, it listed only the lead preparer from the contractor and not each contractor specialist who contributed. Furthermore, it did not provide information about the qualifications or specific role of each contributor. Readers are thus unable to gauge the professional credibility of the technical information presented in the EA and used to base the FONSI.

As specifically required by the Council on Environmental Quality (CEQ) NEPA Regulations, preparation of this EA was a collaborative effort involving not only Coast Guard and contractor staff but also representatives from several interested environmental agencies. Section 6 of the EA lists representatives of several federal, state, and local agencies who contributed technical input into the EA.

This EA, like most EAs, reports consultations with the U.S. Fish & Wildlife Service and state natural resource agencies regarding special-status species and critical habitats and consultations with the state historic preservation agency regarding cultural resources. These initial consultations are sometimes improperly referred to as "Section 7" consultations (referring to Section 7 of the federal Endangered Species Act) and "Section 106" consultations (referring to Section 106 of the National Historic Preservation Act). The proper term for both is informal consultation. The agency responses sometimes indicate that further consultation under Section 7 (or Section 106) is not necessary. Otherwise, formal Section 7 or Section 106 consultations could be required and could require substantial field and library research. A request from an agency for a formal consultation can sometimes indicate that an EIS is required.

This EA, like many EAs completed within the last couple of years, has taken advantage of digital scanning technology to incorporate several photographs into the text. Photographs can convey a large amount of site information more effectively than can text. For example, the photograph in Exhibit 3.2-3 (Page 3-6 of the EA) clearly demonstrates that existing vegetation at the site is sparse and ruderal (weedy), effectively eliminating any concerns that the reader of the EA may have concerning the loss of sensitive natural habitat on the construction site. The contribution of photographs to "concise public documents" such as EAs cannot be underestimated.

Specific Comments

The following specific comments are targeted to specific sections, figures, and pages in the EA.

1. Cover Page: Lists only the lead preparer from the contractor hired to write the EA. There is no List of Preparers at the end of the document. The contractor followed internal Coast Guard procedures for formatting EAs. This EA does not indicate all of the technical individuals responsible for the analysis, the roles played by each individual, or the qualifications of each individual. Decisionmakers are unable to determine the qualifications of the person responsible for any given technical analysis in the document.

2. Section One, Executive Summary: This summary is good in that it summarizes the technical findings of the EA and does not merely state the objectives and scope of the document. However, it would have been easier to read had it been limited to one page.

3. Exhibit 2.1-1: Color reproductions of U.S. Geological Survey maps are much easier to read than black and white reproductions and convey more information (e.g. forest cover, water cover).

4. Exhibit 2.1-2: This map is difficult to interpret. Additional information regarding pavement edges, tree and forest cover, the opposite edge of the river, and surrounding land uses would make the map more useful.

5. Exhibit 2.1-3 (and other photographs): Use of color photographs in an EA can convey a lot of information and help reduce the length of descriptive text. Modern photographic scanning technology has helped increase the use of color photographs in recent EAs.

6. Section Three: This EA is good in that it discusses six alternatives, including a preferred action, no action, and four other alternatives. Many other EAs address only a preferred action and no-action. Such EAs are capable of documenting the insignificance of impacts from the preferred action but are of limited value in overall project planning. Decisionmakers are not provided with the means to analyze alternative courses of action.

7. Exhibit 3.3-1: This summary table provides a concise, "at-a-glance" summary of potential impacts from each alternative using short phrases that convey technical conclusions. It is superior to the "yes/no" summary tables used in many EAs whereby the potential for impacts to each resource is merely indicated using "yes" or "no".

8. Section Four: In accordance with Coast Guard NEPA policy, this EA combines the affected environment and environmental consequences discussions into a single section. However, the EA uses subheadings to break the text under each resource into separate discussions of affected environment and (for each alternative) environmental consequences. The EA thus still allows for a clear distinction between these components. Most importantly, it allows readers to clearly distinguish between the potential environmental consequences for each alternative.

9. Pages 4-11 to 4-13: Affected environment discussion for Hazardous Materials and Hazardous Waste. The preparer of this EA ordered a computerized environmental records database search for properties within a 1-mile radius of the project site. These database searches are typically ordered to support Phase I environmental site assessments (ESAs) prepared to document recognized environmental conditions associated with property undergoing a commercial real estate transaction. But they can also provide a wealth of valuable information, at very little cost, to preparers of NEPA documents.

10. Presentation of the environmental record database search as an appendix may have allowed simplification of the text by eliminating Exhibit 4.3-1. But because the database search companies are private companies who do not make their products publicly accessible, the database search report cannot be incorporated by reference.

Appendix D.2

Environmental Assessment for the Relocation
of Station (Small) Ashtabula, Ashtabula, Ohio

ENVIRONMENTAL ASSESSMENT/FINDING OF NO
SIGNIFICANT IMPACT

for the

RELOCATION OF STATION (SMALL) ASHTABULA
ASHTABULA, OHIO

Responsible Agency:

U.S. Coast Guard – FDCC Atlantic
5505 Robin Hood Road, Suite K
Norfolk, Virginia 23513

Prepared By:
Tetra Tech NUS, Inc.
Gaithersburg, MD

June 1999

U.S. COAST GUARD

ENVIRONMENTAL ASSESSMENT

for the

RELOCATION OF STATION (SMALL) ASHTABULA
ASHTABULA, OHIO

This Coast Guard Environmental Assessment was prepared in accordance with Commandant's Instruction M16475.1C and is in compliance with the National Environmental Policy Act of 1969 (P.L. 91-190) and the Council on Environmental Quality Regulations dated 29 November 1978 (40 CFR 1500-1508).

This Environmental Assessment serves as a concise public document to briefly provide sufficient evidence and analysis for determining the need to prepare an Environmental Impact Statement (EIS) or a Finding of No Significant Impact (FONSI).

This Environmental Assessment concisely describes the proposed action, the need for the proposal, the alternatives, the environmental impacts of the proposal and alternatives, a comparative analysis of the action and alternatives, a statement of environmental significance, and lists the agencies and persons consulted during its preparation.

Date

Preparer: J. Peyton Doub, CEP,
Environmental Scientist, Tetra Tech NUS, Inc.

Date

Environmental Reviewer: K.W. Sykes, PE,
Environmental Support Team Leader,
U.S. Coast Guard, FDCC LANT

Date

Responsible Official: Captain Dale Walker
Commanding Officer, U.S. Coast Guard,
FDCC LANT

U.S. COAST GUARD

FINDING OF NO SIGNIFICANT IMPACT

for the

RELOCATION OF STATION (SMALL) ASHTABULA
ASHTABULA, OHIO

This project has been thoroughly reviewed by the U.S. Coast Guard and it has been determined, by the undersigned, that this project will have no significant effect on the human environment.

This finding of no significant impact is based on the attached U.S. Coast Guard prepared environmental assessment, which has been determined to adequately and accurately discuss the environmental issues and impacts of the proposed action and provides sufficient evidence and analysis for determining that an environmental impact statement is not required.

Date

Environmental Reviewer: K. W. Sykes, PE.
Environmental Support Team Leader,
U.S. Coast Guard, FDCC LANT

Date

Responsible Official: Captain Dale Walker
Commanding Officer, U.S. Coast Guard,
FDCC LANT

TABLE OF CONTENTS

LIST OF EXHIBITS

SECTION ONE   Exective Summary

The following Environmental Assessment (EA) addresses the proposed relocation of facilities associated with U.S. Coast Guard Station (Small) Ashtabula to property leased from the City of Ashtabula. The existing facilities on Station (Small) Ashtabula (referred to through the remainder of the EA as Station Ashtabula) are old, contain asbestos and other potential health hazards, and not properly configured for the current mission. Parking at the station is inadequate. Furthermore, an overhead coal conveyor generates large quantities of wind-borne coal dust that coats the exterior surfaces of station buildings and the cars and other personal belongings of station personnel. Other industrial activity on adjacent property interferes with the station's mission and adversely affects working conditions and the morale of the staff. Additionally, soil and groundwater on the station property have become contaminated by past fuel storage and require environmental remediation. Relocation of ongoing mission-related activities would free up the crowded station property for more expeditious remediation.

Under the Proposed Action, the U.S. Coast Guard would lease from the City of Ashtabula a portion of a municipal dock and a nearby plot of land. The U.S. Coast Guard would establish a 2,050-square foot pre-manufactured modular structure on the land and relocate the station's 41-foot utility boat to the dock. The modular structure would provide modern administrative and living facilities for station staff, thereby replacing the existing station house. The structure would be owned by the U.S. Coast Guard and could be easily moved if the lease were terminated. The U.S. Coast Guard would also place a small storage shed adjacent to the dock for purposes of boat maintenance. The mission of Station Ashtabula would not be changed by the Proposed Action, and response times would be unaffected.

The City of Ashtabula would expand its Ashtabula River dock in order to offset the loss of capacity resulting from the U.S. Coast Guard lease. Although the City of Ashtabula would sponsor this expansion independently of the U.S. Coast Guard, the EA addresses potential environmental impacts from the expansion as secondary impacts resulting from the Proposed Action.

The Proposed Action would not affect ecologically significant natural vegetation, wetlands, hydric soils, prime farmlands, or surface soils likely to contain intact and significant archaeological resources. The potential for impacts to water resources and aquatic biota would be limited to waterfront construction by the City of Ashtabula to expand the city dock. But the proposed shoreline where the expansion would take place is already bulkheaded, and the waterfront construction would involve no disturbance of riparian vegetation and very little disturbance of river sediments. Thus any impacts to water quality or aquatic biota in the river would be very limited.

The Proposed Action would not involve a change in employment numbers or characteristics associated with the station and thus would not substantially affect the local economy or demand for local fire, police, rescue, medical, educational, or recreational facilities. There is no available evidence of environmental contamination in areas proposed for relocation, and the relocation would not require any road improvements other than the proposed new parking spaces. The exterior of the new modular building would be designed to be architecturally compatible with adjacent buildings that are part of the Ashtabula Harbor Historic District.

In addition to the Proposed Action, two other alternatives were addressed in the EA. One is a Renovation Alternative under which the U.S. Coast Guard would remain at the existing station and modernize and upgrade the existing buildings. The other is a No-Action Alternative under which the U.S. Coast Guard would continue to operate at the existing station, without substantial renovation of any building. Neither of these alternatives is superior environmentally to the Proposed Action, and neither would satisfy the purpose and need underlying the Proposed Action.

SECTION TWO   Purpose and Need for Action

Station Ashtabula, as presently operated, is described in Section 2.1. Section 2.2 discusses the need for the changes proposed in the Proposed Action.

2.1   BACKGROUND

U.S. Coast Guard Station Ashtabula occupies approximately 0.23 acre of land on the east bank of the Ashtabula River approximately 3,000 feet south (upriver) from where the river enters Lake Erie (Exhibit 2.1-1). Facilities on the station include a main building (station house), attached three-bay garage, and free-standing boathouse (Exhibit 2.1-2). Exhibits 2.1-3 and 2.1-4 show, respectively, the station house with attached garage and the boathouse. Gravel areas on the side of Front Street, directly east of the garage, provide parking for seven to eight cars (Exhibit 2.1-5).

Station Ashtabula is presently assigned two boats. A 41-foot utility boat (UTB) is moored on the river at the bulkhead adjoining the station house (Exhibit 2.1-6). A 21-foot rigid hull inflatable boat is kept inside the boathouse, with direct access to the river. Fueling facilities formerly located on the station have been removed, and both boats are now fueled at commercial marinas off of the station.

The station is situated in an industrialized part of the City of Ashtabula, Ohio that is termed Ashtabula Harbor. The station property is bounded to the west by the Ashtabula River and to the north, south, and east by heavy industry. To the north are railyards and docks owned by Conrail that are used for loading and unloading coal shipped by rail and freighter. In 1967, Conrail constructed an overhead coal conveyor across the Ashtabula River just 10 feet north of the station perimeter (Exhibit 2.1-7). R. W. Sidley Company, which manufactures concrete, is situated immediately south of the station. Large piles of sand and gravel are stored just south of the station perimeter (Exhibit 2.1-8).

Station Ashtabula operates as a subunit of Station Fairport. Two duty sections of five enlisted personnel each are assigned to Station Ashtabula, with command and control supervision provided by the officer-in-charge of Station Fairport. Station Ashtabula's missions include search and rescue (SAR) (20 percent), recreational boating safety (16 percent), operational training (38 percent), public affairs (6 percent), minor aids to navigation (7 percent), and miscellaneous (13 percent). Station Ashtabula is located near the center of its area of responsibility (AOR), which extends westward along the United States shore of Lake Erie to a point just east of Geneva, Ohio (AOR for Station Fairport) and eastward to Elk Creek, Pennsylvania (AOR for Station Erie). Station Ashtabula's location allows a boat transit time of no more than two hours to any part of its AOR.

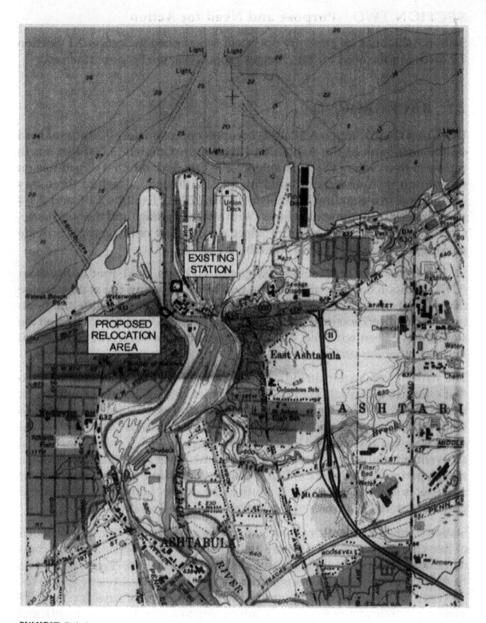

EXHIBIT 2.1-1
Location Map - U.S. Coast Guard Station Ashtabula. Modified from U.S. Geological Survey
7.5-Minute Topographic Map for Ashtabula North, Ohio Quadrangle.

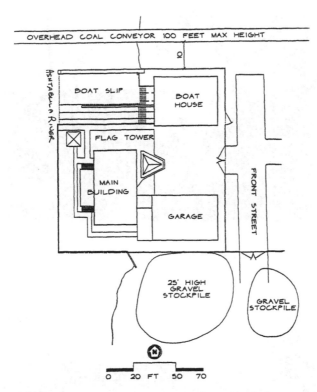

EXHIBIT 2.1-2
Layout of existing buildings on U.S. Coast Guard Station Ashtabula.

EXHIBIT 2.1-3
Existing station house and attached garage on Station Ashtabula.

EXHIBIT 2.1-4
Existing boathouse on Station Ashtabula.

EXHIBIT 2.1-5
View of Front Street, on eastern edge of Station Ashtabula. Station personnel park their cars on unmarked gravel areas on the sides of the street.

2.2   NEED FOR ACTION

The need for the Proposed Action includes the following primary components:

1. a need for safe, modern facilities that are the proper size for the current mission and staffing level;
2. a need for increased and more convenient parking for station personnel, government vehicles, and visitors;

3. a need to escape coal dust generated by the overhead coal conveyor and other interferences caused by adjacent industrial activity (such as noise and dust from other materials); and

4. a need to expedite environmental remediation of soils and ground-water contaminated by past fuel storage at the station.

The U.S. Coast Guard also hopes to reduce maintenance expenses by providing new, modern facilities that are properly sized and configured to the present mission.

EXHIBIT 2.1-6
41-Foot utility boat moored on Astabula River adjacent to existing station house on Station Ashtabula.

EXHIBIT 2.1-7
Footing for overhead coal conveyor that crosses Ashtabula River at northern edge of Station Ashtabula.

EXHIBIT 2.1-8
Large piles of construction materials on R. W. Sidley Company property immediately south of Station Ashtabula.

2.2.1   Provide Safer, Modern Facilities Sized for Current Mission

The station house and attached garage were constructed in 1939 and the boathouse was constructed in 1938. Station Ashtabula was downgraded to a station (small) in 1996 with a reduced manpower complement. The station buildings include a substantial excess of certain categories of floor space, such as kitchen space and other miscellaneous space, but a deficiency in other facilities such as lobby and administrative space, bathroom facilities, and wetrooms. Furthermore, the basement of the station house has flooded during several recent high lake level incidents, and the building contains asbestos-containing material (ACM)

2.2.2   Provide Adequate Parking

The only available parking at the station is provided by gravel areas on the sides of Front Street at the eastern perimeter of the station. None of the parking areas are paved or striped, and cars often must be parked in irregular positions to fit into the available space.

2.2.3   Escape Coal Dust and Other Industrial Interferences

The overhead coal conveyor just north of the station generates large quantities of wind-borne coal dust that falls onto the station property. The coal dust represents a perceived health hazard to station personnel and repeatedly soils the exterior walls of station buildings and the cars and personal property of the staff. Because of a court settlement, Conrail is responsible for cleaning building exteriors on the station once per month. Furthermore, blowing coal dust and other particles from stockpiled materials on adjacent industrial property frequently impairs the vision of the station watchstander.

2.2.4 Expedite Environmental Remediation

Gasoline and diesel fuel used to fuel boats and other U.S. Coast Guard vehicles were formerly stored on the station in two underground storage tanks (USTs) each with a capacity of 1,000 gallons. Both USTs have leaked and contaminated underlying soils and groundwater. The U.S. Coast Guard plans to remediate the contamination in accordance with federal and state law. Vacating ongoing mission activities from the station would reduce interference and thus hasten environmental remediation work.

2.3 PURPOSE OF PROPOSED ACTION

The purpose of the Proposed Action would be to provide safer, modern facilities that are properly sized for Station Ashtabula's mission and that would not be adversely affected by coal dust and other intrusions from adjacent industry. The Proposed Action would reduce maintenance expenses, provide a superior work environment, increase available parking, improve the morale of duty staff, and expedite environmental cleanup of the existing station property.

SECTION THREE Exective Summary

Several alternatives were considered to fulfill the purpose and need. Three alternatives were considered but dismissed from detailed analysis (Section 3.1). Three other alternatives were considered in detail (Section 3.2). Considered in detail include a No-Action Alternative, renovation of the existing station (the Renovation Alternative), and relocation of station facilities to property leased from the City of Ashtabula (the Proposed Action). To be retained for detailed analysis, an alternative had to result in safer, modern facilities; provide adequate parking; and result in reduced maintenance requirements and increased long-term cost savings. The alternative could not adversely affect mission readiness or SAR response times to any part of the Station Ashtabula AOR. The abilities of an alternative to solve problems associated with coal dust and other industrial interferences, and to reduce interference with environmental remediation, were used to evaluate alternatives retained for detailed analysis.

3.1 ALTERNATIVES CONSIDERED AND DISMISSED FROM DETAILED ANALYSIS

The following alternatives were considered but dismissed from detailed analysis.

3.1.1 Replacement of Shore Facilities with an IMARV Facility

Under this alternative, the buildings on Station Ashtabula would be replaced by an Independent Maritime Response Vehicle (IMARV). An IMARV is a 45- to 60-foot vessel with operating characteristics similar to the station's 41-foot UTB but which also provides sleeping, dining, and limited recreational space for a duty crew of four. The station house and other buildings would no longer be necessary and could be demolished or excessed. The IMARV could either be moored on the river at the existing station or at a dock owned by the City of Ashtabula on the western shore of the river.

Mooring the IMARV at the existing station would provide an improved, modern, and properly sized facility for the station and would expedite environmental remediation of the property. But the IMARV would still be exposed to the coal dust and would still be proximate to other industrial activity. Mooring the IMARV at the city dock would provide the same advantages but would also render station operations free from the coal dust and industrial interferences. Additional parking could be made available at either location. However, any alternative involving an IMARV facility had to be dismissed because the U.S. Coast Guard has discontinued the IMARV program for reasons of feasibility.

3.1.2   Construct New Buildings at Existing Station

Under this alternative, the U.S. Coast Guard would raze the existing buildings on the station property, complete necessary environmental remediation, and construct new facilities on the property. This alternative would provide safer, modern facilities that are properly sized and would facilitate environmental remediation. The new buildings could be configured on the property to allow for additional parking space. But the alternative was rejected because it would involve substantial construction expense and still not solve the problems caused by coal dust and industrial interferences. It would also require the use of temporary living and administrative facilities between when the existing buildings are demolished and when the new buildings are completed.

3.1.3   Relocation of Station Ashtabula to Site Outside Ashtabula Harbor

Under this alternative, new station buildings would be constructed at a new location within Station Ashtabula's AOR but outside of Ashtabula Harbor. Sites considered included Redbrook Marina, located approximately five miles west of Ashtabula Harbor; Geneva State Park Marina, located approximately twelve miles west of Ashtabula Harbor; Conneaut Harbor Municipal Marina, located approximately twelve miles east of Ashtabula Harbor; and Lake Shore Park, located approximately one mile east of Ashtabula Harbor. This alternative would provide safer, modern, properly sized facilities and adequate parking in a location distant from industrial activity. But most of the possible locations are not centrally located within the AOR, and many are not as well sheltered against storms as is Ashtabula Harbor. Thus this alternative was rejected.

3.2   ALTERNATIVES RETAINED FOR DETAILED ANALYSIS

The following alternatives were retained for detailed environmental analysis.

3.2.1   No-Action Alternative

Under the No-Action Alternative, Station Ashtabula would continue to operate using the existing facilities on the existing station property. None of the facilities would be renovated, although routine maintenance would continue. Environmental remediation of contaminated soils and groundwater on the station property would continue as planned. The schedule for environmental remediation would not be affected.

3.2.2 Renovation Alternative

Under the Renovation Alternative, the existing structures on Station Ashtabula would be renovated and modernized to better accommodate the station's current mission. Measures would be taken to better prevent flooding of the station house basement during high lake level incidents. Pipe insulation and other building materials containing ACM would be removed. The station's two assigned boats would continue to be moored as at present. Environmental remediation of contaminated soils and groundwater on the station property would continue as planned. The schedule for environmental remediation would not be affected by the No-Action Alternative.

3.2.3 Relocation to Property Leased from City of Ashtabula (Proposed Action)

Under this alternative, which is the Proposed Action, the U.S. Coast Guard would lease a portion of a municipal dock owned by the City of Ashtabula to moor the 41-foot UTB. The U.S. Coast Guard would also lease a nearby vacant lot owned by the City and install a modular building that would provide the functions presently provided by the station house (Exhibit 3.2-1). The U.S. Coast Guard would then vacate the existing station buildings, complete all necessary environmental remediation, and excess the station property.

The city dock (Exhibit 3.2-2) is located on the western shore of the Ashtabula River, approximately 500 feet southwest of the existing Station Ashtabula. The dock presently serves recreational boaters. The City would add shore ties and other dock improvements needed to moor the station's 41-foot UTB or other comparable U.S. Coast Guard boat. The U.S. Coast Guard would place and maintain a small portable shed on City property adjacent to the dock. The shed would provide storage functions presently provided by the boat-house.

It is expected that the City would eventually expand the dock northward along the river to provide additional capacity for recreational boats. The expansion would offset the dock capacity lost in order to accommodate the U.S. Coast Guard. Although the expansion would be conducted by the City of Ashtabula independently of the U.S. Coast Guard, the potential environmental impacts of the expansion have been considered as secondary impacts to the Proposed Action.

The vacant lot (Exhibit 3.2-3) comprises approximately 0.25 acre on the west side of Morton Street, approximately 300 feet west of the city dock. A building formerly used as a garage and warehouse formerly occupied the site and was recently demolished by the City of Ashtabula. The City demolished the building to foster commercial reuse of the site but is willing to execute a 20-year lease of the site to the U.S. Coast Guard. Once leased, the U.S. Coast Guard

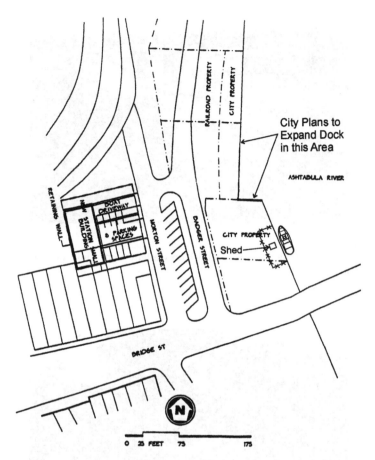

City Plans to
Expand Dock
in this Area

ASHTABULA RIVER

EXHIBIT 3.2-1
Details of proposed relocation area under proposed action. The city-owned dock is located
east of the Badger Street, along the western shore of the Ashtabula River. The proposed new
modular building would be constructed on a vacant lot west of Morton Street.

would purchase a 2,050-square foot pre-manufactured modular structure for
installment on the site. The modular building would include two berthing
(sleeping) rooms, office, kitchenette, dining area, and recreation lounge. The
U.S. Coast Guard would own the modular building and be able to move or
replace it in the future.

A minimum of eight new parking spaces would be constructed either on the
site of the modular building or along the side of Morton Street. The parking
spaces would be paved, striped, and located only a few steps from the modular building. Even if constructed on the side of Morton Street, the new parking spaces would be dedicated to the U.S. Coast Guard rather than open to
the public.

EXHIBIT 3.2-2
Center of photograph shows city-owned dock on western (right) shore of Ashtabula River, just below bridge.

EXHIBIT 3.2-3
Vacant lot that U.S. Coast Guard would lease from City of Ashtabula and install pre-manufactured modular building.

3.3 COMPARISON OF ALTERNATIVES

A comparison of the potential environmental impacts that could result from the Proposed Action, Renovation Alternative, and No-Action Alternative is shown in Exhibit 3.3-1.

EXHIBIT 3.3-1

Comparison of Alternatives Relocation of U.S. Coast Guard Station (Small)
Ashtabula, Ohio

Analysis Items	Alternatives		
	No-Action	Proposed Action	Renovation Alternative
Need for Action			
Provide safer, modern facilities of proper size	Existing buildings are old, pose health hazards, and are not matched to current mission	New modular building would be modern, flexible, safer, and well suited to current mission	Renovated facilities would be safer and better suited to current mission
Provide adequate parking	Inadequate, poorly marked parking	Eight new paved, striped parking spaces, adequate to meet current mission	Renovation could include addition of a few extra parking spaces, but land is limited
Escape coal dust and industrial interferences	Continued exposure to coal dust and other interferences from adjacent industry	Relocated facilities would be across river, away from influence of heavy industry	Continued exposure to coal dust and other interferences from adjacent industry
Expedite environmental remediation	Continued mission use of station property could interfere with planned remediation	Station property would be vacated, thereby facilitating remediation	Continued mission use of station property could interfere with planned remediation
Nonenvironmental Planning Criteria			
Response time	No more than 2 hours to any part of AOR	No more than 2 hours to any part of AOR	No more than 2 hours to any part of AOR
Cost	Maintenance costs would remain very high	Despite costs of relocation, estimated savings of $1,263,000 over life cycle	Would not achieve savings comparable to those under the Proposed Action
Socioeconomic Environment			
Land use	No changes in land use	Consistent with zoning and adjacent land uses, existing station property available for industrial reuse	No changes in land use

(continues)

EXHIBIT 3.3-1

Comparison of Alternatives Relocation of U.S. Coast Guard Station (Small) Ashtabula, Ohio (continued)

	Alternatives		
Analysis Items	No-Action	Proposed Action	Renovation Alternative

Socioeconomic Environment (continued)

Analysis Items	No-Action	Proposed Action	Renovation Alternative
Local economy	No changes to local economy	No permanent changes; minor temporary demand for construction personnel	No permanent changes; minor temporary demand for construction personnel
Housing	Adequate sleeping accommodations in station house for duty personnel	Adequate accommodations would be provided by modular building	Adequate accommodations would be provided by renovated facilities
Community services and medical facilities	City services would be adequate	City services would be adequate	City services would be adequate
Recreational facilities	Adequate facilities off station, no increased demand	Adequate facilities off station, no increased demand	Adequate facilities off station, no increased demand
Fire, rescue, and police services	City services would be adequate	City services would be adequate	City services would be adequate
Schools	No increased school enrollment	No increased school enrollment	No increased school enrollment
Utilities	No increased demand	No increased demand	No increased demand

Transportation

Analysis Items	No-Action	Proposed Action	Renovation Alternative
Transportation	Adequate service provided by local roadways, parking would remain limited	Adequate service provided by local roadways, parking would be adequate	Adequate service provided by local roadways, parking would be increased by remain limited

Physical Environment

Analysis Items	No-Action	Proposed Action	Renovation Alternative
Geology, topography, soils	No potential Impacts	Potential erosion controlled by sediment/erosion plan; no prime farmland or hydric soils affected	No potential impacts

EXHIBIT 3.3-1

Comparison of Alternatives Relocation of U.S. Coast Guard Station (Small) Ashtabula, Ohio (continued)

Analysis Items	Alternatives		
	No-Action	Proposed Action	Renovation Alternative
Physical Environment (continued)			
Climate and air quality	No potential impacts	Very small, temporary emissions from construction equipment, fugitive dust to be controlled	No potential impacts
Water resources	No potential impacts	Limited waterfront construction by City of Ashtabula to expand city dock would not substantially disturb sediment or affect water quality	No potential impacts
Noise	No changes in noise levels	No noticeable changes	No noticeable changes
Hazardous materials and hazardous waste	Continued use of station property for mission could interfere with planned environmental remediation	No evidence of contamination in relocation area, vacating existing station property could facilitate planned environmental remediation	Continued use of station property for mission could interfere with planned environmental remediation
Natural Environment			
Terrestrial environment	No potential impacts	Loss of ruderal (weedy) vegetation only	No potential impacts
Floodplains and coastal zone	No work in mapped floodplain (Zones A or B), episodes of flooding in station house basement would likely continue	No buildings to be located in floodplain, waterfront construction by City of Ashtabula to expand dock would be compliant with Executive Order 11988	No work in mapped floodplain (Zones A or B), renovation could include engineering measures to reduce episodes of flooding in station house.

(continues)

EXHIBIT 3.3-1

Comparison of Alternatives Relocation of U.S. Coast Guard Station (Small)
Ashtabula, Ohio (continued)

	Alternatives		
Analysis Items	**No-Action**	**Proposed Action**	**Renovation Alternative**
Natural Environment (continued)			
Wetlands	No potential impacts	No potential impacts	No potential impacts
Aquatic environment	No potential impacts	Waterfront construction by City of Ashtabula to expand city dock would not substantially affect water quality or aquatic biota	No potential impacts
Threatened and endangered species	No potential impacts	No potential impacts	No potential impacts
Cultural Resources			
Archaeological resources	No potential impacts	Very low potential for impacts because affected soils have long history of urban land use	No potential impacts
Historic Resources	No potential impacts	Exterior of new modular building would be designed to be architecturally compatible with Ashtabula Harbor Historic District	No potential impacts; building exteriors would remain essentially unchanged in appearance

3.4   PREFERRED ALTERNATIVE

The U.S. Coast Guard has selected the Proposed Action, which involves lease from the City of Ashtabula of dock space and land for a new modular building, as the preferred alternative. The modular building would provide modern quarters that are safer, easy to maintain, and properly sized for the mission. Adequate parking would be provided. Both the city dock and the modular building site are over 500 feet distant from the nearest heavy industry and would not experience substantial quantities of wind-borne coal dust. Removing mission activities from the existing station property would reduce interference with, and thus expedite, environmental remediation.

Implementation of the Proposed Action would not compromise the station's mission or increase SAR response times to any part of the AOR. The city dock is only about 500 feet away from the existing station and is thus still at the heart of the AOR. The city dock is still north (lakeward) of all drawbridges and other navigational obstructions on the Ashtabula River. Additionally, the U.S. Coast Guard has estimated that implementation of the Proposed Action would result in a cost savings, relative to the No-Action Alternative, of approximately $1,263,000 over the life cycle of the new facilities.

SECTION FOUR   Affected Environment and Environmental Consequences

The following section describes existing environmental conditions in areas potentially affected by the Proposed Action and alternatives and discusses the potential impacts to specific environmental resources.

4.1   SOCIOECONOMIC ENVIRONMENT

Socioeconomic factors addressed separately below include land use; the local economy; housing; community services and medical facilities; recreational facilities; fire, rescue, and police services; schools; and utilities. Both the existing Station Ashtabula and the proposed new site are located within the incorporated limits of the City of Ashtabula, Ashtabula County, Ohio. The City of Ashtabula is located on the south shore of Lake Erie.

4.1.1   Land Use

Affected Environment: The existing station property occupies approximately 0.23 acre of land on the east side of the lower Ashtabula River. Structures on the property include a 3-story wood-frame station house, a garage with bays for three vehicles, and a boathouse. Gravel areas on the sides of Front Street, directly east of the garage, provide parking for about six cars. The station property is surrounded by industrial development. Conrail's coal loading operation, including an overhead coal conveyor, is located to the north. R. W. Sidley Company, which manufactures concrete, occupies industrial property south of the station.

The City dock and proposed site for the new modular building are located in an area of older residential, commercial, and light industrial development. The proposed modular building site is bounded to the north and west by a steep slope supporting scrub-shrub vegetation. A small city park and a residential neighborhood are located at the top of the slope. An abandoned garage is located just south of the proposed site. A building permit, expiration date 8/17/99, posted on the building indicates that it will be remodeled. A station employee stated that a developer plans to convert the building to a Starbucks restaurant (Suvak, 1999). Several smaller restaurants occupy other older buildings to the south and southwest of the proposed site.

The City of Ashtabula has zoned the station property and adjoining properties east of the Ashtabula River as M-2, Heavy Industrial. The City dock is also zoned M-2. Few limitations are placed on use of land zoned M-2 except that dwellings are not permitted. The proposed modular building site and properties to its south and west are zoned M-1, Light Industrial. Industrial development is permitted on land zoned M-1 but must meet limitations related to the generation of odors, dirt, noise, gas, and vibration. Land north of the proposed modular building site is zoned R-3, Two-Family Residence (City of Ashtabula, 1999).

Environmental Consequences of the Proposed Action: Use of the City dock and establishment of the new modular building would be compatible with adjacent land uses and with zoning established by the City of Ashtabula. The new modular building would be visually compatible with the surrounding neighborhood, and station personnel would benefit from the surrounding shops and restaurants. The sleeping quarters in the new modular building would be permitted under the M-1 zoning assigned to the proposed site. No zoning variances or special exceptions would be required. Excessing the existing station would make available for redevelopment approximately 0.23 acre of waterfront land zoned for heavy industry.

Environmental Consequences of the Renovation Alternative: No land use changes would result from renovation of the existing station facilities. Each facility would continue to be used for its present purpose. It is noted that the sleeping quarters in the existing station house are technically not permitted under the station's M-2 zoning, but U.S. Coast Guard activities at the station are considered by the City to be grandfathered (Perry, 1999).

Environmental Consequences of the No-Action Alternative: No land use changes would result from the No-Action Alternative. It is noted that the sleeping quarters in the existing station house are technically not permitted under the station's M-2 zoning, but U.S. Coast Guard activities at the station are considered by the City to be grandfathered (Perry, 1999).

4.1.2 Local Economy

Affected Environment: The population of Ashtabula County in 1998 was estimated by the Bureau of the Census at 103,300 (Bureau of the Census, 1998), an increase of approximately 3.5 percent over the 1990 population of 99,821 (Bureau of the Census, 1990). Ashtabula County is predominantly rural, with industrial development largely clustered in and around the City of Ashtabula, including Ashtabula Harbor. Nearly half of all jobs in Ashtabula County are in the manufacturing sector (8,300 jobs in 1992). The retail sector accounted for 5,450 jobs in 1992, the services sector accounted for 3,532 jobs in 1992, and the wholesale sector accounted for 666 jobs in 1992 (Bureau of the Census, 1992).

Like many other areas of the Northeastern and Midwestern United States where employment has historically relied heavily on manufacturing, Ashtabula County suffered substantially from the closing and downsizing of manufacturing operations in the 1980s and 1990s. As of the early 1990s, over 700 workers in Ashtabula County had been idled by plant closings, and Ashtabula County's unemployment rate had consistently been between three and five percent higher than that of Ohio or the nation as a whole (Ashtabula County Planning Commission, 1992). County planners have emphasized the establishment of infrastructure to attract industrial development from larger industrial areas to the west (Cleveland, Ohio), south (Youngstown, Ohio), and east (Erie, Pennsylvania) (Ashtabula County Planning Commission, 1991). In response to the reduction in manufacturing jobs, the County has promoted the development of tourism facilities that take advantage of the Lake Erie shoreline and scenic rural areas (Ashtabula County Planning Commission, 1988).

Environmental Consequences of the Proposed Action: The Proposed Action would not result in a change in staff assigned to Station Ashtabula and should have no long-term adverse impact on the local economy. The proposed new facilities would not interfere with the promotion of tourism. Vacating the existing station would make that property available for other industrial development, which could benefit the local economy.

Small numbers of routine construction jobs would be generated on a temporary basis while station facilities are being relocated. Examples include jobs for construction laborers, backhoe operators, carpenters, masons, electricians, plumbers, drywallers, and painters. These demands would be met mostly by the local labor pool, although some workers might commute from other nearby communities. No specialized construction labor demands are anticipated.

Environmental Consequences of the Renovation Alternative: As is true for the Proposed Action, the Renovation Alternative would not result in a change in staff assigned to the station. Renovation of the existing station facilities would generate small numbers of routine construction jobs on a temporary basis.

Environmental Consequences of the No-Action Alternative: The No-Action Alternative would not result in any short-term or long-term changes in employment associated with Station Ashtabula.

4.1.3   Housing

Affected Environment: The existing station house includes four bedrooms, which provide sleeping accommodations for ten staff. Although Ashtabula County is predominantly rural, there is not a serious shortage of housing within a reasonable drive of the station. Ashtabula County included 41,214 housing units in the 1990 Census, of which 4,454 were vacant. Of the vacant housing units, 721 were reported as available for rent (Bureau of the Census, 1990). Several new hotel facilities have recently been constructed, or are still under construction, along Interstate Route 90, which crosses Ashtabula County approximately 6 miles south of the station.

Environmental Consequences of the Proposed Action: The new modular building would provide adequate sleeping and living accommodations for the station's current staff. The Proposed Action would not increase employment levels at the station and thus would not create an increased demand for housing for station personnel. Most construction labor needed to build the new facilities would come from the City of Ashtabula and other communities within commuting distance of the site. Those laborers would not have to find temporary accommodations. A few laborers may come from areas not within commuting distance, but they should be able to find temporary accommodations at nearby hotels or rental units.

Environmental Consequences of the Renovation Alternative: The renovated station facilities would provide adequate sleeping and living accommodations for the station's current staff. There would be no increased employment and thus no increased demand for housing for station personnel. Most construction labor needed to renovate the facilities would come from the City of Ashtabula and other communities within commuting distance of the site. Any construction workers requiring temporary accommodations should be able to find temporary accommodations at nearby hotels or rental units.

Environmental Consequences of the No-Action Alternative: The No-Action Alternative would not result in increased employment at the station and would not generate any requirements for construction labor. It would not result in any increased long-term or short-term demand for housing.

4.1.4 Community Services and Medical Facilities

Affected Environment: Ashtabula County is served by the Ashtabula County Medical Center, located approximately 1 mile south of Station Ashtabula at 2420 Lake Avenue in the City of Ashtabula. The Center is a private, nonprofit, community-oriented facility with a full complement of routine and emergency staff and services (ACMC, 1999). The Ashtabula County District Library, a full service public library, is located approximately 2 miles south of Station Ashtabula at 335 West 44th Street in the City of Ashtabula (NOLA Regional Library System, 1999). Smaller public libraries are located elsewhere in Ashtabula County.

Environmental Consequences of the Proposed Action: The Proposed Action would not result in a change in employment levels associated with Station Ashtabula and thus would not create an increased demand for community services such as hospitals and libraries. The proposed relocation area and the existing station are approximately the same distance from the Ashtabula County Medical Center and the Ashtabula County District Library.

Environmental Consequences of the Renovation Alternative: Renovating the existing station facilities would not result in a change in employment associated with Station Ashtabula and should not generate increased demand for community services.

Environmental Consequences of the No-Action Alternative: The No-Action Alternative would not affect the demand for community services.

4.1.5   Recreational Facilities

Affected Environment: There are no formal exterior recreation facilities on the existing station, although a basketball net has been mounted in an asphalt area between the station house and boathouse. The existing station building includes 240 square feet of interior space designated for "recreation/pantry/physical fitness". This interior space provides lounge facilities and allows for some informal recreation, but station personnel must rely on the surrounding community for formal recreational facilities.

The City of Ashtabula includes a number of parks located only a short distance from the station. The city dock adjacent to the proposed new station house site is the only public boat access on the Ashtabula River. A scenic overlook featuring Ashtabula Harbor is located at the top of the steep slope north of the proposed modular building site. A graded trail leads from Morton Street, which dead-ends at the site, up the slope to the overlook. Two parks are located on the shore of Lake Erie near the mouth of the Ashtabula River. Walnut Beach Park is located in the City of Ashtabula approximately 0.5 mile west of Station Ashtabula, and Lake Shore Park is located approximately 1.1 miles east of Station Ashtabula. The former is located just west of the breakwaters at the mouth of the Ashtabula River, and the latter is located just east of the breakwaters.

Environmental Consequences of the Proposed Action: The proposed new modular building would provide 247 square feet designated for "recreation/pantry/physical fitness", a slight increase over that provided by the existing station house. But station personnel would continue to have to rely mostly on community facilities for recreation. Because no additional personnel would be assigned to the station, there would be no increased demand for the recreation facilities in the surrounding community. Most community recreation facilities would be about the same distance from the new as from the old station facilities. The scenic overlook of Ashtabula Harbor would, however, be substantially closer, less than a five-minute walk. It would thus be readily accessible to station personnel without driving during lunch and other short breaks during the workday.

Use of the City-owned dock to moor a U.S. Coast Guard boat would reduce the capacity of the dock to accommodate privately-owned boats. However, the City of Ashtabula plans to expand the dock northward along the western shore of the river to provide increased capacity. The expanded dock would thus be capable of accommodating at least as many privately-owned boats as at present.

Environmental Consequences of the Renovation Alternative: There would be no change in the amount of space allocated to recreation following renovation of the existing station house. Personnel assigned to the station would continue to have to rely primarily on community facilities for recreation but, because the same number of employees would be assigned to the station, there would be no increased demand for those facilities. U.S. Coast Guard boats would continue to be moored at the existing station and the capacity of the City-owned dock would be unaffected.

Environmental Consequences of the No-Action Alternative: There would be no change in the amount of space on the station allocated to recreation. Personnel assigned to the station would continue to have to rely primarily on community facilities for recreation but, because the same number of employees would be assigned to the station, there would be no increased demand for those facilities. U.S. Coast Guard boats would continue to be moored at the existing station, and the capacity of the City-owned dock would be unaffected.

4.1.6   Fire, Rescue, and Police Services

Affected Environment: Search and rescue (SAR) represents approximately twenty percent of Station Ashtabula's mission. Station personnel represent the most immediate source of aid for boaters and swimmers within Station Ashtabula's AOR. The City of Ashtabula is served by the Ashtabula Fire Department, located at 4326 Main Avenue, Ashtabula, Ohio, approximately two miles south of Station Ashtabula. The Department includes a roster of 26 personnel and two front-line fire trucks acquired in 1991 and 1987 (Ashtabula Fire Department, 1999). The City of Ashtabula has its own police force that is independent from that of Ashtabula County.

Environmental Consequences of the Proposed Action: The proposed relocation area is located almost directly across the Ashtabula River from the existing station. The distance between the City dock and the existing station is only about 500 feet. The dock, like the existing station, is located downstream from the 5th Street drawbridge and thus has navigational access to Lake Erie that is free of drawbridges or other obstructions. Thus, the proposed relocation of the station's facilities and boat would not substantially affect SAR response times. The Proposed Action would not change the number of staff assigned to Station Ashtabula and thus should not increase the demand for community services such as fire and police protection. The new modular building would represent a reduced fire hazard than the existing station because it would be of modern construction and free of the coal dust that has impregnated the exterior walls of the older structure.

Environmental Consequences of the Renovation Alternative: The SAR response time would remain unchanged, as would the demand for fire and police services provided by the community. Renovation of the existing facilities would reduce their fire hazard, but the exterior walls would continue to be exposed to coal dust.

Environmental Consequences of the No-Action Alternative: The SAR response time would remain unchanged, as would the demand for fire and police services provided by the community. The older construction of station facilities and the impregnated coal dust would continue to represent a fire hazard.

4.1.7 Schools

Affected Environment: The Ashtabula City Area School District operates 12 public schools enrolling 5,287 children from grades kindergarten through 12 (based on 1990 Census data). An additional 589 children residing in the area served by the District attend private schools. There are 21 students per teacher in the District, compared to a ratio of 17 students per teacher for the State of Ohio and for the entire United States (SDDB, 1999). These figures suggest a modest shortfall of teachers in schools serving the community surrounding Station Ashtabula.

Environmental Consequences of the Proposed Action: No additional personnel would be assigned to Station Ashtabula as a result of the Proposed Action, which thus would not affect enrollment in area schools. Any teacher shortage in the surrounding communities would not be exacerbated.

Environmental Consequences of the Renovation Alternative: No additional personnel would be assigned to Station Ashtabula as a result of renovating the existing facilities. Thus the Renovation Alternative would not affect enrollment in area schools, and any teacher shortage in the surrounding communities would not be exacerbated.

Environmental Consequences of the No-Action Alternative: No additional personnel would be assigned to Station Ashtabula, and thus enrollment in areas schools would not be affected. Any teacher shortage in the surrounding communities would not be exacerbated.

4.1.8 Utilities

Affected Environment: Electric service is provided to the City of Ashtabula, including the existing station and proposed relocation area, by the Illuminating Company (formerly the Cleveland Electric Illuminating Company), a subsidiary of FirstEnergy Corporation. Formed by the merging of the Illuminating Company, Ohio Edison, Toledo Edison, and Penn Power, FirstEnergy provides electric service to most of northern Ohio and part of northwestern Pennsylvania (FirstEnergy, 1999). Potable water and city sewage are provided by the City of Ashtabula to the existing station and are available to the city dock and to the proposed new modular building site.

Environmental Consequences of the Proposed Action: The relocated facilities would utilize the same sources for electricity, water, and other utilities as does the existing station. The demand for these utilities would be essentially comparable. Slightly reduced electricity usage could result because the floor space in the new station house would be lower than in the existing station house, and because the new building would be better insulated.

Environmental Consequences of the Renovation Alternative: The demand for utilities would remain essentially unchanged, although improved insulation and improved construction in the renovated facilities could slightly reduce electric consumption.

Environmental Consequences of the No-Action Alternative: The demand for utilities would not be affected.

4.2   TRANSPORTATION

Affected Environment: Both the existing station and the proposed relocation area are located on side streets in an urban setting. The nearest arterial road to both areas is State Route 531 (5th Street), which crosses the Ashtabula River in an east-west orientation on a drawbridge constructed in 1925 (USGS, 1960). The drawbridge is manned continuously between March 1 and December 31 but is closed between January 1 and February 28. During this winter period, a 12-hour notice must be submitted to request an opening.

State Route 11, a limited access freeway, provides access to Route 531 approximately 0.75 mile east of the existing station (USGS, 1960). The existing station is accessed from Route 531 by a short segment (approximately 500 feet) of Ferry Street. Most of Front Street, except for a short stretch on the eastern perimeter of the station, has been vacated (abandoned) to provide additional space for industrial development (City of Ashtabula, 1999). The remaining segment of Front Street is used to provide parking for the station, which is inadequate.

The City dock is located on the northeast corner of Route 531 and Petriss Street (formerly Badger Street), on the western shore of the river just north of the drawbridge. The proposed new station house site is located at the northern dead-end of Morton Street, approximately 200 feet northwest of the dock. Twelve perpendicular parking spaces on the east side of Morton Street are used by patrons of businesses on Morton Street and 5th Street.

Environmental Consequences of the Proposed Action: No reconfiguration of existing roadways would be necessary. Eight parking spaces, either on the modular building site or perpendicular to Morton Street would be constructed for the exclusive use of station personnel and visitors. These would include six spaces for duty and supervisory staff, one space for a government vehicle, and one for a visitor. This parking allotment would prevent staff or visitors to the station from having to use the limited street parking in the area.

Environmental Consequences of the Renovation Alternative: Parking would continue to be limited to the edge of Front Street at the eastern perimeter of the station. There would be no increase in parking capacity, which would remain limited.

Environmental Consequences of the No-Action Alternative: Parking would continue to be limited to the edge of Front Street at the eastern perimeter of the station. There would be no increase in parking capacity, which would remain limited.

4.3   PHYSICAL ENVIRONMENT

Elements of the physical environment considered separately below include geology, topography, and soils; climate and air quality; water resources; noise; and hazardous materials and hazardous waste.

4.3.1   Geology, Topography, and Soils

Affected Environment: Surface deposits in the Ashtabula area consist of till and stratified gravel, sand, silt, and clay. The lower course of the Ashtabula River flows through a narrow valley cut through silty shale to a depth of 75 to 100 feet (COE, 1983). The existing Station Ashtabula and the city dock occupy level lands adjoining the Ashtabula River. The proposed new station house site is level but is immediately bounded to the north and west by a steep slope exceeding 4:1 in grade (USGS, 1960). The site appears to have been made level in the past by cutting into the slope.

The Soil Survey for Ashtabula County maps the existing station, city dock, and proposed new station house site as "Made Land", defined as areas of earth, fill, borrow pits, or impervious cover (SCS, 1973). Surface soils in such areas have typically been heavily disturbed, derived from multiple sources, and not possessing uniform or predictable soil properties. Areas mapped as "Made Land" are not regarded as having hydric soils or as being prime or unique farmland. The soil survey maps the slope north and west of the proposed new station house site as "Steep Land, Loamy". The surface soil material is described as loamy and well-drained but possessing a severe erosion hazard. Such soils are also not regarded as either hydric or as prime or unique farmland.

Environmental Consequences of the Proposed Action: Construction of the new modular building, parking, and dock shed would require covering up to 15,000 square feet of disturbed urban soils by impervious surface. Expansion of the city dock by the City of Ashtabula would require replacement of bulkhead along as much as 500 feet of the western shore of the Ashtabula River and covering up to 5,000 additional square feet of urban soil by impervious surfaces. No hydric soils or prime farmland would be affected. The dock construction would be performed in accordance with a soil erosion and sediment control plan designed to prevent the entry of eroded soil into the river.

A retaining wall would be constructed just north and west of the new station house at the edge of the steep slope. The retaining wall would be designed to prevent failure of the slope that could pose a physical risk to the building or a risk of injury to station personnel. Bare areas on the slope above the retaining wall would be vegetatively stabilized with hardy fast-growing grasses to further reduce the potential for erosion.

Environmental Consequences of the Renovation Alternative: No soils would be disturbed by renovating the existing station.

Environmental Consequences of the No-Action Alternative: No soils would be disturbed by the No-Action Alternative.

4.3.2 Climate and Air Quality

Affected Environment: Exposure of station personnel and buildings to coal dust from the overhead Conrail coal conveyor located just south of the station has been a concern since construction of the conveyor in 1967. A cover was installed over the coal chute in 1985 and a sprinkler system was installed in 1987 to minimize coal dust generation. Conrail has also installed a filtered heating, ventilation, and air conditioning (HVAC) system in the station house to prevent the entry of coal dust. Air monitoring surveys conducted in May 1993 indicate no short-exposure health hazards to station personnel, but the exteriors of station buildings must be routinely cleaned of coal dust as part of their maintenance.

Environmental Consequences of the Proposed Action: A beneficial impact of the Proposed Action is that station personnel would be located a greater distance from the coal conveyor and thus would no longer be exposed to airborne coal dust. Air in the new facilities would no longer require special filtration, and routine cleaning of coal dust from the exterior surfaces of the new facilities would not be necessary.

Brief, minor, and localized emissions from construction equipment would take place during site preparation and construction activities. These emissions would be unavoidable and would represent an almost negligible contribution to the regional concentrations of ozone and other air pollutants. Localized releases of fugitive dust could take place until exposed soils at the construction site are stabilized. Temporary actions to control fugitive dust, such as periodic wetting of loose surface soil, would be taken as necessary to prevent visible releases of fugitive dust to areas surrounding the station.

Environmental Consequences of the Renovation Alternative: The station would still be located immediately adjacent to the overhead Conrail coal conveyor and the renovated facilities would thus still require filtered interior air and routine cleaning of coal dust from exterior surfaces.

Environmental Consequences of the No-Action Alternative: The existing station facilities would continue to require filtered interior air and routine cleaning of exterior surfaces to remove coal dust.

4.3.3   Water Resources

Affected Environment: The Ashtabula River drains an area of 137 square miles in northeastern Ohio and northwestern Pennsylvania. The drainage basin is predominantly rural and agricultural, with the City of Ashtabula as the only significant urban center (EPA, 1997).

Water quality in the lower reaches of the Ashtabula River has been degraded by extensive industrial development along the river and certain of its tributaries. The lower two miles of the Ashtabula River, as well as Ashtabula Harbor and associated nearshore waters of Lake Erie have been designated as a Great Lakes Area of Concern (AOC) by the U.S. Environmental Protection Agency (EPA). The United States and Canadian governments have identified 43 Great Lakes AOCs that fail to meet water quality objectives established jointly by the two governments. The Great Lakes Water Quality Agreement, as amended in 1987, directs both federal governments to cooperate with state and provincial governments to develop and implement remedial action plans for each AOC (GLIN, 1999).

That part of the Ashtabula River that passes the existing Station Ashtabula and the City dock is part of the AOC. It is downstream from a variety of riverfront industrial areas, including the Fields Brook Superfund Site, and the sediments have become contaminated by metals, pesticides, nutrients, polychlorinated biphenyls (PCBs), polyaromatic hydrocarbons, and several chlorinated organic compounds (EPA, 1992).

Environmental Consequences of the Proposed Action: Construction of the new modular building and other land-based facilities would not involve disturbance of sediments in the Ashtabula River and thus should not suspend contaminated sediment. Water quality in the Ashtabula River should thus not be affected. Expansion of the City-owned dock by the City of Ashtabula could suspend very small quantities of sediment along the western shore of the river. But because the proposed expansion site is already bulkheaded, the quantity and duration of sediment suspension should be small and brief, and thus should not substantially affect water quality. The potable water demand associated with the new facilities would be comparable to that of the existing facilities.

Environmental Consequences of the Renovation Alternative: Renovation of the existing station would not require disturbance of sediment or banks along the Ashtabula River. It would thus have no potential to affect water quality. The demand by the station for potable water would not substantially change.

Environmental Consequences of the No-Action Alternative: The No-Action Alternative would not involve disturbance of sediment or banks along the Ashtabula River. It would have no potential to affect water quality. The demand by the station for potable water would not substantially change.

4.3.4 Noise

Affected Environment: Noises in the Ashtabula Harbor area are dominated by the sounds of heavy equipment; ship, truck, and rail traffic; and other industrial sounds (COE, 1983). Considerable noise is also generated by operation of the coal conveyor and the horns of craft approaching the drawbridge just south of the city dock.

Environmental Consequences of the Proposed Action: Temporary noise generated by new construction activities would blend almost indistinguishably into the overall industrial noise of the surrounding area. No permanent new noise sources would be generated.

Environmental Consequences of the Renovation Alternative: Temporary noise generated by renovation of existing facilities would blend almost indistinguishably into the overall industrial noise of the surrounding area. No permanent new noise sources would be generated.

Environmental Consequences of the No-Action Alternative: No additional noise would be attributed to the No-Action Alternative.

4.3.5 Hazardous Materials and Hazardous Waste

Affected Environment: Hazardous materials and petroleum products currently stored or handled on Station Ashtabula are limited to small quantities of oils, antifreezes, and paints necessary for routine maintenance of the station's structures and boats. Drums of lube oil used on the boats and waste oil collected from the boats are stored in the boathouse. Although the floor of the boathouse is not bermed, it is concrete in apparently good condition. Several small cans of oil, antifreeze, deicer, and other liquids used on the boats are also kept in the boathouse. Several house-hold sized containers of spray paint, mineral spirits, polyurethane, caulking, WD-40 oil, and other paints and cleaners are stored in a labeled cabinet for flammable materials in the garage (USCG, 1998).

Boats are presently fueled at commercial marinas rather than at the station. On-station fueling facilities which were decommissioned in 1986 (USCG,

1986) included a 1,000-gallon UST that stored gasoline and a 1,000-gallon UST that stored diesel fuel. Both tanks were of steel construction and lacked containment, lining, or leak detection features (USCG, 1985). Both tanks have been removed but have been confirmed to have leaked to underlying soil. Both are formally listed as leaking UST (LUST) incidents, and the U.S. Coast Guard is proceeding with corrective action (EDR, 1999a).

Federal and state environmental databases were searched, using a computerized record search service, for Station Ashtabula and surrounding areas within one mile. The specific databases searched are listed and described in Exhibit 4.3-1. The database search revealed the two LUST incidents for the former fueling USTs on the station. It did not reveal any records of environmental contamination for the proposed new station house site or for the City-owned dock property. The database search did identify the Fields Brook Project, a National Priorities List (NPL) site focussed on an upstream tributary of the Ashtabula River (EDR, 1999a). As noted in the discussion of water quality, metals, PCBs, and organic contaminants released into the tributary by bordering industrial sites have accumulated in the sediment in that part of the river that passes the existing station and the City-owned dock.

Several other LUST incidents have been recorded in the databases for sites close to the existing station and the proposed relocation area (EDR, 1999a). The frequency of these incidents suggest a potential for groundwater contamination in the area but do not suggest potential surface-level contamination that could affect relocation and future use of the subject facilities.

Environmental Consequences of the Proposed Action: Relocation of the station facilities would not delay or interfere with ongoing investigation of, and implementation of any necessary corrective action for, the LUST sites on the existing station. The U.S. Coast Guard would prepare an environmental baseline survey and finding of suitability to transfer (FOST), lease (FOSL), or assign (FOSA) for the existing station property before allocating the property for reuse. No fuel storage or fueling facilities would be constructed at the City dock. The U.S. Coast Guard would continue to fuel its boats at commercial marinas. The U.S. Coast Guard would construct a new storage shed adjacent to the dock to store tools and small containers of lube oil, waste oil, and other materials associated with boat maintenance. The shed would include a bermed concrete floor that would not allow exterior leakage of oils or other liquids used in boat maintenance.

Environmental Consequences of the Renovation Alternative: Renovation of the station facilities would not delay or interfere with ongoing investigation of, and implementation of any necessary corrective action for, the LUST sites on the station. The U.S. Coast Guard would continue to fuel its boats at commercial marinas.

EXHIBIT 4.3-1

Environmental Databases Searched Station (Small) Ashtabula
and One-Mile Radius May 1999

Database	Administering Agency	Issues Addressed
Federal Records		
Comprehensive Environmental Response, Compensation, and Liability Information System (CERCLIS)	EPA	Sites on, proposed for, or undergoing screening for inclusion on NPL
Emergency Response Notification System	EPA/NTIS	Reported releases of oil and hazardous substances
National Priority List sites (NPL)	EPA	Sites identified by EPA for priority cleanup under Superfund program
Treatment, Storage, and Disposal Facilities (RCRIS-TSD)	EPA/NTIS	Facilities permitted under RCRA for treatment, storage, and disposal of hazardous waste
Small Quantity Generators (RCRIS-SQG)	EPA/NTIS	Small quantity generators of hazardous waste permitted under RCRA
Large Quantity Generators (RCRIS-LQG)	EPA/NTIS	Large quantity generators of hazardous waste permitted under RCRA
Corrective Action Report	EPA	RCRA-permitted hazardous waste handlers with corrective action activity
Biennial Reporting System (BRS)	EPA/NTIS	Data on the generation and management of hazardous waste
Superfund Consent Decrees	EPA	Legal settlements that establish responsibility and standards for cleanup of NPL sites
Facility Index System	EPA/NTIS	Pointers to other possible sources of environmental information
Hazardous Materials Information Reporting System (HMIRS)	DOT	Hazardous material spill incidents reported to U.S. Department of Transportation
Material Licensing Tracking System	NRC	Sites which possess or use radioactive materials and are licensed by NRC
NPL Liens	EPA	Federal Superfund Liens
PCB Activity Database System	EPA	EPA lists of generators, transporters, commercial storers, brokers, and disposers of PCBs

EXHIBIT 4.3-1

Environmental Databases Searched Station (Small) Ashtabula
and One-Mile Radius May 1999 (continued)

Database	Administering Agency	Issues Addressed
Federal Records (continued)		
RCRA Administrative Action Tracking System	EPA	Enforcement actions issued under RCRA for major violators
Records of Decision (ROD)	NTIS	RODs issued for NPL sites
Toxic Chemical Release Inventory System (TRIS)	EPA/NTIS	Facilities which release toxic chemicals to air, water, or land in reportable quantities under SARA Title III, Section 313
Toxic Substances Control Act (TSCA)	EPA	Manufacturers and importers of chemicals on TSCA inventory
State Records		
Leaking Underground Storage Tanks (LUST)	Ohio Department of Commerce	Reported leaking underground storage tank incidents
Master Sites List (SHWS)	Ohio Environmental Protection Agency	Ohio's equivalent to CERCLIS
Licensed Solid Waste Facilities (LF)	Ohio Environmental Protection Agency	Solid waste disposal facilities and landfills
Underground Storage Tanks (UST)	Ohio Department of Commerce	Underground storage tanks registered with Michigan DEQ
Emergency Response Database (SPILLS)	Ohio Environmental Protection Agency	Spills and other releases to the environment

Source: EDR, 1999a
EPA: U.S. Environmental Protection Agency
NTIS: National Technical Information Service
DOT: U.S. Department of Transportation
NRC: U.S. Nuclear Regulatory Commission
RCRA: Resource Conservation and Recovery Act

Environmental Consequences of the No-Action Alternative: Renovation of the station facilities would not delay or interfere with ongoing investigation of, and implementation of any necessary corrective action for, the LUST sites on the station. The U.S. Coast Guard would continue to fuel its boats at commercial marinas.

4.4   NATURAL ENVIRONMENT

Elements of the natural environment considered separately below include the terrestrial environment; floodplains and the coastal zone; wetlands; the aquatic environment; and threatened and endangered species.

4.4.1 Terrestrial Environment

Affected Environment: Vegetation on the existing station is limited to small areas of mowed lawn and a bed of small shrubs facing Front Street. Vegetation at the city dock comprises mowed grass. Vegetation at the proposed new station house site is limited to a sparse cover of ruderal (weedy) herbs such as tall fescue, dandelion, stinging nettle, dames rocket, black medick, dock, blackberry, and goldenrod. The steep slope north and west of the site supports similar vegetation but also includes multiflora rose and scattered seedlings and saplings of successional hardwood trees such as ailanthus, black locust, and boxelder.

Environmental Consequences of the Proposed Action: Construction of the new modular building would result in the conversion of up to 10,000 square feet of ruderal (weedy) vegetation to impervious surface or mowed lawn.

Environmental Consequences of the Renovation Alternative: Renovation of the existing station would not affect any vegetation.

Environmental Consequences of the No-Action Alternative: The No-Action Alternative would not affect any vegetation.

4.4.2 Floodplains and Coastal Zone

Affected Environment: The Flood Insurance Rate Map for the City of Ashtabula (HUD, 1980) shows that the 100-year floodplain associated with the lower reach of the Ashtabula River is limited to the open waters of the river. The existing station, city dock, and proposed new modular building site are shown as located in Zone C, defined as areas of minimal flooding (i.e. outside of the 100-year and 500-year floodplains). Even though the existing station house is located in Zone C, it has experienced several incidents of basement flooding during periods of high lake levels. A copy of the Flood Insurance Rate Map is provided in Appendix A.

Environmental Consequences of the Proposed Action: Construction of the new modular building and other land-based facilities would not involve work within either the 100-year or 500-year floodplain. If the City of Ashtabula extends the city dock, work performed at the edge of the river would be in the 100-year floodplain. Because the riverbank at that point already comprises a concrete retaining wall, installation of a dock would not affect flood patterns. The Proposed Action would thus comply with Executive Order 11988 (Floodplain Management).

Environmental Consequences of the Renovation Alternative: Renovation of the existing station would not involve work within either the 100-year or 500-year floodplain. The Renovation Alternative would comply with Executive Order 11988 (Floodplain Management). Engineering measures would be necessary as part of the renovation to prevent future incidents of basement flooding in the renovated station house.

Environmental Consequences of the No-Action Alternative: Renovation of the existing station would not involve work within either the 100-year or 500-year floodplain. The Renovation Alternative would comply with Executive Order 11988 (Floodplain Management). The station house would likely continue to experience frequent incidents of basement flooding.

4.4.3   Wetlands

Affected Environment: National Wetland Inventory maps prepared by the Fish & Wildlife Service (FWS, 1977) and wetland inventory maps produced as part of the Ohio Wetland Inventory (Ohio DNR, 1999) show no wetlands on the existing station, city dock property, or proposed new station house site. The Ashtabula River is mapped as a Lacustrine limnetic open water downstream from the existing station and as a Riverine lower perennial open water upstream from the station, including where it passes the city dock. Both sides of the river, extending upstream from its mouth past the city dock, have been reinforced with concrete retaining walls and lack fringing wetlands.

Environmental Consequences of the Proposed Action: No wetlands would be affected by the Proposed Action. The Proposed Action would be in compliance with Executive Order 11990 (Protection of Wetlands).

Environmental Consequences of the Renovation Alternative: No wetlands would be affected by the Renovation Alternative. The Renovation Alternative would be in compliance with Executive Order 11990 (Protection of Wetlands).

Environmental Consequences of the No-Action Alternative: No wetlands would be affected by the Renovation Alternative.

4.4.4   Aquatic Environment

Affected Environment: Species of fish in Ashtabula Harbor have been described as typical of warmwater fish communities found in river mouths at Lake Erie. Protected areas of water, such as the lower river reach at Station Ashtabula and the city dock, are described as containing relatively large numbers of yellow perch (*Perca flavescens*), white bass (*Morone chrysops*), pumpkinseed (*Lepomis gibbosus*), white crappie (*Pomoxis annularis*), goldfish (various species), and emerald shiner (*Notropis atherinoides*). In the spring months, walleye (*Stizostedion vitreum*) and smallmouth bass (*Micropterus dolomieu*) are reported to migrate into the Ashtabula River from Lake Erie to spawn (EPA, 1992).

Fish species reported from the lower Ashtabula River in a 1983 EIS for dredging include carp (*Cyprinus carpio*), black bullhead (*Ameiurus melas*), largemouth bass (*Micropterus salmoides*), gizzard shad (*Dorosoma cepedianum*), and goldfish (COE, 1983). Fish species collected in large numbers (at least 10 or more individuals from Ashtabula Harbor in a 1984 survey include brown bullhead (*Ameiurus nebulosus*), white crappie, black crappie (*Pomoxis nigromaculatus*), pumpkinseed, bluegill (*Lepomis macrochirus*), rock bass (*Ambloplites rupestris*), and yellow perch (COE, 1990). The 1984 survey included collection rounds in May, July, and August of that year.

Benthic macroinvertebrates in the Ashtabula Harbor area are reported in the EIS as dominated by Oligochaetes (Family Tubificidae) (COE, 1983). Dominance by this taxon is commonly associated with warm, stagnant water that is polluted. The Ohio Department of Health and Ohio EPA issued an advisory against the consumption of fish caught from the Ashtabula River downstream from the 24th Street Bridge, including the reach at the city dock and existing station (EPA, 1992).

Environmental Consequences of the Proposed Action: Waterfront construction work resulting from the Proposed Action would be limited to the modest extension of the city dock. The dock would be expanded in an area that is already bulkheaded by a concrete retaining wall. No wetlands or other shallow-water aquatic habitat would be lost or disturbed. No dredging would taken place.

Environmental Consequences of the Renovation Alternative: Aquatic biota and its habitat would not be affected by renovation of the existing station. No alterations to the waterfront would take place.

Environmental Consequences of the No-Action Alternative: No impacts to aquatic biota would take place under the No-Action Alternative.

4.4.5   Threatened or Endangered Species

Affected Environment: Letters requesting information on the possible presence of federal and state threatened, endangered, or other special status species and critical habitats were submitted to the Ohio Department of Natural Resources (DNR), Division of Natural Areas and Preserves and to the U.S. Fish and Wildlife Service (FWS). The Ohio DNR responded that their Natural Heritage maps include several records of state listed species within a one-mile radius of Station Ashtabula (Woischke, 1999). But the maps provided with the letter, included in Appendix B, do not show any sites of record on or immediately adjacent to the existing station, the City-owned dock, or the proposed new station house site. A copy of the letter is provided in Appendix B.

The Ohio DNR letter indicates that the bigeye chub (*Notropis amblops*), a state-listed fish species, has been sighted in the Ashtabula River approximately 0.9 mile upriver from the City-owned dock. If the bigeye chub occurs in that part of the river, it also likely occurs in the lower part of the river where it passes the existing station and proposed relocation area. Another state-listed fish species, the burbot (*Lota lota*), has been recorded in the open waters of Ashtabula

Harbor, just inside the breakwaters to Lake Erie. A breeding population of the ring-necked gull (*Larus delawarensis*) has been recorded at the Pinney Dock, approximately 0.5-mile northeast of the existing station. Most of the other sightings are plants recorded along the Lake Erie waterfront, especially in the sand dunes and wetlands in Walnut Beach Park, located over 1,000 feet from the existing station and proposed relocation area (Woischke, 1999).

A written response has not yet been received from the FWS but will be included in Appendix B following receipt. Four federally-listed threatened and endangered species are thought to occur in Ashtabula County. They include the Indiana bat (*Myotis sodalis*), the bald eagle (*Haliaeetus leucocephalus*), the piping plover (*Charadrius melodus*), and the clubshell (*Pleurobema clava*) (FWS, 1997). The Indiana bat and bald eagle favor forested habitats and thus would not likely visit or nest close to the existing station or proposed relocation area. The piping plover favors sandy beaches as habitat and thus could occur along the beaches of Lake Erie. But it would not likely visit the bulkheaded riverfront at the existing station or City-owned dock.

The clubshell is a mussel that inhabits soft sediment in rivers and is thought to potentially inhabit the Ashtabula River. It could perhaps inhabit sediment in the river where it passes the existing station and proposed relocation site. But any populations there have likely been reduced or eliminated by the history of industrial pollution of the sediments. Sensitivity to industrial pollution is recognized as a key reason as to why the species is endangered (FWS, 1997).

Environmental Consequences of the Proposed Action: Establishment of the proposed new modular building would not affect any habitat potentially inhabited by state or federally listed threatened, endangered, or special status species. Expansion of the city dock would involve a small amount of disturbance of sediment in the Ashtabula River at a point on the western shore that is already bulkheaded. This disturbance would be too localized and temporary to adversely affect either the bigeye chub or the clubshell, if they inhabit that part of the river.

Environmental Consequences of the Renovation Alternative: Renovation of the existing station would not affect any habitat potentially inhabited by state or federally listed threatened, endangered, or special status species.

Environmental Consequences of the No-Action Alternative: The No-Action Alternative would not affect any habitat potentially inhabited by state or federally listed threatened, endangered, or special status species.

4.5 CULTURAL RESOURCES

Cultural resources considered separately below include archaeological and historic resources.

4.5.1 Archaeological Resources

Affected Environment: The Ashtabula River is thought to have formed a boundary between lands controlled by the Iroquois and Algonquin Indian nations. The name "Ashtabula" is thought to have been derived from a now obscure Indian term for "river of many fish", and early settlers reported an abundance of game and fish in the area (Ashtabula Chamber of Commerce, 1990). Although people from one or both of the Iroquois and Algonquin tribes, or from other tribes, likely inhabited the lands bordering the lower Ashtabula River, the potential for significant artifacts is very low for either the existing station or proposed relocation area. That is because soils in both areas have been highly disturbed by urban development in the nineteenth and twentieth centuries.

Environmental Consequences of the Proposed Action: The Proposed Action would involve disturbance only of surface soils that have a history of recent heavy physical disturbance. It thus would have little potential for disturbance of significant archaeological resources. A letter has been submitted to the Ohio Historic Preservation Office, which houses the State Historic Preservation Officer (SHPO) for Ohio, requesting a review of the existing station property and proposed relocation area. Because of the disturbed soils, the U.S. Coast Guard expects that the SHPO will not require additional investigation to comply with Section 106 of the National Historic Preservation Act. It is possible, however, that the SHPO will request a Phase I archaeological survey of the affected areas before approving the Proposed Action.

Environmental Consequences of the Renovation Alternative: Renovation of the existing station facilities would not involve disturbance of surface soils and thus has no potential for disturbance of significant archaeological resources.

Environmental Consequences of the No-Action Alternative: The No-Action Alternative would not involve disturbance of surface soils and thus has no potential for disturbance of significant archaeological resources.

4.5.2 Historical Resources

Affected Environment: European settlement of Ashtabula County is thought to have originated with the construction of a log cabin in 1801 by Thomas Hamilton at the mouth of the Ashtabula River. Steamboats were making commercial use of Ashtabula Harbor by 1837. The Lake Shore and Michigan Southern Railroad arrived at Ashtabula in 1852; and the Pittsburgh, Youngstown, and Ashtabula Railroad arrived in 1873. By the end of the 19th century, Ashtabula Harbor was becoming one of the largest ore and coal ports in the world. The City of Ashtabula was incorporated in 1892, at which time it included a hospital, electric rapid transit system, and a telephone company (Ashtabula Chamber of Commerce, 1990).

The Ashtabula Harbor Commercial District, which includes both sides of West 5th Street on the west side of the Ashtabula River, is listed on the National Register of Historic Places (NRHP). The city dock and proposed site for the new station house are located in the District, and the existing station is visible from the District. The District was listed because of its architectural significance and history. The West 5th Street Drawbridge, which crosses the Ashtabula River just north of the city dock, is also listed on the NRHP. The drawbridge is also visible from the existing station and the proposed new station house site (EDR, 1999b).

Environmental Consequences of the Proposed Action: The new station house would be constructed within the Ashtabula Harbor Commercial District, which is listed on the NRHP. The proposed site is presently vacant, as the City of Ashtabula recently demolished a garage-type building. The exterior of the new station house would be designed to be architecturally compatible with the surrounding 19th Century buildings in the District.

Environmental Consequences of the Renovation Alternative: The existing station buildings, which were constructed in the late 1930s, are not listed on the NRHP. The overall exterior appearance of the buildings would not be changed.

Environmental Consequences of the No-Action Alternative: No impacts to features listed on the NRHP or areas visible from listed features would occur under the No-Action Alternative.

SECTION FIVE   Regulatory Requirements

The following regulatory requirements are anticipated to apply to the Proposed Action. Some but not all of these requirements could apply to the Renovation Alternative. No regulatory requirements would apply to the No-Action Alternative.

5.1   LOCAL

Even though no building permits would be necessary for the Proposed Action or any of the other alternatives, the U.S. Coast Guard would furnish the City of Ashtabula with copies of plans and specifications before initiating any construction activities.

5.2   STATE

No permits from the State of Ohio would be necessary for the Proposed Action, Renovation Alternative, or No-Action Alternative. Because the City of Ashtabula would be the party expanding the city dock, any state permits necessary for waterfront construction activities would have to be acquired by the city.

5.3   FEDERAL

No federal permits would be necessary for the Proposed Action, Renovation Alternative, or No-Action Alternative. Because the City of Ashtabula would be the party expanding the city dock, any permits necessary for waterfront construction activities would have to be acquired by the city.

SECTION SIX Agencies and Persons Contacted

The following agencies were contacted by letter, telephone, or in-person during the preparation of this document.

Federal	Nathan Paskey Ashtabula Soil and Water Conservation District 39 Wall Street Jefferson, Ohio 44047
	U.S. Department of the Interior Fish and Wildlife Service Region 3 Ecological Services Office BHW Federal Building 1 Federal Drive Fort Snelling, Minnesota 55111
	Nickolas Suvak U.S. Coast Guard Station Ashtabula 1 Front Street Ashtabula, Ohio 44004
	Frank Blaha U.S. Coast Guard Civil Engineering Unit Cleveland 1240 East Ninth Street Cleveland, Ohio 44199
State	Deborah Woischke Ohio Department of Natural Resources Division of Natural Areas and Preserves 1889 Fountain Square Columbus, Ohio 43224
	Ohio Historic Preservation Office (State Historic Preservation Officer) 567 East Hudson Street Columbus, Ohio 43211
Local	Joseph Perry Building Commissioner City of Ashtabula Municipal Building 4400 Main Avenue Ashtabula, Ohio 44004
	Ashtabula County District Library 335 West 44th Street Ashtabula, Ohio 44004

SECTION SEVEN   Glossary of Terms

ACM	Asbestos-containing Material
AOC	Area of Concern
AOR	Area of Responsibility
DNR	Department of Natural Resources
CERCLIS	Comprehensive Environmental Response, Compensation, and Liability Information System
EA	Environmental Assessment
EIS	Environmental Impact Statement
EPA	U.S. Environmental Protection Agency
FONSI	Finding of No Significant Impact
FOSA	Finding of Suitability to Assign
FOSL	Finding of Suitability to Lease
FOST	Finding of Suitability to Transfer
FWS	Fish and Wildlife Service
GLIN	Great Lakes Information Network
HUD	Department of Housing and Urban Development
IMARV	Independent Maritime Response Vehicle
LQG	Large Quantity Generator
LUST	Leaking Underground Storage Tank
NPL	National Priorities List
NRHP	National Register of Historic Places
PCB	Polychlorinated Biphenyl
RCRA	Resource Conservation and Recovery Act
RCRIS	Resource Conservation and Recovery Information System
ROD	Record of Decision
SAR	Search and Rescue
SHPO	State Historic Preservation Officer
SQG	Small Quantity Generator
TSCA	Toxic Substances Control Act
TSD	Treatment, Storage, and Disposal
UST	Underground Storage Tank
UTB	Utility Boat

SECTION EIGHT   References

Ashtabula Chamber of Commerce. 1990. History of Ashtabula County, by the Ashtabula Chamber of Commerce in 1990. http://www.infinet.com/~dzimmerm/achs1.html

ACMC (Ashtabula County Medical Center). 1999. ACMC Homepage. http://www.acmchealth.org/ACMC/acmc.html

Ashtabula County Planning Commission. 1988. Ashtabula County Tourism Plan.

Ashtabula County Planning Commission. 1991. Ashtabula County Major Thoroughfare Plan.

Ashtabula County Planning Commission. 1992. Ashtabula County Overall Economic Development Program.

Ashtabula Fire Department. 1999. Ashtabula Fire Department Homepage. http://www.aimanagers.com/AFD

Bureau of the Census. 1990. 1990 U.S. Census Data, Database C90STF1A, Summary Level: State-County, for Ashtabula County, Ohio. http://venus.census.gov/cdrom/lookup/927037211

Bureau of the Census. 1992. 1992 Economic Census – Area Profile for Ashtabula County, Ohio. http://www.census.gov/epcd/www/92profiles/county/39007.TXT

Bureau of the Census. 1998. County Population Estimates for July 1, 1998 and Population Change for July 1, 1997 to July 1, 1998. http://www.census.gov/population/estimates/county/co-98-1/98C1_39.txt

City of Ashtabula. 1999. Zoning Index Map. Up to date as of May 1999.

COE (U.S. Army Corps of Engineers). 1983. Ashtabula Harbor Draft Environmental Impact Statement.

COE (U.S. Army Corps of Engineers). 1990. Information Summary, Area of Concern: Ashtabula River, Ohio. Miscellaneous Paper EL-90-22. Environmental Laboratory, Waterways Experiment Station, Vicksburg, Mississippi.

EDR (Environmental Data Resources). 1999a. The EDR Radius Map for U.S. Coast Guard Station Ashtabula, 1 Front Street, Ashtabula, Ohio 44004. Inquiry Number 374692.1s. May 28, 1999.

EDR (Environmental Data Resources). 1999b. The EDR-NEPA Report with Detail Map for U.S. Coast Guard Station Ashtabula, 1 Front Street, Ashtabula, Ohio 44004. Inquiry Number: 374692.3s. May 27, 1999.

EPA (US Environmental Protection Agency). 1992. Assessment and Remediation of Contaminated Sediments (ARCS) Program, Baseline Human Health Risk Assessment: Ashtabula River, Ohio Area of Concern. Great Lakes National Program Office. EPA 905-R92-007. December 1992.

EPA (U.S. Environmental Protection Agency). 1997. Ashtabula River Area of Concern. Compiled October 27, 1997. http://www.epa.gov/glnpo/aoc/ashtabula.html

FirstEnergy. 1999. Corporate Profile. http://www.firstenergycorp.com/profile/

FWS (US Fish and Wildlife Service). 1977. National Wetlands Inventory Map, Ashtabula North Quadrangle.

FWS (U.S. Fish and Wildlife Service). 1997. Ohio – County List of Federal Threatened (T), Endangered (E) and Proposed (P) Species, December 1997. http://www.fws.gov/r3pao/eco serv/endangrd/lists/ohio.html

GLIN (Great Lakes Information Network). 1999. Great Lakes Areas of Concerns. Revised March 24, 1999. http://www.great-lakes.net/places/aoc/aoc.html

HUD (Department of Housing & Urban Development). 1980. Flood Insurance Rate Map for City of Ashtabula, Ohio. Community-Panel Number 390011 0005 B.

NOLA Regional Library System. 1999. Ashtabula County District Library. http://www.nolanet.org/spotlite/spot01-99.htm

Ohio DNR (Department of Natural Resources). 1999. Ohio Wetland Inventory Map, Ashtabula North Quadrangle.

Perry, J. 1999. Personal communication dated May 19, 1999 between Joseph Perry, Building Commissioner for City of Ashtabula, Ohio and Peyton Doub of Tetra Tech NUS.

SCS (Soil Conservation Service). 1973. Soil Survey of Ashtabula County, Ohio.

SDDB (School District Data Book). 1999. School District Data Book Profile for Ashtabula City Area School District, Ohio. http://govinfo.kerr.orst.edu/cgi-bin/sddb-state?stateis=Ohio

Suvak, N. 1999. Personal communication dated May 19, 1999 between Nickolas Suvak of the U.S. Coast Guard, Station Ashtabula, and Peyton Doub of Tetra Tech NUS.

USCG (U.S. Coast Guard). 1985. Memorandum from Officer in Charge, USCG Station Ashtabula, Ohio to Commander, Ninth Coast Guard District, Cleveland, Ohio. Subject: Underground Tank Inventory. December 18, 1995.

USCG (U.S. Coast Guard). 1986. Memorandum from Commander, US Coast Guard Group Buffalo to Commander, Ninth Coast Guard District. Subject: Assessment of Operational Need for Fueling Facilities at Group Buffalo Units. June 23, 1986.

USCG (U.S. Coast Guard). 1998. Hazardous Material Inventory – Flammable Locker, CG Station Ashtabula.

USGS (US Geological Survey). 1960. Ashtabula North, Ohio 7.5-Minute Topographic Quadrangle. Photorevised 1970, Photoinspected 1988.

Appendix E

Appendix E

Appendix E.1

Example 2:   Critique of the Sitewide Environmental Assessment for Continued Development of Naval Petroleum Reserve No. 3

Agency: U.S. Department of Energy
Preparer: Tetra Tech NUS, Gaithersburg, Maryland
Date: June 1995
Action: Continued oil extraction activities at active production facility for 5-year period

Overview

This EA addresses ongoing petroleum extraction actions and support activities at an active oilfield of 9,481 acres. At the time, the oilfield was owned by the US Navy but managed by a contractor to the Department of Energy (DOE). Alternatives considered in detail include

1. Continued oil production at NPR-3 using conventional and newer oil recovery technologies (the Proposed Action),

2. Continued oil production at NPR-3 but emphasizing newer oil recovery technologies,

3. Divestiture (sale) of NPR-3 by the Navy,

4. Decommissioning of NPR-3 (cessation of oil extraction activities followed by any necessary environmental restoration of the site), and

5. A No-Action Alternative, under which oil production would continue using only conventional technologies.

The primary difference between most of the alternatives is the degree of emphasis upon conventional oil production technology versus newer methods designed to extract additional oil from formations incapable of yielding further oil using conventional technology. The No-Action Alternative is to continue production using only conventional technology until that technology is no longer productive, and then to cease operations and decommission. The Proposed Action is to use conventional technology where still productive and then use newer technology to extract additional oil. The second alternative is to phase out conventional technology, even where still potentially production, and focus mainly on the newer technologies.

Readers should note that the No-Action Alternative is not defined as cessation of all production activity on the installation. Rather, such a scenario is considered as the Decommissioning Alternative. No-Action in the context of National Environmental Policy Act (NEPA) is to maintain the current baseline without substantial changes. Conventional oil production activities were already in place for nearly 20 years prior to initiation of the EA. The affected environment section describes environmental conditions as already affected by those conventional oil production activities.

The DOE prepares "sitewide" NEPA documents to broadly consider the potential individual and cumulative impacts of ongoing and reasonable foreseeable future actions for an installation. Sitewide NEPA documents are essentially programmatic documents, prepared to address alternative approaches to site management, from which activity-specific documentation can be tiered in accordance with 40 CFR 1508.28. The DOE typically prepares Sitewide Environmental Impact Statements (EIS) (essentially Programmatic EISs targeted to a specific installation) for controversial and/or complex installations such as National Laboratories or sites within the nuclear weapons complex. Because NPR-3 was a small, relatively non-controversial site with a relatively simple mission focused almost exclusively on oil production, DOE decided to prepare an EA to determine whether a FONSI, rather than an EIS, could support sitewide management decisions.

The subject document is thus an example of what is essentially a Programmatic EA. Historically, most programmatic documents prepared under NEPA have been Programmatic EISs, as most agencße decided that programmatic decisions are typically too complex to be adequately addressed through a FONSI. As agencies continue to increase their reliance on EAs and FONSIs to document increasingly complex and controversial decisions, a greater number of Programmatic EAs can be expected in the future.

As with Programmatic EISs, the heart of a Programmatic EA is the ability to consider broadly the potential environmental consequences of overall management alternatives and save the consideration of individual actions until a time closer to implementation. The individual actions will then be better designed and thus more "ripe" for a detailed analysis through a tiered document. Most NEPA documentation tiered off of a programmatic EA will, of course, comprise EAs or categorical exclusions (CATXs). If a component activity under a management alternative broadly addressed in a programmatic document were suspected of potentially resulting in significant environmental impacts, the proper course of action would be to prepare a Programmatic EIS rather than EA. However, it is conceptually possible that changes in circumstances or in the availability of supporting information could result in a situation where the potential for impacts from a component activity is greater than originally anticipated. An agency might therefore decide to prepare an activity-specific EIS tiered off of a Programmatic EA.

One particularly good feature of the NPR-3 Sitewide EA is that its Chapter 2 (Description of the Proposed Action and Alternatives) provides a detailed tabular listing of each action anticipated under each alternative. Such a list clearly demonstrates that the preparers carefully considered the types of specific actions that would be taken under each alternative. The list also provides the detailed information necessary for a reader or reviewer of the document to fully understand each alternative. Furthermore, reviewers can compare the listed actions against the DOE's lists of activities typically covered by CATXs, EAs, and EISs (Appendices A through D of 10 CFR 1021). Many programmatic NEPA documents address their alternatives too vaguely to be of real value to decisionmakers. Such programmatic documents often degrade into almost meaningless writing exercises that do little more than demonstrate compliance with the mechanics of NEPA.

SPECIFIC COMMENTS

The following specific comments are targeted to specific sections and figures in the EA.

1. Executive Summary: The Executive Summary would be enhanced by a bulleted or numbered list of each alternative considered in detail.

2. Executive Summary: The "Issues Tracking Matrix" is an excellent pointer to parts of the EA addressing specific issues. This is particularly useful to reviewers with specialized agendas (e.g. agencies charged with protection of threatened or endangered species).

3. Section 2: The detailed discussion of each alternative, and especially the tabularized listing of component activities, is one of the strengths of this document. Although simplification should always be aggressively pursued when writing descriptions of alternatives in NEPA documents, vague descriptions will defeat the entire analysis.

4. Section 2: Another strength of this document is the fact that it considers several alternatives in addition to a Proposed Action and a No-Action Alternative. Furthermore, the alternatives differ with respect to more than one "thread". That is, some alternatives differ with respect to technology and some alternatives differ with respect to overall approach (e.g. the Decommissioning Alternative). NEPA documents whose alternatives differ only with respect to a single "thread", e.g. differ only with respect to construction footprint

location, sometimes overlook reasonable alternatives that should be considered.

5. Section 4, Mitigation Measures subsections: When originally drafted, this EA discussed mitigation measures in a separate section following Section 4. The DOE editors combined the mitigation measures subsections into the impact analysis text. Such an approach is sometimes superior, as many FONSIs must rely on the implementation of mitigation measures to avoid potentially significant impacts (i.e., triggering preparation of an EIS).

6. Section 5, List of Preparers: There is some debate as to whether Lists of Preparers should be included in an EA. The Council on Environmental Quality (CEQ) Regulations specifically mention that the preparers should be listed for an EIS. The opinion of this book is that the list should be included in EAs as well. Reviewers and decisionmakers need to know the sources of expertise used in developing technical supporting data. The need for the list becomes even greater as larger and more complex projects are being increasingly addressed by EAs rather than EISs.

7. Section 7, Bibliography: This lengthy bibliography is suggestive of the very thorough research underlying this document. Readers encountering poorly or skimpily referenced EAs, particularly when the proposal involves complex issues, should be suspicious of the conclusions presented in those EAs. An EA that fails to demonstrate that a scientifically thorough approach was pursued is sometimes evidence of a weakly investigated analysis and may indicate that a FONSI is not appropriate.

Appendix E.2

Environmental Assessment for Continued Development of Naval Petroleum Reserve No. 3

EA-1008; Final Sitewide Environmental Assessment and FONSI EA-1008 for Continued Development of Naval Petroleum Reserve No. 3 (NPR-3)

TABLE OF CONTENTS

LIST OF TABLES

LIST OF FIGURES

ABBREVIATIONS AND ACRONYMS

AQCR	Air Quality Control Region
ASP	Alkaline-Surfactant-Polymer (flood)
AUM	Animal Unit-Month
CAEDA	Casper Area Economic Development Alliance, Inc.
CERCLA	Comprehensive Environmental Response, Compensation & Liability Act
COE	U.S. Army Corp of Engineers
CO_2	Carbon Dioxide Gas
CX	Categorical Exclusion
DOE	U.S. Department of Energy
EA	Environmental Assessment
EOR	Enhanced Oil Recovery
FD	Fluor Daniel (NPOSR), Inc.
FIRM	Flood Insurance Rate Map
FONSI	Finding of No Significant Impact
FWS	U.S. Fish & Wildlife Service
H_2S	Hydrogen Sulfide
JBEC	John Brown E&C, Inc. (previous M&O contractor)
LPG	Liquified Petroleum Gas
MEOR	Microbial Enhanced Oil Recovery
MER	Maximum Efficient Rate
M&O	Management and Operation
NEPA	National Environmental Policy Act
NORM	Naturally Occurring Radioactive Material
NOx	Nitrogen Oxides
NOSR	Naval Oil Shale Reserves
NPDES	National Pollutant Discharge Elimination System
NPOSR-CUW	Naval Petroleum and Oil Shale Reserves in Colorado, Utah and Wyoming
NPR-3	Naval Petroleum Reserve No. 3
NTCHS	National Technical Committee for Hydric Soils

NWI	National Wetland Inventory
OSHA	Occupational Safety & Health Administration
PCB	Polychlorinated Biphenyl
RCRA	Resource Conservation & Recovery Act
RMOTC	Rocky Mountain Oilfield Testing Center
SARA	Superfund Amendment Reauthorization Act
SCS	U.S. Soil Conservation Service
SHPO	State Historic Preservation Officer
T&E	Threatened and Endangered
TDS	Total Dissolved Solids
TPQ	Threshold Planning Quantities
TSD	Treatment, Storage and Disposal
TSP	Total Suspended Particulates
UIC	Underground Injection Control
USDA	U.S. Department of Agriculture
USDW	Underground Sources of Drinking Water
USGS	U.S. Geological Survey
UST	Underground Storage Tank
VPD/ADT	Vehicles Per Day/Average Daily Totals
WGFD	Wyoming Game and Fish Department
WNDDB	Wyoming Natural Diversity Data Base
WYDEQ	Wyoming Department of Environmental Quality

EXECUTIVE SUMMARY

This Sitewide Environmental Assessment (EA) has been prepared for the United States Department of Energy to address the Proposed Continued Development of Naval Petroleum Reserve No. 3 (NPR-3) over the next five years. NPR-3, or Teapot Dome, is a 9,481-acre (3,837 ha) oilfield located in Natrona County, Wyoming, approximately 35 miles (56 km) north of the City of Casper. The United States Department of Energy (DOE) has managed NPR-3 for oil recovery at the "Maximum Efficient Rate" (MER) since 1976. The Sitewide EA has been prepared for the DOE in order to comply with the National Environmental Policy Act of 1969 (NEPA), (42 USC 4321, et. seq.), the DOE's implementing regulations for NEPA (10 CFR 1021) and the DOE's NPOSR-CUW NEPA Guidance Manual (DOE, 1992a).

The Proposed Action is the continued development of NPR-3 for the next five years. Continued development includes all activities typically required to profitably manage a mature stripper oilfield, such as NPR-3, at the MER. Continued development comprises four general categories of activity: continued development drilling utilizing conventional oil recovery technologies; continued and expanded use of Enhanced Oil Recovery (EOR) techniques that are necessary for continued oil production from reservoirs after primary or secondary recovery; continuation of general operations and support activities; and full implementation of the Rocky Mountain Oilfield Testing Center.

Continued development activities either have no potential to result in adverse environmental impacts or would only result in adverse impacts that could be readily mitigated. This Sitewide EA summarizes the potentially affected environment at NPR-3 as of 1994, discusses all potentially adverse environmental impacts, and proposes specific mitigation measures that offset each identified adverse impact. Resource types discussed in detail include land resources, air quality and acoustics, water resources, geology and soils, biological resources, cultural resources, socioeconomics, and waste management.

Continued development of NPR-3, as outlined in the Proposed Action, would not substantially alter the character of existing operations and would be consistent with NPR-3's historic role as an oilfield. Continued development is not expected to result in major changes in the types and quantities of air emissions and wastewater discharges already generated by existing operations at NPR-3. Continued development, especially where it involves expansion of EOR activities, would result in small areas of new land disturbance at several locations on NPR-3, especially in the already intensively developed central area.

Alternatives to the Proposed Action that were reviewed include: other chemical and thermal EOR technology alternatives to maintain oil and gas production, divestiture of NPR-3 by the Federal government, a no-action alternative of continuing operation of NPR-3, but without further development, and the immediate decommissioning of the project.

Table i-1 ISSUES TRACKING MATRIX

Issue	Executive Summary	Section 1.0 Purpose & Need	Section 2.0 Alternatives	Section 3.0 Affected Environment	Section 4.0 Environmental Consequences
Land Resources	i	1-4	2-2, 2-6, 2-12, 2-15, 2-16	3-1, 3-4	4-1, 4-2, 4-3
Air Quality	i	1-4	2-11, 2-15, 2-16	3-4, 3-5, 3-6	4-4, 4-5, 4-6
Water Resources	i	1-4, 1-5	2-12, 2-13, 2-15, 2-16	3-6, 3-7, 3-8, 3-9	4-6, 4-7, 4-8, 4-9, 4-10, 4-11
Geology & Soils	i	1-4	2-2, 2-6, 2-7	3-10, 3-13, 3-17	4-11, 4-12, 4-13
Biological Resources	i	1-4, 1-5	2-12, 2-13, 2-16	3-17, 3-18, 3-20, 3-21, 3-22, 3-25, 3-26, 3-27	4-13, 4-14, 4-15, 4-16, 4-17, 4-18, 4-19, 4-20
Cultural Resources	i	1-4	2-1, 2-15, 2-16	3-28, 3-29	4-20, 4-21
Socio-economics	i	1-4	2-15, 2-16	3-29, 3-30, 3-31	4-22, 4-23, 4-24

Waste Management	i	1-4	2-13	3-32, 3-33, 3-34, 3-35, 3-36	4-24, 4-25

1.0 PURPOSE OF AND NEED FOR ACTION

1.1 Introduction

This Sitewide Environmental Assessment (EA) is prepared to address the Proposed Continued Development of Naval Petroleum Reserve No. 3 (NPR-3, or Teapot Dome), a 9,481-acre (3,837 ha) oilfield owned by the U. S. Department of Energy (DOE) in Natrona County, Wyoming (Figure 1-1). NPR-3 is operated under a Management and Operation (M&O) contract by Fluor Daniel (NPOSR) Inc., hereinafter referred to as FD. The Sitewide EA has been prepared for the DOE in order to comply with the National Environmental Policy Act of 1969 (NEPA), (42 USC 4321, et. seq.), the DOE's implementing regulations for NEPA (10 CFR 1021) and the DOE's NPOSR-CUW NEPA Guidance Manual (DOE, 1992a).

NPR-3 was created by Executive Order of President Wilson in 1915 as an emergency source of liquid fuels for the military. Production began in the 1920s during a time of substantial exploration and production, when leases were issued by the Interior Department under the Mineral Leasing Act. Production was discontinued after 1927 and renewed between 1959 and 1976 in a limited program to prevent the loss of U.S. Government oil to privately-owned wells on adjacent land.

In response to the Arab oil embargo of 1973-74, which demonstrated the nation's vulnerability to oil supply interruptions, Congress authorized and directed, in 1974, that the Naval Petroleum Reserves be explored and developed to their full economic and productive potentials. In 1976, Congress formally passed the Naval Petroleum Reserves Production Act (Public Law 94-258), which required that the Naval Petroleum Reserves be produced at their maximum efficient rate (MER), consistent with sound engineering practices, for a period of six years. The law also provided that at the conclusion of the initial 6-year production period, the President (with the approval of Congress) could extend production in increments of up to three years each, if continued production was found to be in the national interest. The President has authorized five 3-year extensions since 1982, extending production continuously through April 5, 1997.

This Sitewide EA is prepared to address continued development activities at NPR-3 for the next five years. Substantial changes are currently proposed to the scope and character of existing production activities at NPR-3 that necessitate new NEPA documentation beyond that approved in 1990. This Sitewide EA serves both to update the 1990 EA to reflect 1994 conditions and to revise the 1990 EA to reflect the changes in production strategy that have occurred since that time.

An Environmental Impact Statement (EIS) was approved in 1976 for the initial development of NPR-3 (U.S. Navy, 1976). A subsequent EA for continued development of NPR-3 was approved in 1990, under which present operations at the Reserve are covered (DOE, 1990). In addition, DOE prepared an EA (DOE/EA-0334) in 1988 that analyzed the difference in the environmental and socioeconomic impacts of the development and operation of NPR-3 that would be caused by changing ownership from the public to the private sector. A Finding of No Significant Impact (FONSI) was issued for the proposal to sell NPR-3, although no further consideration has been given to the proposal to sell the

Federal Government's ownership interest in NPR-3.

1.2 Decisions needed

Decisions that must be made regarding the material in this document include:

- Whether any significant issues have been raised by the Proposed Action or any of the alternatives;
- Whether the Proposed Action or any of the alternatives would result in significant impact to the environment; and
- Whether the DOE would prepare an Environmental Impact Statement (EIS) or a Finding of No Significant Impact (FONSI) in response to this Environmental Assessment.

1.3 Scoping Summary

1.3.1 Internal Scoping

Meetings were held between the DOE, its Management and Operation (M&O) Contractor - FD and the consulting firm Halliburton NUS. DOE and Contractor staff determined the probable level of activity over the next five-year period and supplied the necessary background information. Halliburton NUS conducted site surveys, reviewed available background information, and recommended the general scope of the EA. DOE and FD adopted their proposed scope and it appears in Sections 3.0 and 4.0.

1.3.2 External Scoping

Several meetings were held with local, state and Federal agencies to provide them with the opportunity to present key areas of concern that should be addressed in the document. Governmental Agencies that were contacted include:

- U. S. Fish & Wildlife Service
- State Office of Historic Preservation
- U. S. Department of Agriculture
- Natrona County Planning Department
- Soil Conservation Service
- Wyoming Game and Fish Department
- Natrona County School District
- Wyoming Transportation Department

1.4 Discussion of Major Issues

Two major issues have been determined as a result of consultations with governmental agencies. These issues are:

From the U. S. Fish and Wildlife Service:

> Additional area would be disturbed for new wells, roads, pipelines and production facilities. As a result, there may be some impact to the local biological community because the total area available for vegetation and

wildlife would also decrease by a corresponding amount. Additional development may otherwise stress wildlife including big game, raptors, migratory species, and Threatened and Endangered (T&E) species by creating additional traffic, oilfield pits, power poles, and other hazards to wildlife.

From the Wyoming State Office of Historic Preservation:

Previous archaeological and cultural resource surveys no longer meet current standards. SHPO wants to ensure that all sites have been identified, recorded, collected or preserved. Additional surface disturbance may inadvertently impact those sites.

The remainder of the agencies contacted indicated that they could not foresee any major issues resulting from the Proposed Action or any of the Alternatives.

1.5 Summary of Federal Permits, Licenses, and Entitlements

Table 1-1 presents information regarding environmental permits at NPR-3. Most of the permits presented in this table are for federal programs for which the State of Wyoming has obtained primacy. For example, the Wyoming Department of Environmental Quality (WYDEQ) regulates and permits wastewater discharges under the National Pollutant Discharge Elimination System (NPDES), as described in the Clean Water Act. The Department of Energy generally holds the permits, except that the Contractor (Fluor Daniel) obtains routine permits from the Wyoming Oil and Gas Conservation Commission. In Table 1-1, permits for Underground Injection Control fall into this group.

In addition to the current permits, it is believed that several new air quality permits would be required in order to comply with the provisions of the Clean Air Act Amendments of 1990. The need for an Operating Permit under Title V of the Clean Air Act has been identified and the application is currently being prepared by a consulting firm.

Second, it is possible that some construction projects may disturb an area greater than 5 acres. In this case, a stormwater discharge permit would be obtained from WYDEQ, Water Quality Division.

Also, it is envisioned that the number of active NPDES permits would be substantially reduced, since many of the permitted facilities no longer discharge. One research project proposed for RMOTC involves the creation of a biological treatment area designed to use halophytic (salt-loving) plant species to bind chlorides in produced water and lower its toxicity. If successful, biological treatment and surface discharge of produced water would be preferable to underground injection.

Underground Injection Control (UIC) permits for oilfield water injection in Class II wells would remain relatively stable although the specific wells would change as areas of the field are depleted and other areas are brought under injection.

1.6 Preview of Remaining Chapters

Five alternatives, including the Proposed Action are considered in this Sitewide EA and are discussed in Section 2.0. They include:

1) The Proposed Action, which is composed of four principal components:

Continued infill and development drilling of NPR-3 utilizing conventional oil recovery technologies.

Continuation and expansion of the use of Enhanced Oil Recovery (EOR) techniques required to profitably extract additional oil from oil-bearing geological strata (reservoirs). Specific EOR technologies considered under the Proposed Action are also discussed in Section 2.0.

Continuation of general operations and support activities at NPR-3, including the continued use and expansion of the existing infrastructure comprising oil transport pipelines, water treatment facilities, warehouses, office facilities, roads, and electric distribution and transmission lines.

The development of the Rocky Mountain Oilfield Testing Center at NPR-3, whose purpose would be to provide facilities and necessary support to government and private industry for testing and evaluating new oilfield and environmental technologies, and to transfer these results to the petroleum industry through seminars and publications.

2) An Additional EOR Technology Alternative under which one or more EOR technologies, other than those considered under the Proposed Action, would be implemented as a substitute for drilling activity. General operations and support activities would continue as needed to supply and support the changed focus.

3) A Divestiture Alternative under which the DOE would sell or lease NPR-3 to one or more private concerns, effectively privatizing oil development on the Reserve.

Table 1-1 Federal Permits in Effect at NPR-3

Item	Permit No.	Facility
Air Quality (Stack Permits)	CT-360	Gas Plant Heat Transfer Fluid Heater
	CT-361A	Gas Plant Smokeless Flare
	CT-361A-2	Steam Generator No. 1
	CT-778	Steam Generator No. 2
	CT-850	Steam Generator No. 3

	CT-874	Steam Generator No. 4
	CT-937	Steam Generator No. 5
Water Quality (NPDES Permits)	WY-0028894	B-1-3 Tank Battery
	WY-0028908	B-1-10 Tank Battery
	WY-0028932	B-2-10 Tank Battery
	WY-0028274	B-TP-10 Tank Battery
	WY-0028916	B-1-28 Tank Battery
	WY-0028924	B-1-33 Tank Battery
	WY-0031895	North Waterflood
	WY-0032115	Water Disposal Facility
	WY-0034029	Steam Generator No. 2
	WY-0034495	Steam Generator No. 3
	WY-0035076	Steam Generator No. 4
	WY-0035297	Steam Generator No. 5
	WY-0034037	Water Treatment Facility

	WY-0034126	North Waterflood Floor Drains
Solid Waste	NPR-Ind #2	Operation of NPR-3 Industrial Landfill
	1-2 permits per year	Application of crude oil sludge to NPR-3 Roads
Ground Water Appropriation	UW-60713	B-1-3 Tank Battery
	UW-60714	B-1-10 Tank Battery
	UW-60715	B-2-10 Tank Battery
	UW-60716	B-TP-10 Tank Battery
	UW-60717	B-1-14 Tank Battery
	UW-60718	B-1-20 Tank Battery
	UW-60719	B-1-28 Tank Battery
	UW-60720	B-2-28 Tank Battery
	UW-60721	B-1-33 Tank Battery
	UW-60722	B-1-35 Tank Battery
	UW-43810	17-WX-21 Madison Water Well
	UW-85156	57-WX-3 Madison Water Well
Underground Injection Control	No permit number issued	124 Water Injection Wells

	No permit number issued	34, 51 & 74-CMX-10 for Oilfield Brine Disposal
	No permit number issued	86-LX-10, 25-LX-11, 14-LX-28
Underground Storage Tanks	963-1	Diesel Storage Tank
	963-2	Unleaded Gasoline Storage Tank
	963-3	Unleaded Gasoline Storage Tank
EPA Hazardous Waste ID No.	WY 4890090042	Hazardous Waste Disposal ID for NPR-3 (Also amended for PCB activity)

4) A No-Action Alternative, under which NPR-3 would continue to be produced using present conventional and enhanced oil recovery technologies, but whereby no new development activities would be implemented. Petroleum production would begin to decline to the economic limit of the project, but the RMOTC would provide a purpose for continuing limited operations at NPR-3 after that time. General operations and support activities would continue as needed to support the limited activity.

5) A Decommissioning Alternative in which the DOE would promptly cease commercial operation of NPR-3 and begin environmental restoration.

The affected environment on and surrounding NPR-3 is characterized in Section 3.0. This characterization has been updated from the earlier characterizations provided in the 1976 and 1990 NEPA documents to reflect present conditions at NPR-3. Environmental consequences potentially resulting from the Proposed Action and each alternative are discussed in Section 4.0, which also details the mitigation measures necessary to offset any potential adverse environmental consequences identified for the Proposed Action. A discussion of potential cumulative impacts from the Proposed Action is also provided in Section 4.0, as are the potential impacts from the Alternatives to the Proposed Action. Sections 5.0, 6.0 and 7.0 provide a list of preparers, agencies and persons consulted, and bibliography, respectively.

2.0 DESCRIPTION OF THE PROPOSED ACTION AND ALTERNATIVES

Elements of the Proposed Action for continued development of NPR-3 are described below (Section 2.1). This is followed by a discussion of alternatives to the Proposed Action (Section 2.2), including Additional EOR Technology Alternatives (Section 2.2.1), Divestiture of NPR-3 (Section 2.2.2), the No-Action Alternative (Section 2.2.3), and the Decommissioning Alternative (2.2.4).

2.1 Proposed Action

The DOE has developed a number of continued development projects which could be implemented to continue maximum efficient rate (MER) production at NPR-3 for the next five years. For a mature stripper field such as NPR-3, MER corresponds to the maximum economic rate of withdrawal, which is highly dependent upon the price of petroleum and associated products on the open market. Since 1986, wide swings in petroleum prices have been experienced, with prices ranging from $11 per barrel to over $40 per barrel. The Proposed Action has therefore been designed to encompass several projected ongoing and new projects. While some of these new projects may not be economically feasible under current oil prices, their inclusion in the Proposed Action allows for greater flexibility in planning NPR-3 activities, providing contingencies for changing market conditions. The Proposed Action addresses the specific NPR-3 activities identified in Table 2-1 .

Continued development activities under the Proposed Action would include the drilling of approximately 250 oil production and injection (gas, water, and steam) wells, the construction of between 25 and 30 miles of associated gas, water, and steam pipelines, the installation of several production and support facilities, and the construction of between 15 and 20 miles of access roads. This work would be performed over the next five years. Since excavation and construction are important parts of this alternative, one could expect further impact to land resources, biological resources, and cultural resources.

In addition to the continued development of oil and gas resources to support production at the MER, it is proposed to fully develop the Rocky Mountain Oilfield Testing Center (RMOTC). The mission of RMOTC is to provide facilities and necessary support to government and private industry for testing and evaluating new oilfield and environmental technologies, and to transfer these results to the petroleum industry through seminars, training, and publications. This project is already partially at work at NPR-3, providing assistance on small projects that fall under DOE Categorical Exclusions (CXs). The goal would be to improve the economics of oil production at NPR-3 and other stripper oilfields. Since much of the country's domestic oil supply remains in older, marginally economic fields, RMOTC would provide research and development (R&D) benefits to the oil producers most in need of technological assistance. Construction requirements are included in the totals discussed in the previous paragraph.

As noted in Section 1.0, activities under the Proposed Action generally correspond to four major program elements: 1) Continued Drilling Activity; 2) Enhanced Oil Recovery (EOR); 3) General Operations and Support Activities; and 4) Development and Operation of RMOTC. Proposed developments associated with these program elements are discussed below. Specific activities which are included within these program elements are presented in Tables 2-2 through 2-4.

2.1.1 Continued Drilling Activity

Several programs are planned to continue drilling at NPR-3. The geology of NPR-3 and the surrounding area is very complicated; there are at least 11 different geologic formations that have yielded oil. Numerous faults further divide each oil producing zone into many separate reservoirs. Geologists and petroleum engineers may review available geologic data and find areas of the field that are not being adequately drained by existing wells. The technical staff may also find that there is oil production possible outside the previously defined boundary of the petroleum reservoir. Drilling activity designed to exploit those areas is called *development drilling*.

NPR-3 would usually employ conventional oilfield technology to drill vertical wells. In general, a rotary drilling rig would drill a 12- to 15-inch diameter hole deep enough to protect surface waters and potential groundwater resources. While the well is being drilled, compressed air or drilling mud would be pumped down the inside of the drill pipe and circulated back up the outside of the pipe in order to carry the rock chips out of the well. *Drilling mud* is actually a very special fluid containing sodium bentonite (a type of clay) and other additives. Steel casing would then be run into the well and cemented in place. This is called the *surface casing*. The drilling tools would then be run inside the casing to the bottom of the well and begin drilling again where the first drilling phase ended. Drilling would continue until the objective is reached. The drilling tools would then be removed from the well and electronic instruments (*logging tools*) would be lowered into the well to measure various properties of the rock formations that were penetrated. Those instruments may contain radioactive sources. The next string of steel casing (production casing) would be run to the bottom of the well and cement would be pumped into the annular space between the outside of the casing and the rock wall of the well. After this phase is completed, the drilling rig would be moved off of the well and completion operations would begin.

Directional drilling is a modification of routine vertical drilling in which the well would start vertically, and then is slowly curved in order to reach a particular geographic location within the oil zone. The process of drilling would be the same as for vertical drilling, but special tools would also be used in order to steer the well on a specific compass heading. Usually this is done because the surface location directly above the target is inaccessible or otherwise unavailable for the drilling rig. Alternatively, several wells could be drilled from one larger location in order to minimize surface occupancy or impact.

Table 2-1 List of Continued Development Projects Under the Proposed Action

Maintenance and installation of fences.
Siting, construction and maintenance of buildings.
Emergency response and fire training exercises.
Maintenance of roads and locations.
Maintenance, construction or modification of pits, boxes or tanks including bird netting and liner installation.
Environmental sampling and monitoring as required by Federal, state and local regulations and permits.
Development of a program to evaluate Naturally Occurring Radioactive Materials (NORM) issues.
Dismantling of electrical distribution lines to abandoned wells and reclaimed locations.
Tapping and installing electrical tap lines for new installations.
Electrical transmission line repair, modification, relocation or expansion.
Electrical substation construction or modifications.

Communications and electronic equipment installation, repair and maintenance.
Hazardous material clean-up, storage and disposal.
Operation, maintenance and modification of hazardous waste accumulation areas.
Emergency planning and evacuation routes.
Data gathering and process sampling.
Relocation of existing equipment.
Polychlorinated Biphenyls (PCB) removal, handling and disposal.
Routine maintenance activities as defined in DOE NEPA regulations.
Recompletions of existing wells in other reservoir intervals.
Recompletions of existing wells in current reservoir intervals.
Pump and piping configuration modifications.
Pipeline construction, maintenance, repair or replacement.
Construction of new pipelines and related facilities.
Waste collection and waste treatment facility construction, operation and maintenance.
Leasing and oversight of grazing activities.
Research and technology demonstration projects of limited scale and impact run by RMOTC.
Water disposal activities including: a) Drilling additional Class II disposal wells in the Crow Mountain formation, as well as any other formation that may be approved; b) Installing flowlines to disposal wells; c) Adding pump(s) or pump capacity to water disposal facility, and production facilities to ship increased water volumes; d) Constructing a new water disposal facility; e) Converting existing water disposal facility to a chase waterflood facility; f) Siting, design and installation of equipment such as tanks, pumps,

separators and pipelines, to treat produced water so that it could be recycled in chase water, waterflood supply, or for steam generator feed water;

g) Adding tanks, separators, filters, or other water treatment equipment at production facilities and the water disposal facility, for treatment and storage of produced water;

h) Constructing new production test facilities to support new wells and steamflood development, or expanding the capacity of existing facilities by adding equipment such as pumps, tanks, piping, flowlines, separators, electrical equipment and buildings; and

i) Constructing a biotreatment facility to treat oil and grease in produced water and discharge under an NPDES permit.

Crude oil sludge and contaminated soils handling activities including:

a) Storage;

b) Transportation;

c) Centrifuge;

d) Road application;

e) Construction of a new landfarm or expand the existing landfarm;

f) Using biological and chemical accelerators to speed up the decomposition of petroleum contaminated materials;

g) Composting; and

h) Off-site disposal.

Chase water injection activities designed to inject water into oil-bearing formations behind steam injection; including:

a) Converting wells from production to injection (or the reverse);

b) Installing flowlines; and

c) Constructing injection facilities, including pumps, tanks, electrical installations, and buildings.

Hydrogen sulfide (H_2S) treatment and control activities including:

    a) Testing;

    b) Building and operation of a pilot facility;

    c) Installation of two-phase separators at test facilities;

    d) Install low pressure gas gathering system and construct pipeline to existing gas processing plant;

    e) Installation of amine H_2S removal system at existing gas processing plant;

    f) Installation of amine H_2S removal system at existing test facilities;

    g) Disposal or recycling of spent amine;

    h) Regeneration of spent amine on-site;

    i) Installation of additional natural gas compressors; and

    j) Chemical, microbial or biocide treatment of wells for H_2S control.

Design, permitting, and construction of a new solid waste disposal facility and landfill.

Procurement and transportation of drinking water from a regulated municipal water source.

Cathodic protection projects including:

    a) Drilling deep bed anode holes; and

    b) Installing transformers and electrical lines necessary for operation of an impressed current cathodic protection system.

Use of gas tracers, foamers, polymers, gels and surfactants in the steam drive patterns to aid in mobility control.

Gas processing plant expansion or modification as required to meet demands. May consist of adding, moving or resizing equipment.

Reclamation of right-of-ways, pits, production facilities, well pads, and other abandonment activities.

Construction or enlargement of production, workover, or evaporative pits, both for temporary and long-term application.
Installation of a vehicle wash facility with water recycling capability.
Installation, operation and maintenance of air quality and meteorological monitoring stations.
Injection of air or natural gas for reservoir pressure maintenance and/or gas storage.
Implementation of Shannon formation waterflood.
Participation with adjoining mineral rights holders for drilling wells under a cooperative agreement.
Use of polymers and gel conformance treatments to improve reservoir conformance.
Drilling and completion of new production and injection wells using vertical, directional or horizontal drilling technology and techniques.
Relocation sections of existing injection lines from "old" to "new" injectors.
Modification of existing equipment to reduce emissions.
Gas huff 'n puff (cyclic) injection.
Steam huff 'n puff (cyclic) injection.
Conducting emergency response and fire training exercises.
Emergency planning.
Data gathering and process sampling.

Horizontal drilling is a modification of directional drilling in which the well starts vertically, and then is slowly curved all the way to horizontal within the oil zone. The well may continue for several thousand feet horizontally before drilling is stopped. Surface occupancy is virtually identical to that of a vertical well, except that larger production equipment is required. This is because one horizontal well may tap the same geographic area as two or more vertical wells. The increased efficiency of one horizontal well would tend to reduce the number of well locations required.

The *completion* phase of operations involves establishing oil and/or natural gas production from the well. At the end of drilling, the well would still be completely sealed from the petroleum producing rock by the well casing and cement. In order to establish production from the well, a truck-mounted well servicing rig would be set up over the well. This rig is also known as a *pulling unit* or *workover rig*. Additional logging tools would be lowered into the well that measure the quality of the cement between the production casing and the rock. This is important because the cement must be completely impermeable and must prevent fluids from migrating up and down within the annular space. If the well logs showed a poor cement bond, additional cementing would be required. Assuming a good reading, shaped explosive charges called *perforating charges* would then be lowered into the well and detonated inside the casing, adjacent to the oil producing formation. Each charge would punch a 1/4 inch diameter hole through the casing, cement, and approximately 24 inches into the rock. This would allow oil and gas to flow into the well if it is to become a producer, or it would provide a way for water or gas to be injected into the targeted formation.

In most cases, however, the flow rate into the well would be too slow and it would be *stimulated* in order to achieve an economic production rate. Two alternatives are available to improve production: chemical treatment and hydraulic fracturing. Chemical treatment may include pumping hydrochloric and/or hydrofluoric acid into the rock formation to improve flow into the well; or it may include other chemicals depending on the circumstances.

If the circumstances indicated its use, an alternative would be *hydraulic fracturing*, which involves pumping fluids into the well under sufficient pressure to cause the oil-producing formation to fracture vertically. The fluid would force the fracture open while coarse sand is pumped into it. When the pressure is released, the fracture would close partially, but would be held open by the coarse sand. This sand-filled fracture would form a highly permeable conduit for fluid flow toward the well from the surrounding rock. A fracture 100 feet long and 1/8-inch wide would be expected to double the production or injection rate. Radioactive tracers may be used in the fracturing fluid in order to determine the vertical extent of the fracture.

After completion, various surface facilities would be constructed. A pipeline would be constructed to a central production facility that would separate produced crude oil, natural gas and water. All materials would be measured and the oil, water, and gas delivered into existing pipelines for further handling. Oil would be transported to a commercial pipeline for transportation to a nearby refinery. Natural gas would be sent to the gas plant for processing and re-injection as part of one of the secondary recovery projects. Water would be collected and disposed of by underground injection.

The previous description is by necessity a simplification. Actual circumstances may justify minor deviation from this summary of well drilling and completion operations.

As stated previously, this project would result in drilling between 250 oil production and injection (gas, water, and steam) wells, one water supply well, the construction of between 25 and 30 miles of associated gas, water, and steam pipelines, the installation of several production and support facilities, and the construction of between 15 and 20 miles of access roads.

In addition, the potential also exists for the communitization of wells at the site perimeters to prevent drainage of NPR-3 reserves. A detailed listing of specific activities associated with conventional in-fill development is presented in Table 2-2 .

2.1.2 Enhanced Oil Recovery

Enhanced oil recovery (EOR) technologies, which utilize fluid injection techniques to maintain reservoir pressure and displace oil, are currently utilized in production of three of the nine currently producing zones: Second Wall Creek, Shannon, and Muddy. Pressure maintenance commenced in the Second Wall Creek Sand in 1979, followed by testing of EOR techniques in the Shannon Sand in 1981, and finally pressure maintenance in the Muddy in 1985. No other productive zones currently offer economic EOR opportunities, although testing the potential for introducing cyclic gas injection into the Lakota reservoir is planned. Though no other projects are currently planned, other similar injection programs may be considered in these and other formations in the future. A detailed listing of specific activities associated with EOR development is presented in Table 2-3 . The following specific EOR activities are included as part of continued development of NPR-3:

- Shannon Reservoir Steam Drive (Steamflood) EOR Development: The largest remaining reserve potential on NPR-3 is from the Shannon EOR steam drive, or steamflood. The Shannon sandstone contains large quantities of oil at very shallow depths. The oil lacks any natural reservoir pressure to push oil into wells and production rates have historically been disappointing. Numerous techniques have been attempted over the years to push oil to producing wells but results have been poor. Steam injection was tried because it had three beneficial effects on the trapped crude oil:

  It provided the pressure necessary to get oil flowing into wells.

  It heated the oil and reduced its viscosity, permitting the oil to flow faster.

  It distilled some oil and created solvent that thins the oil.

Since it began in 1985, the project has expanded to five steam generators, which are large gas-fired boilers. Each steam generator is designed to inject steam into five different well patterns for a period of 3 to 4 years each, usually followed by water injection to scavenge heat from the area immediately surrounding the injection well. This last procedure is called *chase water injection*.

The Proposed Action includes continuation of the current operations without notable changes. Associated with each steam generator would be a well pattern consisting usually of 10 steam injection wells, 15 oil production wells, a steam distribution pipeline, and a fuel gas supply system. A water treatment facility, which provides high quality water for steam production and a water disposal system would continue to serve the five generators and associated wells in common. Approximately 3,000 feet (900 m) of surface steam lines would be required for each new steam injection pattern. After steam injection into a pattern is completed, injection would be relocated to a new pattern or to portions of other patterns.

- Reservoir Microbial Treatment: This project would utilize microbial enhanced oil recovery (MEOR) for increasing oil production. Other strains of bacteria may be used to inoculate existing wells for the treatment of hydrogen sulfide (H_2S), carbonate or sulfate scale, and paraffin problems associated with production. Testing of bacteria that could tolerate high temperatures ($100°$ to $110°$ C, or $212°$ to $230°$ F) is of particular interest in the steam-injection area.

- Second Wall Creek Alkaline-Surfactant-Polymer (ASP) EOR Pilot Project: The Northern Second Wall Creek Reservoir is highly faulted and fractured. Despite the ongoing pressure maintenance program via gas cap injection and pattern waterflood, a sharp decline in oil production has been observed. For this reason, a field pilot test to determine the feasibility of an alkaline-surfactant-polymer (ASP) flood is being considered. A mix of alkaline, surfactant, and polymer agents with Madison water would be injected into the Second Wall Creek to reduce interfacial tension between oil and water. If the pilot test proves successful technically and economically, an expanded program could be implemented to recover additional oil from the northern part of the Second Wall Creek Reservoir and other reservoirs if the technique is found to be applicable.

Table 2-2 List of Continued Drilling Activities (2.1.1)

Pump and piping configuration modification.
Pipeline construction, maintenance, repair or replacement.
Construction or enlargement of production, workover, and emergency pits, both for temporary and long-term application.
Drilling additional Class II disposal wells in the Crow Mountain formation, as well as any other formation that may be approved.
Drilling of additional water supply wells in the Madison formation.
Drilling and completion of new production and injection wells using vertical, directional or horizontal drilling technology and techniques, including: a) The use of rotary drilling rigs common to the industry;   b) The use of air or drilling mud as a medium for cleaning cuttings from the well;   c) The use of steel or fiberglass tubular goods of various sizes for casing and tubing;   d) The use of cement blends to seal the annular space between the well and the hole;   e) The use of electronic well logs - including those with radioactive sources - to measure the properties of the rock surrounding the well bore;   f) The use of concentrated salt brines as completion fluids;   g) The use of shaped explosive charges to penetrate well casing, cement, and the outer rock for the establishing of production or injection;   h) The use of polymer fluids and/or compressed gasses in hydraulic fracturing treatments;   i) The use of hydrochloric and hydrofluoric acid mixtures to clean out

wells and restore production or injection; and

j) The use of various commercial oilfield treatment chemicals to prevent the deposition of paraffin and carbonate/sulfate scale and to break oil/water emulsions.

Tapping and installing electrical tap lines for new installations.

Electrical substation construction or modifications.

Construction of new production facilities including:

a) Pipeline headers and manifolds;

.b) Pipeline pig launching stations;

c) Production test satellites with fluid separators, tanks, chemical injection equipment, metering equipment, and facility operator office; and

d) Tank batteries with fluid separators, tanks, chemical injection equipment, metering equipment, tankage for oil and water storage, and facility operator office.

Table 2-3 List of Enhanced Oil Recovery Activities (2.1.2)

Pump and piping configuration modification.

Pipeline construction, maintenance, repair or replacement.

Water disposal activities including:

a) Drilling additional Class II disposal wells in the Crow Mountain formation, as well as any other formation that may be approved;

b) Installing flowlines to disposal wells;

c) Adding pump(s) or pump capacity to water disposal facility, and production facilities to ship increased water volumes;

d) Constructing a new water disposal facility;

e) Converting existing water disposal facility to a chase waterflood facility;

f) Siting, design and installation of equipment such as tanks, pumps, separators and pipelines to treat produced water so that it could be recycled in chase water, waterflood supply or for steam generator feed water;

g) Adding tanks, separators, filters or other water treatment equipment at production facilities and the water disposal facility, for treatment and storage of produced water; and

h) Constructing new production test facilities to support new wells and steamflood development, or expand the capacity of existing facilities by adding equipment such as pumps, tanks, piping, flowlines, separators, electrical equipment and buildings.

Chase water injection activities designed to inject water into oil-bearing formations behind steam injection including:

a) Converting wells from production to injection (or the reverse);

b) Installing flowlines; and

c) Constructing injection facilities, including pumps, tanks, electrical installations, and buildings.

Use of gas tracers, foamers, polymers, gels and surfactants in the steam drive patterns and waterfloods to aid in mobility control.

Construction or enlargement of production, workover, or evaporative pits, both for temporary and long-term application.

Relocate sections of existing injection lines from "old" to "new" injectors.

Huff 'n Puff (cyclic injection) using natural gas and steam.

Injection of air and natural gas for reservoir pressure maintenance.

Hydrogen sulfide (H_2S) treatment and control activities including:

a) Testing;

b) Building and operation of a pilot facility;

c) Installation of two-phase separators at test facilities;

d) Installation of low pressure gas gathering system and construction of pipeline to existing gas processing plant;

e) Installation of amine H_2S removal system at existing gas processing plant;

f) Installation of amine H_2S removal system at existing test facilities;

g) Disposal or recycling of spent amine;

h) Regeneration of spent amine on-site;

i) Installation of additional natural gas compressors;

j) Chemical, microbial or biocide treatment of wells for H_2S control; and

k) Construction of flares for the combustion of H_2S contaminated natural gas.

Implementation of waterfloods in conducive reservoirs.

Drilling and completion of new production and injection wells using vertical, directional or horizontal drilling technology and techniques.

Installation of tanks at facilities to increase pump head pressure.

- Huff 'n Puff Treatment: Several Huff 'n Puff techniques (cyclic injection) would be employed in the Shannon and Lakota zones. Under this project, existing Shannon wells would be injected with steam or natural gas, and Lakota wells would be injected with carbon dioxide. These fluids would be allowed to "soak" into the oil in the producing zone, which would then be pumped back to produce a short-term increase in oil flow. This recovery technology has been applied previously to several wells at NPR-3 with encouraging results. Again, usage of the treatment would be expanded to other formations and areas of the field if information became available that suggested its feasibility.

- Shannon Waterflood: This project would involve the construction of a water injection pipeline and the drilling of water injection wells in the Shannon sandstone. The injected water would spread out radially away from the injection wells and displace the existing crude oil toward producing wells. If the pilot project were to be deemed successful, waterflooding would be applied to some portions of the Shannon reservoir outside the steamflood area, possibly on the east or southern sides of the field. Produced water would increase, resulting in increased power usage and disposal capacity at the Water Disposal Facility. Water would be disposed via UIC-permitted injection into the three existing disposal wells. Additional surface disturbance would result from the associated pump stations, pipelines, and injection well pads.

With the installation and ongoing operation of Steam Generators Nos. 1, 2, 3, 4 and 5, effective utilization of produced water for steam make up and effective produced water disposal has been increasingly important. Under the current water handling scheme, all National Pollutant Discharge Elimination System (NPDES) regulations are satisfactorily met. Three Class 2 disposal wells, 74-CMX-10, 34-CMX-10, and 51-CMX-10 have UIC permits to handle produced water as well as brine from the water treatment facility. Several additional disposal wells and related surface equipment are anticipated to be installed as part of continued EOR activities.

2.1.3 General Operations and Support Activities

Implementation of the Proposed Action would require the operation, maintenance, and continued development of support facilities and programs. Items under this program element correspond to activities and infrastructure requirements necessary to support ongoing and projected day-to- day production and operations at NPR-3. Included are all support facilities used for processing crude oil and wet gas, as well as solid waste and waste water disposal operations, field management activities, general maintenance activities, environmental monitoring programs, and health and safety programs. Additional support functions include electrical power distribution systems, potable water and sewer systems, and cathodic protection systems. A detailed listing of specific activities associated with general operations and support is presented in Table 2-4 .

One additional program under general operations is the oversight and leasing of grazing activities. A grazing program at NPR-3 would include leases to one or more individuals, for a total of no more than 450 Animal Unit Months (AUM).

Finally, NPR-3 would use its gas reservoirs to seasonally store natural gas for use in the steam generators and for use by other Federal agencies. Gas would be purchased in the summer, when prices were low, and be injected into the gas reservoirs. The gas would be produced in the winter when priced increase and demand is high. Since the gas plant compressors are electric, the incremental impact of this program would be negligible.

2.2 Alternatives to the Proposed Action

2.2.1 Other EOR Technology Alternatives

Instead of drilling, additional chemical and thermal EOR technology alternatives, other than those in the Proposed Action, would be considered to maintain oil and gas production at the MER. These other processes include chemical injection, carbon dioxide gas (CO_2) or nitrogen gas (N_2) flooding (as opposed to Huff 'n Puff in the Proposed Action), in-situ fireflooding, and oil mining. Such technologies would not be viable alternatives under today's economic conditions due to their high cost and low process performance, but all have technical merit. Serious operational problems would also be associated with these techniques, such as wet CO_2 (carbonic acid) corrosion. In-situ combustion, which has been tried as a pilot project at NPR-3, would present safety problems associated with operation of pressurized oxygen or air systems.

Under this scenario, general operations and support activities would continue as needed to support these different projects. RMOTC would also have a role that would be changed little from that in the Proposed Action.

Table 2-4 List of General Operations and Support Activities (2.1.3)

Maintenance and installation of fences.
Siting, construction and maintenance of buildings.
Emergency response and fire training exercises.
Maintenance of roads and locations.
Maintenance, construction or modification of pits, boxes or tanks including bird netting and liner installation.
Environmental sampling and monitoring as required by Federal, state and local regulations and permits.
Development of a program to evaluate Naturally Occurring Radioactive Materials (NORM) issues.
Dismantling of electrical distribution lines.
Tapping and installing electric tap line for new installations.
Electrical transmission line repair, modification, relocation or expansion.
Production and recompletions in existing well bores.
Hazardous material clean-up, storage and disposal.
Operation, maintenance and modification of hazardous waste accumulation areas.
Emergency planning and evacuation routes.
Data gathering and process sampling.
Relocation of existing equipment.
Polychlorinated Biphenyls (PCB) removal, handling and disposal.
Routine maintenance activity as defined by DOE NEPA regulations.
Communications and electronic equipment installation, repair and maintenance.
Electrical substation construction or modification.
Electrical transmission line relocation.
Leasing and oversight of grazing activities.

Waste collection and waste treatment facility construction, operation and maintenance.

Crude oil sludge and contaminated soils handling activities including:
 	
    a) Storage;
 	
    b) Transportation;
 	
    c) Centrifuge;
 	
    d) Road application;
 	
    e) Construct a new landfarm or expand the existing landfarm;
 	
    f) Using biological and chemical accelerators to speed up the decomposition of petroleum contaminated materials;
 	
    g) Composting; and
 	
    h) Off-site disposal.

Closure of solid waste disposal facility and NPR-3 industrial landfill, with off-site disposal of solid waste.
Procurement and transportation of drinking water from a regulated municipal water source.
Gas processing plant expansion or modification as required to meet demands. May consist of adding, moving or resizing equipment.
Reclamation of right-of-ways, pits, production facilities, and other abandonment activities.
Installation of a vehicle wash facility with water recycling capability.

Cathodic protection projects including:
 	
    a) Drilling deep bed anode holes; and
 	
    b) Installing transformers and electrical lines necessary for operation of an impressed current cathodic protection system.

Installation, operation and maintenance of air quality and meteorological monitoring stations.
Modify existing equipment to reduce emissions.

As outlined in Section 3.2, other EOR techniques that have been considered include gas injection (of carbon dioxide or nitrogen), chemical flooding (surfactants, polymers), thermal methods (in-situ fireflooding), and oil mining. None are viable economic alternatives at present due to high cost and low process performance. Serious operational problems associated with some of these techniques could be encountered. Examples include carbon dioxide corrosion; inert atmospheric safety considerations associated with nitrogen; and air pollutant emissions from high pressure compressors needed to inject carbon dioxide or nitrogen. In-situ combustion, which has been tried as a pilot project at NPR-3, presents safety considerations, including operation of pressurized oxygen and nitrogen injection systems. Implementation of any of these EOR technology alternatives would require the same amount of land disturbance (and cultural resources impact) as would implementation of the EOR technologies under the Proposed Action.

2.2.2 Divestiture of NPR-3

All Naval Petroleum Reserves are the property of the United States and, pursuant to Public Law 95-91, are operated under authority delegated to DOE. Therefore, any change in ownership or management must be specifically authorized by Congress through legislation. In 1988, the Administration submitted legislation to Congress requesting authorization to sell the government's ownership interest in NPR-3 (and also the larger Naval Petroleum Reserve No. 1 in California). DOE prepared an EA (DOE/EA-0334) that analyzed the difference in the environmental and socioeconomic impacts of the development and operation of NPR-3 that would be caused by changing ownership from the public to the private sector. A Finding of No Significant Impact (FONSI) was prepared for the proposal to sell NPR-3.

Subsequent to publication and distribution of the aforementioned EA, no further consideration had been given to the proposal to sell the Federal government's ownership interest in NPR-3. Public ownership and management of NPR-3 are expected to continue for the next few years because the NEPA review of such a proposal would need to be repeated and none of the required legislation has been proposed.

If NPR-3 were sold to a private interest, it would likely be managed as an oilfield in a manner similar to that used by the DOE under the Proposed Action. However, an independent operator may choose to operate NPR-3 as a stripper oilfield and minimize new investment. The potential environmental impacts would basically be similar to, or less than, those under the Proposed Action. On the other hand, an independent operator may be less attentive to environmental protection than DOE, and the net impact is therefore difficult to quantify.

2.2.3 No-Action Alternative

The No-Action Alternative assumes that none of the actions outlined in the Proposed Action would be initiated. Existing wells and facilities would continue to be operated on a well-by-well basis until the costs to lift a barrel of oil exceed the revenue gained. Implementation of the No-Action Alternative would not be consistent with the statutory mandate to produce NPR-3 at MER.

Plugging and abandonment of wells, an on-going project, and shutdown and rehabilitation of battery sites would be accelerated under the No-Action Alternative. There would also be a reduction in work force as the project changes pace from an aggressive production mode, to a remediation mode, and finally, to a caretaker.

Impact on cultural resources would be minimal, since no new development requiring construction or excavation would occur.

At some point after steam is stopped, produced water would become minimal and the Water Treatment Facility would be scaled down. Also, some of the three Crow Mountain disposal wells would no longer be needed and would be plugged. Steam generators would also be phased out gradually, decommissioned and salvaged.

With the resultant decline in production, the economics of operating NPR-3 would necessitate a substantial reduction of the current DOE and operating contractor staffs. Area socioeconomics would be adversely impacted since many DOE, operating contractor, and support group employees and their families would have to leave Natrona County to find work. There would be no additional new disturbed acreage, resulting in slightly lower levels of fugitive dust and less disturbance of natural habitat. Roads and facilities would be reclaimed to natural habitat as wells became uneconomical to continue production.

2.2.4 Decommissioning Alternative

Under this alternative, NPR-3 would cease production and begin environmental restoration. The abandonment of the project while it is still economic to operate would result in negative socioeconomic impact to DOE staff, contractor staff, and to Natrona County.

The level of activity would remain relatively high for several years while restoration and decommissioning occurs, but would cease at the completion of remedial action

Negative impact to wetlands would be substantial, since most wetland areas at NPR-3 would dry up as produced water discharge ceased.

Finally, this alternative would result in the least impact to land and cultural resources because no new disturbance or construction would occur.

3.0 AFFECTED ENVIRONMENT

3.1 Land Resources

3.1.1 Land Use

The principal land use of Natrona County (5,300 square miles or 13,700 square km) is sheep and cattle ranching. Areas adjacent to the NPR-3 are utilized primarily for oil production, with limited livestock grazing. Under the Zoning Ordinance of Natrona County, these lands are zoned RF (Ranching and Farming) although mineral extraction activities are exempt from the Zoning Resolution (Natrona County, 1978). No residential development is currently present or proposed for the immediate area surrounding NPR-3 (Halliburton NUS, 1993), especially because of the lack of potable water.

Land at NPR-3 is utilized primarily for oil production. The land surface is characterized by prairie with occasional sagebrush, severely cut ravines, and sandstone bluffs. Although formerly utilized for livestock grazing, leasing of NPR-3 lands for grazing had been discontinued. This practice would be resumed under the Proposed Action. Developed features in NPR-3 include gravel and dirt roads, wellheads and pumping units, oil and gas production facilities and equipment, storage areas, and an office complex. Existing well locations, shown in Figure 3-1, are concentrated in a 2,500-acre (1,000 ha) area located in the center of NPR-3, with substantially less development taking place in the northern and southern portions of the site. Most wells are located within the basin and at a considerable distance from the surrounding bluffs. Several wells in the extreme southern portion of NPR-3 are located near steeper slopes. Existing roads and facility locations, similarly concentrated in the center of NPR-3, are depicted in Figure 3-2.

Construction of facilities and supporting infrastructure requirements from 1915 to 1989 has resulted in the disturbance of approximately 1,623 acres (657 ha), approximately 17% of the total acreage of NPR-3. As of 1990, approximately 939 of these disturbed acres (380 ha) had been reclaimed (revegetated) and the other 684 acres (277 ha) were required to support ongoing production operations (DOE, 1990). Between 1990 and the present, additional construction of wells, roads and pipelines have disturbed approximately 100 additional acres, although 80 acres of previous wellsites and roads have been reclaimed.

3.1.2 Aesthetics

NPR-3 is typical of much of the central portion of Wyoming. It consists of rolling terrain covered with native grass and sagebrush, and is fragmented by numerous small gullies. NPR-3 is surrounded by a rim of sandstone bluffs. Although portions of NPR-3 operations are visible from the north along Wyoming Route 259, bluffs to the south, east and west generally isolate NPR-3 visually from the public (Halliburton NUS, 1993). The southern-most end of this rim does provide a scenic view of the entire project, although this viewpoint is limited to NPR-3 employees and a few local ranchers (DOE, 1990). Oilfield structures and activities associated with NPR-3 operations are aesthetically consistent and a common visual feature of offsite conditions.

Much of the area inside the sandstone bluffs at NPR-3 has already been altered to some degree by installation of facilities and service roads since operations first began in the 1920s, and since full scale development (at MER) was ordered in 1976.

3.1.3 Recreation

There are no public recreation facilities in the immediate vicinity of NPR-3, and no areas within NPR-3 are open to public recreation (Halliburton NUS, 1993). The nearest public recreation facility to NPR-3 is the Moses Ballfield, located approximately 7 miles (11 km) north near the town of Midwest. Additional recreational facilities maintained within Natrona County include several county parks, reservoirs, and recreation areas. These offer a large variety of activities including picnicking, camping, fishing, boating, swimming, and hiking (Natrona County, 1978).

3.2 Air Quality and Acoustics

3.2.1 Meteorology and Climate

The climate of NPR-3 is characterized as semi-arid with approximately 9-12 inches (23 - 30 cm) of precipitation annually. Precipitation is seldom sufficiently abundant and evenly distributed to keep the soil moist throughout the entire summer. Typical high temperatures in the summer are 80-85°F (27-30°C), and low temperatures in the winter are around 0°F (-18°C). However, extreme temperatures could reach 100°F (38°C) in summer and -40°F (-40°C) in winter. Winds are usually westerly or southwesterly and are most predominant during the late fall and spring months. (FD Services, 1992a)

3.2.2 Air Quality

NPR-3 is located in Natrona County, Wyoming, which is part of the Casper Intrastate Air Quality Control Region (AQCR)(40 CFR 81.213), designated as being in attainment by the EPA for all criteria pollutants (40 CFR 81.351). An ambient air quality monitoring program at NPR-3 was established to monitor air quality parameters set forth by the Wyoming Department of Environmental Quality (WYDEQ), Division of Air Quality, and as recommended by the June 1989 Environmental Survey Team. Ambient air quality meets State of Wyoming standards at the perimeter of the property (FD Services, 1992a). The air quality program includes ambient air monitoring for H_2S, nitrogen oxides (NOx) and hydrocarbons. In order to address worker health and safety, H_2S sampling has been conducted in the areas of highest potential concentrations (FD Services, 1992a). The primary areas associated with elevated H_2S levels include facilities in the steamflood patterns, the main ones being T-5-3, T-5-10, and B-3-3/T-4-3 tank batteries (FD Services, 1992b).

Prior to the NPR-3 studies, ambient air quality data for Natrona County generally, and NPR-3 specifically, were limited. Data prior to 1976 indicate that background levels of suspended particulates in the area ranged from 20 to 30 mg/m^3. No values for hydrocarbons were available for Natrona County. However, hydrocarbon sampling done in Converse County (adjacent to Natrona County) revealed that background levels there were apparently exceeding current state standards. Levels of H_2S measured on NPR-3 in June 1976 were less than 4 ppm.

From July 1 through December 31, 1981, ambient air monitoring for total suspended particulates (TSP), sulfur dioxide (SO_2), nitrogen dioxide (NO_2) and hydrogen sulfide (H_2S) was done to establish background levels of the above parameters and to monitor emissions associated with the Fireflood Pilot Project which was initiated at NPR-3 in 1982. During this period, the sampling results for TSP, SO_2, NO_2 and H_2S were lower than the annual regulated standard. Additional ambient air monitoring for TSP, SO_2, H_2S, and NO_2 was also conducted between July 1982 and March 1983. During this period the sampling results for hydrocarbons, TSP, NO_2, and SO_2 were also less than the annual standard. (DOE, 1990)

In August 1986 the annuli between the casing and tubing on various steamflood wells were sampled for H_2S. Prior to steam injection these wells did not produce H_2S. As the steam front spread through the formation, the growth of anaerobic sulfate-reducing bacteria was stimulated, resulting in the formation of the gas. H_2S levels were stabilized by means of chemical treatment of the wells with biocides. (DOE, 1990)

Further sampling of ambient H_2S, ozone, PM-10 and hydrocarbons occurred in 1989. Again, sampling results, indicated that PM-10, ozone and H_2S levels were less than the standard. (DOE, 1990)

Table 3-1 lists the NPR-3 facilities currently operating under air quality permits issued by the Wyoming Department of Environmental Quality and their respective emission inventories for calendar year 1993.

The permitting and operation of Steam Generator #5 in January 1993 increased the potential emissions of nitrogen oxides (NO_x) (including NO_2) at NPR-3 to levels exceeding the 100 tons (102 metric tons) per year threshold for a major source. (Khatib, 1993a) Because of NPR-3's major source status, a Title V Operating Permit application is currently being prepared for NPR-3 by a consulting firm.

3.2.3 Acoustics

The major noise sources within NPR-3 include various facilities, equipment and machines (steam generators, engines, pumps, drilling rigs, vehicles, etc.). Buildings associated with the North Waterflood, Water Disposal Facility, and all steam generators have been identified as having inside noise levels exceeding 85 decibels, and hearing protection is required for workers within these areas (FD Services, 1992b). Although sound-level monitoring of ambient acoustic conditions at NPR-3 has not been conducted, the contribution from NPR-3 operations to ambient noise levels beyond the Reserve boundary is estimated to be minimal, and no residences are located within audible range of general operations.

Table 3-1 Permitted Air Quality Emission Sources at NPR-3

		1993	Emissions	Data	
Source	Permit Number	Particulate Matter	Sulfur Dioxide	Nitrogen Oxide	Carbon Monoxide
Gas Plant Heater	CT-360	0.32×10^{-4} lb/hr 1.4×10^{-4} tpy[a]	3.9×10^{-6} lb/hr 1.7×10^{-5} tpy	0.64×10^{-3} lb/hr 2.8×10^{-3} tpy	1.3×10^{-4} lb/hr 5.6×10^{-4} tpy
Gas Plant Smokeless Flare	CT-361A	b	b	b	b
Steam Generator No. 1	CT-361A-2	0.219 lb/hr 0.94 tpy	0.013 lb/hr 0.05 tpy	1.50 lb/hr 6.42 tpy	b
Steam Generator No. 2	CT-778	0.208 lb/hr 0.878 tpy	0.030 lb/hr 0.12 tpy	10.13 lb/hr 21.46 tpy	b
Steam Generator No. 3	CT-850	b	b	b	b
Steam Generator No. 4	CT-874	b	b	4.49 lb/hr 19.66 tpy	b

Steam Generator No. 5	CT-937	c	c	4.49 lb/hr 19.66 tpy	c

 a tpy = Metric tons per year Source: 1993 Emissions Inventory Report for Criteria Pollutants

[a] tpy = Metric tons per year Source: 1993 Emissions Inventory Report for Criteria Pollutants

[b] Facility was not tested at NPR-3, submitted by FD to WYDEQ

[c] Began operation in 1993 on 3-28-94.

3.3 Water Resources

3.3.1 Surface Water Quantity

NPR-3 is drained by a series of ephemeral or intermittent stream channels that flow through steep topographic swales, locally referred to as draws. Little Teapot Creek originates in the highlands south of NPR-3 and enters NPR-3 in a northerly direction across the southern boundary as an intermittent stream. Teapot Creek originates approximately 15 miles (24 km) southwest of NPR-3 and enters NPR-3 in an easterly direction across the northwestern boundary as an intermittent stream. All other ephemeral and intermittent streams on NPR-3 drain into Little Teapot or Teapot Creeks. Little Teapot and Teapot Creeks merge immediately south of NPR-3's northern boundary and exit NPR-3 in a northerly direction. The merged stream flows into Salt Creek less than 1 mile (1.6 km) north of NPR-3, which flows to the Powder River, approximately 25 miles (40 km) to the north. (USGS, 1974)

Several small impoundments, none larger than 10 acres (4 ha) in surface area, had been constructed in the draws to serve as reservoirs during earlier operations on NPR-3 in the 1920s (Halliburton NUS, 1993). The remains of several of these impoundments still exist, but the basins only support wetlands.

Produced water obtained from the Tensleep and Madison formations is discharged to Little Teapot Creek and its tributaries through 14 outfalls. Discharges through each outfall are regulated under NPDES permits issued by WYDEQ, Water Quality Division. Although many of the outfalls are presently inactive, and discharges through some outfalls are only sporadic, discharge through other outfalls is continuous, resulting in perennial flow in Little Teapot Creek. (DOE, 1990) Discharge under any necessary general stormwater discharge permits would not be expected to make a contribution to surface flows.

Current operations at NPR-3 do not involve the withdrawal of any surface water from the streams or ponds.

3.3.2 Ground Water Quantity

There are no high quality fresh water aquifers in the strata underlying NPR-3. Those strata that produce fluids either produce water with excessive levels of total dissolved solids (TDS) or a mixture of hydrocarbons and water. The Steele Shale formation occupies the interval from the surface to an approximate depth of 2,000 feet (610 m). There are two porous and permeable sandstone formations within the Steele Shale. The Sussex sandstone outcrops in a ring near the center of the Teapot Dome structure, but does not appear to contain an aquifer. The second sandstone body is the Shannon sandstone which is an oil reservoir in much of the field. A fault separates the oil reservoir from the

Shannon outcrop at Salt Creek to the north. Groundwater is encountered in the Shannon in some areas north of the fault, but the concentration of Total Dissolved Solids exceeds 10,000 mg/l. No Underground Sources of Drinking Water (USDWs) or other shallow fresh water aquifers have been detected in the 795 wells drilled since 1976.

It should be noted that there is a strong distinction at NPR-3 between "fresh water aquifers" and "USDWs". Exempted aquifers are not USDW's under the Safe Drinking Water Act, which permits aquifer exemptions for fresh water aquifers being used for Class II injection. Several such aquifer exemptions exist at NPR-3. In addition, aquifers that contain crude oil, natural gas, or other contaminants that make it undesirable for a water supply could also be exempted. Several other aquifers at NPR-3 qualify for exemption under this criteria, although the actual exemption has not been pursued with the Wyoming Oil & Gas Conservation Commission. Produced water from oil and gas production is put to beneficial use for livestock and wildlife at NPR-3, but there would be no intention to protect it as a source of municipal water supply.

The Madison formation, which could be a high yield, fresh water aquifer, lies below the deepest producing geologic unit within NPR-3 at a depth of below 6,000 feet (1,800 m) but yields water of only fair quality, with a TDS level of approximately 3000 mg/L. (DOE, 1990) The Madison could be considered a USDW, but activities at NPR-3 are not likely to impact this aquifer.

Although not suitable as drinking water, water from the Madison and Tensleep formations (at a depth approximately 5400 feet or 1,600 m from the surface) is utilized to supply make-up water for existing steamflooding and waterflooding EOR activities at NPR-3. (Fosdick, 1992b)

3.3.3 Surface Water Quality

The effluent limits from each National Pollutant Discharge Elimination System (NPDES) permit under which water is discharged to the draws at NPR-3 are listed in Table 3-2. The DOE submits semi-annual Discharge Monitoring Reports to the WYDEQ. Samples are taken bimonthly to monitor discharge water quality. (DOE, 1990; Dunn, 1993)

Water is discharged in large quantities only from the Tensleep Battery (B-TP-10) (NPDES Permit WY-0028274). The other NPDES permits listed in Table 3-2 are either inactive, represent highly occasional discharges, or represent discharges of very small quantities of effluent. Water discharged from the Tensleep Battery is formation water produced with the Tensleep oil. Although the natural temperature of water at the time of withdrawal from Tensleep formation is 180°F (82°C), temperatures of the effluent are typically under 100°F (38°C) (Doyle, 1993). Because the streams are generally less than 1 foot (0.3 m) deep, the elevated temperatures at the point-of-discharge rapidly diminish to ambient levels through atmospheric cooling.

The WYDEQ has determined that the streams at NPR-3 are all Category IV streams (Doyle, 1993). Category IV streams are defined in the Wyoming Water Standards as "surface waters, other than those classified as Class I, which are determined by the Wyoming Game and Fish Department not to have the hydrologic or natural water quality potential to support fish". Thermal effluent limits are not established by the WYDEQ for NPDES Permits for discharges to Class IV streams.

3.3.4 Ground Water Quality

Groundwater produced with crude oil and natural gas is disposed underground by injection into the

Crow Mountain formation. The water treatment plant softener regeneration water is also injected into a disposal well. These wells are permitted through EPA's Underground Injection Control (UIC) program, which is managed by the Wyoming Oil and Gas Conservation Commission. Geologic formations that receive injected water also have an aquifer exemption authorized by the Oil and Gas Conservation Commission, which has primacy for regulating class II injection wells under the Safe Drinking Water Act.

3.3.5 Potable Water

Because there are no potable water wells in the vicinity of NPR-3, all potable water must be trucked to NPR-3 from either the city of Casper or the town of Midwest. Both supplies are community water systems and have been approved by the EPA as drinking water systems. Drinking water samples are taken quarterly at NPR-3 to monitor for coliform and confluent bacteria. Samples are analyzed by the Natrona County Health Department. A copy of the analytical results is retained by the Contractor's Environmental Department and a copy is sent by the Natrona County Health Department to the EPA Region VIII (DOE, 1990). Sampling is also conducted for lead and copper levels as required by the Lead and Copper Rule.

Table 3-2 Summary of NPDES Permit Limits

Permit Number	Name of Source	Oil and Grease[1]	Specific Conductance[2]	COD[3]
WY-0028274	B-TP-10 Tank Battery	10	N/A	N/A
WY-0034126	North Waterflood Floor Drains	10	7500	100
WY-0031895	North Waterflood	10	N/A	N/A
WY-0028894	Tank Battery B-1-3	10	N/A	N/A
WY-0028908	Tank Battery B-1-10	10	N/A	N/A
WY-0028932	Tank Battery B-2-10	10	N/A	N/A
WY-0028916	Tank Battery B-1-28	10	N/A	N/A
WY-0028924	Tank Battery B-1-33	10	N/A	N/A
WY-0034037	Water Treatment Facility	10	7500	100
WY-0032115	Water Disposal Facility	10	N/A	N/A
WY-0034029	Steam Generator 2	10	7500	100
WY-0034495	Steam Generator 3	10	7500	100
WY-0035076	Steam Generator 4	10	7500	100

WY-0035297	Steam Generator 5[4]	10	7500	100

[1] In mg/l, daily maximum

[2] In umhos/cm, daily maximum

[3] In mg/l, daily maximum

[4] Additional limits are set for pH (minimum of 6.5 and maximum of 8.5) and flow in conduit or through treatment plant (30 mgd, as daily maximum or 30-day average)

3.4 Geology, Soils, and Prime and Unique Farmlands

3.4.1 Geology

NPR-3 is centered over the crestal axis of an asymmetrical doubly-plunging anticline called the Teapot Dome, which is the southern extension of the much larger Salt Creek anticline. The Salt Creek anticline underlies the prolific Salt Creek Oilfield, located to the north of NPR-3. (DOE, 1990)

The geologic column for the Teapot Dome is shown in Figure 3-3. The oil productive horizons are the Shannon, Steele Shale, Niobrara Shale, Second Wall Creek, Third Wall Creek, Muddy, Dakota, Lakota, and Tensleep formations. Formations currently undergoing enhanced oil recovery (EOR) operations include the Shannon and Second Wall Creek sands and the Muddy formation.

The topography of the region surrounding NPR-3 is characterized by rolling plains interspersed with ridges and isolated bluffs. The central part of NPR-3 consists of a large plain, dissected by ravines (draws), that is encircled to the east, west, and south by a rim of sandstone (U.S. Navy, 1976). The area surrounding NPR-3 is not known to be seismically active (Halliburton NUS, 1993).

3.4.2 Soils

The USDA Soil Conservation Service (SCS) has completed a Class III soil survey of portions of Natrona County, including NPR-3 and surrounding lands. Soil survey mapping units covering NPR-3 are outlined in Figure 3-4. Map pages from the soil survey covering NPR-3 are provided in Table 3-3. Soils throughout NPR-3 are largely derived from sodic (alkaline) parent materials and are highly alkaline and saline. The high salinity of soils on NPR-3 is limiting to plant growth. All soils on NPR-3 are well drained. Most soils on NPR-3 are highly or moderately susceptible to erosion caused by heavy downpours (Davis, 1993a).

Most upland soils throughout all parts of NPR-3 other than the peripheral ridges are mapped as Cadoma-Renohill-Samday clay loams. The Cadoma soil series is typically found on hillsides of 3 to 12 percent slope, the Renohill soil series is typically found in swales of 3 to 6 percent slope, and the Samday soil series is typically found on ridges of 3 to 12 percent slopes. These soils are derived from slopewash alluvium and residuum derived dominantly from sodic (alkaline) shale. The Cadoma and Renohill soils are moderately deep and well drained, while the Samday soils are shallow and well drained. All of these soils are highly susceptible to water erosion. (Davis, 1993a)

Scattered areas of upland soils are mapped under other names and comprise soils mapped in other soil series. Most of these other upland soils are also derived from sodic (alkaline) materials. All are well drained but differ widely in their susceptibility to water erosion (Davis, 1993a). Soils in the major draws on NPR-3 are mapped in the Haverdad-Clarkelen complex, a mosaic of soils in the Haverdad series (Haverdad loam) and the Clarkelen series (Clarkelen sandy loam). The Haverdad and Clarkelen soils are very deep and well drained, and they are only slightly susceptible to water erosion. (Davis, 1993a)

Table 3-3 Soil Survey Mapping Units

Map Unit 112: Arvada-Absted-Slickspots complex, 0 to 6 percent slopes
Location on NPR-3: Scattered upland areas throughout all parts of the reserve except for the bluffs.
Composition: 35% Arvada clay loam; 30% Absted clay loam; and 15% Slickspots.
Origin: Alluvium derived dominantly from sodic shale (Arvada and Absted soils).
Drainage: Well drained (Arvada and Absted soils).
Hazard of Water Erosion: Slight (Arvada and Absted).
Capability Subclass: VIs (Arvada and Absted soils)
Map Unit 113: Arvada, runon-Slickspots complex, 0 to 3 percent slopes
Location on NPR-3: Isolated upland area in the northern part of the reserve.
Composition: 60% Arvada loam, overflow and 25% Slickspots.
Origin: Alluvium derived dominantly from sodic shale (Arvada soil).
Drainage: Well drained (Arvada soil).
Hazard of Water Erosion: Slight (Arvada soil).
Capability Subclass: VIs (Arvada soil).

Map Unit 125: Blackdraw-Lolite-Gullied land complex, 3 to 20 percent slopes

Location on NPR-3: Scattered upland areas in the northern part of the reserve.

Composition: 45% Blackdraw clay loam; 20% Lolite clay loam; and 20% gullied land.

Origin: Slopewash alluvium and residuum derived dominantly from noncalcareous sodic shale (Blackdraw soil); residuum derived dominantly from noncalcareous sodic shale (Lolite soil).

Drainage: Well drained (Blackdraw and Lolite soils).

Hazard of Water Erosion: Severe (Blackdraw and Lolite soils)

Capability Subclass: VIe (Blackdraw soil); VIIe (Lolite soil).

Map Unit 134: Bowbac-Taluce-Terro complex, 6 to 20 percent slopes

Location on NPR-3: Scattered upland areas in the northern part of the reserve.

Composition: 40% Bowbac sandy loam; 25% Taluce sandy loam; and 15% Terro fine sandy loam.

Origin: Slopewash alluvium and residuum derived dominantly from sandstone (Bowbac soil); residuum derived dominantly from sandstone (Taluce soil); alluvium derived dominantly from sandstone (Terro soil).

Drainage: Well drained.

Hazard of Water Erosion: Moderate (Bowbac and Terro soils); High (Taluce soil)

Capability Subclass: IVe (Bowbac and Terro soils); VIIe (Taluce soil).

Map Unit 140: Cadoma-Renohill-Samday clay loams, 3 to 12 percent slopes

Location on NPR-3: Characteristic soil on the uplands throughout all parts of the reserve except for the bluffs.

Composition: 40% Cadoma clay loam; 25% Renohill clay loam; and 25% Samday clay loam.

Origin: Slopewash alluvium and residuum derived dominantly from sodic shale (Cadoma and Renohill soils).

Drainage: Well drained.

Hazard of Water Erosion: Severe.

Capability Subclass: VIe (Cadoma soil); IVe (Renohill soil); VIIe (Samday soil).

Map Unit 195: Haverdad-Clarkelen complex, saline, 0 to 3 percent slopes

Location on NPR-3: Characteristic soil within the larger draws throughout all parts of the reserve.

Composition: 50% Haverdad loam, saline and 35% Clarkelen sandy loam, saline

Origin: Stratified alluvium from mixed sources.

Drainage: Well drained.

Hazard of Water Erosion: Slight.

Capability Subclass: IVs - irrigated; VIs - nonirrigated.

Map Unit 208: Kayner sandy clay loam, 3 to 10 percent slopes

Location on NPR-3: Characteristic soil on the high ground at the foot of the bluffs near the eastern, western, and southern boundaries.

Composition: Over 80% of this map unit is Kayner sandy clay loam.

Origin: Alluvium derived dominantly from sodic sandstone and shale.

Drainage: Well drained.

Hazard of Water Erosion: Moderate.

Capability Subclass: VIe.

Map Unit 209: Keyner-Absted-Slickspots complex, 0 to 6 percent slopes

Location on NPR-3: Small, isolated area of uplands near the western boundary.

Composition: 50% Keyner sandy loam; 20% Absted sandy clay loam; and 15% slickspots.

Origin: Alkaline alluvium derived from mixed sources (Keyner soil); alluvium derived dominantly from sodic shale (Absted soil).

Drainage: Well drained.

Hazard of Water Erosion: Slight (Keyner and Absted soils).

Capability Subclass: No information.

Map Unit 214: Lolite-Rock outcrop complex, 10 to 40 percent slopes

Location on NPR-3: Small, scattered areas of uplands in the northern part of the reserve.

Composition: 60% Lolite clay and 20% Rock outcrop.

Origin: Residuum derived dominantly from sodic shale (Lolite soil).

Drainage: Well Drained.

Hazard of Water Erosion: Severe (Lolite soil).

Capability Subclass: VIIe.

Map Unit 215: Lolite, dry-Rock outcrop, 5 to 50 percent slopes

Location on NPR-3: Isolated area of uplands near the interior of NPR-3.

Composition: 50% Lolite clay, dry and 30% Rock outcrop.

Origin: Residuum derived dominantly from noncalcareous, sodic shale (Lolite soil).

Drainage: Well drained (Lolite soil).

Hazard of Water Erosion: High (Lolite soil).

Capability subclass: VIIe (Lolite soil).

Map Unit 256: Rock outcrop-Ustic torriorthents, shallow-Rubble land complex, 30 to 100 percent slopes

Location on NPR-3: Characteristic soil on the bluffs near the eastern, western, and southern boundaries.

Composition: 40% Rock outcrop; 25% Ustic torriorthents, shallow; and 15% Rubble land

Drainage: Well to excessively well drained (Ustic torriorthents).

Hazard of Water Erosion: Moderate to severe. (Ustic torriorthents)

Capability Subclass: VIII.

Map Unit 278: Silhouette-Petrie clay loams, 1 to 6 percent slopes

Location on NPR-3: Small upland area in northwestern corner.

Composition: 50% Silhouette clay loam and 30% Petrie clay loam

Origin: Alluvium derived dominantly from shale (Silhouette soil); alluvium derived dominantly from sodic shale (Petrie soil).

Drainage: Well drained.

Hazard of Water Erosion: Moderate.

Capability Subclass: VIII.

Map Unit 283: Theedle-Shingle-Kishona complex, 6 to 40 percent slopes, gullied

Location on NPR-3: Small area on extreme west-central periphery

Composition: 30% Theedle clay loam, 25% Single loam, and 20% Kishona clay loam

Origin: Slopewash alluvium and residuum derived dominantly from sedimentary rocks

Drainage: Well drained.

Hazard of Water Erosion: High (Theedle and Single soils); Moderate (Kishona soil)

Capability Subclass: VIe (Theedle and Kishona soils); VIIe (Shingle soil)

Higher elevation lands approaching the peripheral ridges are mapped as Keyner sandy clay loam. These soils are deep and well drained. The hazard of water erosion is moderate. Soils on and immediately at the base of the bluffs are mapped in the Rock outcrop-Ustic Torriorthents, shallow-Rubble land complex. These areas are characterized by exposed rock, colluvial boulders, and shallow soil. (Davis, 1993a)

3.4.3 Prime and Unique Farmlands

The SCS does not presently recognize any prime or unique farmlands or farmlands of local importance within the boundaries of NPR-3 (Davis, 1993b). All soils on NPR-3 are mapped in Capability Classes IV or higher, and the majority are mapped in Capability Classes VI and higher (Davis, 1993a). The SCS defines Class IV soils as soils that have very severe limitations that reduce the choice of plants or that require very careful management, or both. The SCS defines Class VI soils as soils having severe limitations that make them unsuitable for cultivation. In general, soils in the highest numbered Capability Classes are less suitable for cultivation than soils in the lowest numbered Capability Classes.

3.5 Biological Resources

3.5.1 Aquatic Biology

Aquatic habitats at NPR-3 are limited to intermittent streams within the draws, shallow perennial streams fed primarily by produced water discharged under NPDES permits, and man-made ponds. Fish have not previously been reported in the draws on NPR-3 (DOE, 1990). The Wyoming Game and Fish Department (WGFD) stocked fingerling (5 to 6 inch/14 cm) rainbow trout in two of the abandoned impoundments at NPR-3 between 1987 and 1989. Water in one of the impoundments comprises run-off from snow melt and rain, and water in the other comprises produced water originating from the Madison formation on an adjoining privately owned oilfield. One year later, the trout in the second pond had grown to 11-14 inches (28-36 cm) in length, while the first pond dried up. The following year, they had reached a length of approximately 18 inches (46 cm) (DOE, 1990).

A fish survey of the surface waters on NPR-3 has not been conducted. NPR-3 lies within the geographic range of approximately 17 fish species. Although only a few of these species (such as creek chub or killifish) would be expected in streams onsite, NPR-3 is within the watershed of the Powder River, which may contain most of these species (Page and Burr, 1991).

3.5.2 Terrestrial Vegetation

NPR-3 is located in part of North America where vegetation is characterized by shortgrass prairie. The last vegetation survey of NPR-3, performed prior to intensive development of the Reserve by the DOE in 1978, identified six major vegetation associations (Figure 3-5). These include three rangeland associations on the upland plains, two riparian associations in the bottoms of the draws, and a pine-juniper association on the peripheral ridges. (U.S. Navy, 1976)

Much of the rangeland vegetation has been physically disturbed by construction of wells, drill pads, access roads, and other DOE activity since 1978. Disturbance is generally continuous throughout certain areas of intensive activity in the center of the Reserve east of the office and warehouse complexes. Disturbance elsewhere is generally localized around scattered wells and other work areas. The pine-juniper vegetation on the peripheral ridges has not generally been disturbed by DOE operations since 1978. Except at a few road crossings, riparian vegetation in the draws has not generally been physically disturbed by DOE operations. However, riparian vegetation downstream of NPDES-permitted points of discharge has experienced increased water flows and increased water temperatures. (Halliburton NUS, 1993)

The DOE reclaims and reseeds drill pads, flowline rights-of-way, and abandoned well sites on NPR-3, using guidelines provided by the SCS (SCS, 1992). The reseeded areas provide browse for the larger mammals, habitat for smaller animals, and reduce water and wind erosion.

The DOE does not presently lease any of the rangeland within NPR-3 for grazing, although this would be part of the Proposed Action. The last grazing lease terminated in 1986 (Doyle, 1993). Prior to that time, rangeland within NPR-3 was overgrazed (Young, 1986; Watson, 1987). Between 1981 and 1986, grazing on NPR-3 exceeded 2,000 animal unit-months (AUM), whereas the Soil Conservation Service had recommended in 1965 that grazing on NPR-3 not exceed 1,185 AUM (Watson, 1987).

Trees at NPR-3 are largely limited to piZon pine, ponderosa pine, and juniper within small zones of pine-juniper forests on the peripheral ridges, and to a few cottonwood trees among the riparian vegetation in the draws (DOE, 1990). Except for the peripheral ridges, uplands throughout NPR-3 lack trees. No land on NPR-3 is managed for timber production (Doyle, 1993).

During the summer of 1987, and spring of 1988, a pilot project was initiated to introduce narrow leaf cottonwood (*Populus angustifolia*) and Russian olive (*Eleagnus angustifolia*) trees to NPR-3. Both species are hardy and were expected to adapt to the dry summers and cold winters. Four hundred and fifty cottonwood trees, Russian olive trees, and willow (*Salix* sp.) shrubs were planted along streams and ponds on the Reserve. Due to drought conditions that occurred during these years and damage done by wildlife, few of the trees survived (DOE, 1990). This project may be tried again, but using indigenous species to increase the probability of success.

3.5.3 Terrestrial Wildlife

The Wyoming Game and Fish Department (WGFD) maintains a database (Wildlife Observation System) of wildlife sightings throughout the state by township, range, and section. A list of species recorded in the database for those townships and ranges in the immediate vicinity of NPR-3 is provided in Table 3-4. This list also includes several other species which have been observed over the years on NPR-3 by the DOE staff and its contractors (US Navy, 1976; Stark, 1993). This does not represent a systematic inventory of terrestrial wildlife known to occur on NPR-3. According to a bird and mammal distributive study for Wyoming, approximately 222 bird species and 49 mammal species have been observed in the region containing the NPR-3 site (WGFD, 1991). NPR-3 lies within the geographic range with at least 6 amphibians and 9 reptile species (Stebbins,1985). Table 3-4 indicates recorded observations of 3 amphibian, 4 reptile, 61 bird, and 20 mammal species at NPR-3.

Pronghorn antelope and mule deer are the principal big game mammals seen at NPR-3 (DOE, 1990). The DOE does not presently allow any hunting on NPR-3 (Doyle, 1993). NPR-3 does not contain any Critical Winter Range for either antelope or deer. Range within NPR-3 is classified by the WGFD as Winter Year-Long Range for both species. The range is utilized by both species throughout the year but is not depended upon during the winter by transient deer or antelope populations that reside elsewhere during the growing season (Thiele, 1993).

Other characteristic mammal species of NPR-3 include: raccoons, striped skunk, porcupine, badger, fox, bobcat, prairie dog (three known colonies), cotton-tail rabbit, and deer mouse. Apparently common species among the variety of birds found at NPR-3 are the red-tailed hawk, American kestrel, golden eagle, horned lark, western meadowlark, Brewer's blackbird, vesper sparrow, Brewer's sparrow, lark bunting, and sage thrasher. Characteristic amphibians and reptiles found on NPR-3 include: toad species, sagebrush lizard, short-horned lizard, garter snake, and western rattlesnake (DOE, 1990; WGFD, 1991; WGFD, 1993).

3.5.4 Threatened and Endangered Species

The offices of the U.S. Fish and Wildlife Service (FWS) and the WGFD, both in Cheyenne, Wyoming, and the Nature Conservancy in Laramie, Wyoming, were consulted to determine which federally and/or state listed threatened, endangered, or candidate species or critical habitats could potentially occur at NPR-3.

In a letter dated January 14, 1993, (attached) the FWS indicated that several of the species shown in Table 3-5 could be present in the area of NPR-3. According to the FWS, the black-footed ferret (Federally-listed endangered) could inhabit prairie dog towns in the vicinity of NPR-3 (Davis, C. P., 1993). Three prairie dog colonies, each less than 100 acres (40 ha) in area, are known to occur near the eastern and southern boundaries of NPR-3 on rangeland that is undisturbed by present oil drilling operations (Stark, 1993). Two of these colonies are large enough to potentially support the black-footed ferret. No evidence of the black-footed ferret was found during an earlier survey conducted in 1986 (DOE, 1990).

The FWS also indicated that the bald eagle (Federally-listed endangered) could be a winter resident or a migrant to the area of NPR-3 and that the peregrine falcon (Federally-listed endangered) could be a migrant to the area (Davis, C. P., 1993). An adult bald eagle has been observed perched on the bluffs immediately west of the administration building on NPR-3 (Soehn, 1993). There are no known bald eagle or peregrine falcon nests in the vicinity of NPR-3. The closest known bald eagle nests to NPR-3 are on the Platte River east of Glenrock and in Ednes Kimball Wilkens Park in Casper (Thiele, 1993).

The FWS identified several federal candidate species which potentially occur in the vicinity of NPR-3 (Table 3-5). The FWS is especially interested in the narrow-foot hygrotus diving beetle, which is currently known only from Dugout, Cloud, and Dead Horse Creeks, all intermittent streams in draws within a 25-mile (40-km) radius of NPR-3. (Davis, C. P., 1993; Leech, 1966)

The loggerhead shrike (Category 2) has been observed at NPR-3 and is a known breeder in the region. The ferruginous hawk (Category 2) is also a known breeder and year-round resident to the region. Suitable habitat exists at NPR-3, but there are no documented occurrences. The white faced ibis and black tern (both Category 2) have been observed within the region, but there is very little suitable habitat at NPR-3 to attract these species. The mountain plover (Category 1) has also been observed in the region but it is not known to breed in the region. Although suitable habitat exists NPR-3, this species has not been observed. (WGFD, 1992)

There are no known threatened, endangered or other special status fish species known to occur at NPR-3. The Powder River provides important habitat for the sturgeon chub (Category 2) and the shovelnose sturgeon, both considered to be "Sensitive Species" in Wyoming. (Collins, 1993)

The Nature Conservancy maintains the Wyoming Natural Diversity Data Base (WNDDB), a data base of species sightings recorded by township, range, and section. The WNDDB has no records of threatened or endangered species within Townships 37 - 40N or Ranges 77 - 79W; which constitute the area within and immediately surrounding the NPR-3 site (Neighbours, 1993). The WNDDB does contain two records of a plant species, Barr's Milkvetch (Category 2), in the area surrounding NPR-3. However, this species has been recommended for downlisting to Category 3 (not appropriate for listing as threatened or endangered) because it has been found to be more common than originally believed. The Barr's Milkvetch generally grows where vegetative cover is sparse, and is thought to prefer a whitish, sandy-silty soil that may be calcareous. (Neighbors, 1993).

Table 3-4 List of Species Recorded in the Vicinity of NPR-3

Common Name	Scientific Name
AMPHIBIANS	
Boreal chorus frog	*Pseudacris triseriata malculata*
Tiger salamander[b]	*Ambystomia tigrinum*
Toad sp.[c]	*Bufo sp.*
REPTILES	
Sagebrush lizard[b]	*Sceloporus graciosus*
Short-horned lizard[b]	*Phrynosoma douglassi*
Western terrestrial garter snake[bc]	*Thamnophis elegans*
Western rattlesnake[bc]	*Crotalis viridis*
FISH	
Minnow sp.[c]	*Undetermined species*
BIRDS	
American robin[a]	*Turdus migratorius*
American kestrel[ab]	*Falco sparverius*
American wigeon[ab]	*Anas americana*
American avocet[ac]	*Recurvirostra americana*
Bald eagle[ac]	*Haliaeetus leucocephalus*
Black-billed magpie[abc]	*Pica pica*
Blue-winged teal[ab]	*Anas discors*
Brewer's blackbird[a]	*Euphagus cyanocephalus*
Brewer's sparrow[ab]	*Spizella breweri*
Chukar[a]	*Alectoris chukar*
Cliff swallow[a]	*Hirundo pyrrhonota*
Common poorwill[ab]	*Phalaenoptilus nuttaillii*
Common nighthawk[a]	*Chordeiles minor*
Common snipe[b]	*Capella gallinago*
Double-crested cormorant[c]	*Phalacrocorax auritus*
European starling[a]	*Sturnus vulgaris*
Gadwall[ab]	*Anas strepera*

Golden eagle[abc]	*Aquila chrysaetos*
Great horned owl[ab]	*Bubo virginianus*
Green-winged teal[ab]	*Anas crecca*
Horned lark[ab]	*Eremophila alpestris*
House wren[ab]	*Troglodytes aedon*
Killdeer[ab]	*Charadrius vociferus*
Lark bunting[a]	*Calamospiza melanocorys*
Lark sparrow[b]	*Chondestes grammacus*
Lesser yellowlegs[b]	*Tringa flavipes*
Loggerhead shrike[abc]	*Lanius ludovicianus*
Mallard[bc]	*Anas platyrhyndios*
McCown's longspur[a]	*Calcarius mccownii*
Mountain bluebird[ab]	*Sialia currucoides*
Mourning dove[ab]	*Zenaidura macroura*
Northern shrike[a]	*Lanius excubitor*
Northern (red-shafted) flicker[a]	*Colaptes (cafer) auratus*
Northern (yellow-shafted) flicker[ab]	*Colaptes auratus*
Northern harrier[ab]	*Circus cyaneus*
Northern rough-winged swallow[b]	*Stelgidopteryx serripennis*
Pectoral sandpiper[b]	*Calidris melanotos*
Pintail[b]	*Anas acuta*
Pinyon jay[b]	*Gymnorhinus cyanocephalus*
Plover sp.[c]	*Charadrius sp.*
Prairie falcon[ab]	*Falco mexicanus*
Red-tailed hawk[abc]	*Buteo jamaicensis*
Red-winged blackbird[ab]	*Agelaius phoeniceus*
Rock wren[ab]	*Salpinctes obsoletus*
Rough-legged hawk[c]	*Buteo lagopus*
Sage sparrow[ab]	*Amphispiza belli*
Sage grouse[ab]	*Centrocercus urophasianns*
Sage thrasher[ab]	*Oreoscoptes montanus*
Say's phoebe[ab]	*Sayornis saya*
Sharp-shinned hawk[a]	*Accipiter striatus*
Short-eared owl[a]	*Asio flammeus*

Spotted sandpiper[a]	*Actitis macularia*
Swainson's hawk[ab]	*Buteo swainsoni*
Turkey vulture[a]	*Cathartes aura*
Vesper sparrow[ab]	*Pooecetes gramineus*
Violet-green swallow[b]	*Tachycineta thalassina*
Western grebe[c]	*Aechmophorus occidentalis*
Western meadowlark[abc]	*Sturnella neglecta*
Western kingbird[ab]	*Tyrannus verticalis*
White-throated swift[a]	*Aeronautes saxatalis*
Wilson's phalarope[a]	*Phalaropus tricolor*

MAMMALS

Black-tailed prairie dog[a]	*Cynomys ludovicanus*
Bobcat[ac]	*Lynx rufus*
Brush-tailed woodrat[b]	*Neotoma cinerea*
Coyote[bc]	*Canus latrans*
Deer mouse[b]	*Peromyscus maniculatus*
Desert cottontail[b]	*Sylvilagus auduboni*
Eastern cottontail[a]	*Sylvilagus floridanus*
Least chipmunk[b]	*Eutamias minimus*
Mountain lion[a]	*Felis concolor*
Mountain cottontail[a]	*Sylvilagus nuttallii*
Mule deer[ac]	*Odocoileus hemionus*
Muskrat[c]	*Ondatra zibethica*
Northern pocket gopher[b]	*Thomomys talpoides*
Porcupine[bc]	*Erethizon dorsatum*
Pronghorn[ac]	*Antilocapra americana*
Raccoon[c]	*Procyon lotor*
Red fox[ac]	*Vulpes vulpes*
Striped skunk[bc]	*Mephitis mephitis*
Swift fox[b]	*Vulpes velox*
Wyoming pocket mouse[b]	*Perognathus fasciatus*

Source: WGFD, 1993; US Navy, 1976; Stark, 1993; Soehn, 1993.

[a] Species observed within Township T 38-39N, Range R78W (on or in the vicinity of NPR-3).

[b] Species observed during survey of NPR-3, August 1975 (US Navy, 1976).
[c] Species observed by FD staff.

Table 3-5 Threatened, Endangered or Other Special Status Species Potentially in the Vicinity of NPR-3

Common Name	Scientific Name	Status[a] Federal	State	Habitat/Location
Plants Barr's milkvetch	Astragalus barrii	C2	NL	Whitish sandy silty soil with sparse vegetative cover.
Invertebrate Narrow-foot hygrotus diving beetle	Hygrotus diversipes	C2	NL	Natrona County, near Midwest, Wyoming
Fish Sturgeon chub Shovelnose sturgeon	Hybopsis gelida Scaphirhynchus platorynchus	C2 NL	S S	Powder River drainage Powder River drainage
Birds White-faced ibis Bald eagle[b] Peregrin falcon Ferruginous hawk Mountain plover Black tern Loggerhead shrike[b]	Plegadis chihi Haliaeetus leucocephalus Falco peregrinus Buteo regalis Charadrins montanus Chlidonias niger Lanius ludovicianus	C2 E E C2 C1 C2 C2	PI E E NL NL PII NL	Wetland Winter resident/migrant Migrant Grassland Grassland Wetland Woodland/shrubland

Mammal				
Black-footed ferret	Mustela nigripes	E	E	Potential resident in prairie dog colonies
Plains (eastern) spotted skunk	Spilogale putorius interrupta	C2	NL	East of Bighorn and Laramie Mts.

Sources: Collins, 1993; Davis C. P., 1993; Neighbors, 1993; Soehn, 1993.

ª Status Code: C1 = Federal candidate - Category 1 (appropriate for listing)
C2 = Federal candidate - Category 2 (possibly appropriate for listing)
E = Endangered
NL = Not Listed
PI = Priority I - Species in need of immediate attention
PII = Wyoming Priority II - species in need of additional study
S = Sensitive Species in Wyoming
ᵇ Species observed on NPR-3.

3.5.5 Floodplains and Wetlands

Although Flood Insurance Rate Maps (FIRM's) are available for certain parts of Natrona County, none have been prepared for the area around NPR-3 (Keller, 1993a). The FWS prepared National Wetland Inventory (NWI) Maps for the area surrounding NPR-3 in February 1993, which document the many impoundments and reservoirs within NPR-3. Some portions of the major stream beds are also classified as wetlands.

The topography of NPR-3, characterized by gently rolling uplands punctuated by narrow draws with steep embankments, suggests that floodplains are limited to lands within the embankments of the draws (Figure 3-6). It is likely that the areal extent of floodplains on NPR-3 roughly corresponds to Map Unit 195 in the soil survey in Table 3-3. The low permeability of the sodic soils which predominate in much of the watershed of the draws (Davis, 1993a) suggests that brief but very intense floods could occur following infrequent downpours.

Wetlands and other areas at NPR-3 that are regulated under Section 404 of the Clean Water Act appear to be limited to man-made ponds, stream channels, and to certain areas within the embankments of the draws. The basins of several small impoundments constructed in the larger draws on NPR-3 during the 1920's to create reservoirs to support early oil drilling efforts (Doyle, 1993) are likely to be wetlands. No soils on the list of hydric soils compiled by the SCS for Natrona County (Davis, 1993c) or Hydric Soils of the United States (NTCHS, 1991) appear on the soil survey for areas at NPR-3 outside of the draws.

The channels of perennial and intermittent streams within the draws are regulated under Section 404 of the Clean Water Act, even if they lack vegetation and therefore do not technically meet the definition of wetlands. Available information suggests that some portions of the draw bottoms are wetlands, although further study would be required to determine exactly how much. Draw bottom areas are mapped in Figure 3-5. Areas with the Flowing and Impounded (Wet) Riparian Vegetation Association, which is dominated by sedges (*Carex* sp. and *Cyperus* sp.), rushes (*Juncus* sp.), and cattails (*Typha* sp.), were likely to have met the definition of wetlands at the time that the figure was generated. Areas mapped with the Upland (Dry) Riparian Vegetation Association, which is characterized by thistle (*Cirsium flodmanii*), yarrow (Achillea *lanulosa*), goldenrod (*Solidago* sp.) and occasional grasses and grass-like species, were likely not to have met the definition of wetlands (US Navy, 1976). The distribution of riparian vegetation may have changed since 1976 in draw bottoms downstream of NPDES-permitted points of discharge.

The partial extent of wetlands within the draw bottoms is also supported by soil survey data. The soil survey mapping unit which encompasses the draw bottoms (Figure 3-3) is primarily comprised of soils in the Haverdad and Clarkelen soil series, which are not listed as hydric by the National Technical Committee for Hydric Soils (NTCHS, 1991). However, the SCS notes that inclusions of other soil series which are hydric are known to occur within Map Unit 195. (Davis, 1993c)

The FWS has developed a system to classify wetlands and other waters of the United States (Cowardin, 1979). The man-made ponds discussed in Section 3.5.1 could be classified as Palustrine Open Water (POW) wetlands. The intermittent stream channels could be classified by the FWS as Riverine Intermittent Streambeds (R4SB). The perennial stream channels could be classified as Riverine, Upper Perennial Streambeds (R3SB). Areas within the draw bottoms but outside of the channels could be classified as Palustrine Emergent (PEM) or Palustrine Scrub-Shrub (PSS) Wetlands.

3.6 Cultural Resources

Shoshoni and Sioux tribes lived on the Wyoming Plains until the 1840's, when westward movement brought settlers on their way to Oregon via the Oregon Trail. The Oregon Trail followed a portion of the North Platte River Valley through Fort Laramie, Fort Caspar, and Fort Bridger. The land on which NPR-3 is located was used as hunting grounds by Native American tribes in the area. (Halliburton NUS, 1993)

Surveys of NPR-3 which were conducted in 1976 were unable to identify specific tribal groups which may have used the property. Six areas were identified as having a concentration of flakes and/or artifacts. Only one of these areas was recommended for additional survey work in 1976, and the remaining areas were determined to be of no importance. The one area identified for additional work is located in the southeast part of NPR-3. This area was classified as lithic, ceramic scatter, with possible rock shelters. The area contained a large number of scattered tools and ceramic shards, suggesting that the area could have been occupied on a seasonal basis. All of the artifacts collected during the survey were estimated to date back to AD 400. (U.S. Navy, 1976)

During the comment period for this EA, the Wyoming State Historic Preservation Office (SHPO) requested that additional surveys be done to locate cultural resources at NPR-3. The resulting Class III cultural resource inventory was completed in June 1995. The inventory identified 17 prehistoric sites, 13 isolated artifacts, and one historic site. Two of the 17 prehistoric sites are recommended for additional survey work and are considered eligible for listing on the National Register of Historic Places. Both of these sites contain hearth and rock shelter features which could provide additional information.

Petroleum development has shaped the history of NPR-3 and its immediate surroundings since the turn of the century. NPR-3 was established in 1915 in the wake of a national emphasis toward mineral resource conservation. Public versus private use of petroleum resources on these lands was a hotly contested political issue in the early 1900's, culminating in the "Teapot Dome Scandal" of 1924 (US Navy, 1976). Oil production at NPR-3 was discontinued in 1927 and did not resume again until 1959.

From 1959 until 1976, oil production operations were established at NPR-3 in order to prevent the loss of oil to adjacent lands (Lawrence Allison, 1987; Halliburton NUS, 1993). In response to the oil shortages of the mid-1970's, President Carter authorized the development of NPR-3 to the maximum efficient rate (MER). Since that time, oil has been continuously pumped from NPR-3.

Teapot Dome Oil Field (Site 48NA831) has been determined to be eligible for inclusion in the National Register of Historic Places. This was confirmed by the 1995 inventory. In addition, three cultural resources sites located within NPR-3 (48NA2180, 48NA2181, and 48NA2182) were not evaluated for inclusion during the 1976 inventory. These sites were not relocated during the 1995 inventory.

Several other sites which are eligible for listing or are listed on the National Register are located close to NPR-3. These include: Casper Buffalo Trap, Casper (6/25/74); Fort Casper, Casper (8/12/71 and 7/19/76); Independence Rock, Casper (10/15/66); Martin's Cove, Casper (3/8/77); Midwest Oils Company Hotel, Casper (11/17/83); South Wolcott Street Historic District, Casper (11/23/88); Stone Ranch Stage Station, Casper (11/01/82), Teapot Rock, 6 miles SW of NPR-3 (12/30/74); and Townsend Hotel, Casper (12/25/83). (U.S. National Park Service, 1991)

3.7 Socioeconomics

3.7.1 Population and Employment

The socioeconomic study area is defined for the purposes of this EA as Natrona County (including the City of Casper and other incorporated municipalities). The estimated 1990 population of Natrona County was 61,226 (CAEDA, 1993). The estimated 1990 population of the City of Casper was 46,742, which accounted for 76.3 percent of the total population of Natrona County (CAEDA, 1993). Population in Natrona County dropped by 14.8 percent between 1980 and 1990, from 71,856 to 61,226 (CAEDA, 1993). This population loss reflected a statewide trend brought about by declining oil prices and subsequent decreasing employment in the early 1980's.

Population growth in the county is expected to occur at a slow but steady rate over the next five years, with the population projected to reach 64,926 in 1998. This is a projected 6 percent increase over the 1990 total population, but is still less than the peak 1980 population of 71,856 (State of Wyoming, 1992a). This growth rate is approximately the same as that projected for the entire state, which is also expected to grow by about 6 percent over the same period (State of Wyoming, 1992a). The majority of Natrona County's population growth is expected to occur in and around the City of Casper.

Total employment in Natrona County was 36,637 in 1990, an increase of 3.6 percent over 1989. Unemployment in Natrona County in 1990 was 5.8 percent, down from 6.9 percent in 1989, and slightly higher than the statewide average of 5.4 percent (CAEDA, 1993). The largest employment sectors in the county (for non-proprietary employees) are in services (32%), retail trade (23%), and government and government enterprises (17%), which together in 1990 employed 72 percent of all workers in the study area (State of Wyoming, 1992a). On a statewide level, these sectors accounted for about 62.3 percent of all jobs in 1990 (State of Wyoming, 1992b).

Per capita income in Natrona County was $13,446 in 1990, slightly higher than the statewide average of $12,008 (Rand McNally, 1992).

3.7.2 Housing

Natrona County has approximately 29,082 housing units, of which approximately 69 percent are owner-occupied and approximately 31 percent renter-occupied. Within the City of Casper, the ratio is 66 percent owner-occupied to 34 percent renter-occupied (Morris, 1993). Eighteen percent of all housing units in Natrona County were vacant in 1990, compared to 14.7 percent in Casper that same year (Morris, 1993). The median home value in Natrona County in 1990 was $53,100, approximately 16 percent lower than the median value of $61,600 for the state of Wyoming. For the renter-occupied housing units, the median rent in 1990 was $252, compared to the statewide average of $270 (Wyoming State Data Center, 1992). New construction in Natrona County (as indicated by the number of building permits issued) decreased by 43 percent between 1980 and 1990, from 1,343 to 764 (CAEDA, 1992).

3.7.3 Transportation

Interstate Highway 25 provides the major north-south access through much of Natrona County, and is located approximately 8 miles (13 km) west of the NPR-3 site. Interstate 25 is a four-lane interstate highway with a median and narrow shoulders. Wyoming Route 259 is a two-lane secondary road with no median and narrow shoulders, which runs in a general north-south direction, connecting Interstate 25 with Wyoming Route 387. The NPR-3 site is accessed by a gravel road which is entered from Route 259, approximately 5 miles (8 km) south of the town of Midwest.

In 1991, the estimated Vehicles Per Day/Average Daily Totals (VPD/ADT) for Interstate 25 at the north Casper city limit was 3,710 (both directions). The VPD/ADT for Interstate 25 at Ormsby Road was also 3,710, and the VPD/ADT for Interstate 25 at Wyoming Route 259 was 3,270 in 1991. Wyoming Route 259 had an estimated VPD/ADT of 1,490 in 1991 (Leek, 1993). VPD/ADT totals show the current level of service on these road segments to be well below their carrying capacity. Traffic conditions on these roads, therefore, could be characterized as free-flowing with no congestion (Leek, 1993).

A road construction project on approximately 10 miles (16 km) of Interstate 25 north of the Casper city limit was recently completed. No other major construction projects are anticipated for roads in the vicinity of NPR-3 (Leek, 1993).

Air transportation services in Natrona County are provided at the Natrona County International Airport in Casper. The airport offers both freight and passenger services. Private airstrips are likely to exist in the county, although information concerning their exact number and location is not available (Keller, 1993b).

Rail transportation services are provided by the Burlington Northern Railroad and the Chicago and Northwestern Railroad. Both railroads run in a northwest-southeast direction and are located approximately 35 miles (56 km) south of NPR-3. Both railroads provide freight service only (no passenger service) to the Casper area.

3.7.4 Community Services

Public education in Natrona County is provided by the Natrona County School District No. 1, which has jurisdiction over the entire county. Total enrollment during the current school year (1992-1993) is

12,975, and the total number of certified teachers is 904 (Cadwell, 1993; Catellier, 1993). The district operates a total of 39 schools, including 26 elementary schools, 3 high schools, 4 junior high schools, 1 correctional school, and 6 rural schools. Attendance in these schools is generally below capacity (Cadwell, 1993).

Health services in Natrona County are provided by the Wyoming Medical Center in Casper, which has a maximum capacity of 232 beds.

Police protection in Natrona County is provided by the Natrona County Sheriff's Office, which has one police station and approximately 70 sworn officers (Calder, 1993). The City of Casper also maintains a police force, consisting of one station and approximately 68 sworn officers (Honeycutt, 1993).

Fire protection services in the county are provided by the Natrona County Fire Department, which has one fire station and 13 full-time firefighters. Additional fire protection is provided by 6 volunteer fire departments, which are located throughout the county. Fire protection services for NPR-3 are provided by the Midwest and Edgerton volunteer fire departments, approximately 15-20 minutes away (Sullivan, 1993). The City of Casper Fire Department consists of 5 stations and 67 firefighters (Loomis, 1993).

The chief provider of electric service in Natrona County is the Pacific Power & Light Company. Gas service is provided by Northern Gas of Wyoming (CAEDA, 1992).

Municipal water for the city of Casper is derived from the North Platte River and local wells, and is treated locally by chlorination. Total capacity is 40 million gal (151,000 m^3)/day, with a storage capacity of 21.5 million gallons (81,400 m^3). Peak demand is 28 million gal (106,000 m^3)/day (CAEDA, 1992). The town of Midwest receives its potable water from Casper through an underground pipeline, and Edgerton has three main wells which supply the town with water (U.S. Navy, 1976).

The Casper sewage treatment system serves the Casper metropolitan area. The system consists of primary and secondary treatment, chlorination and chlorine removal. The current capacity is 12.8 million gal (48,400 m^3)/day and the current load is 7.5 million gal (28,400 m^3)/day. (Hill, 1993)

Residential garbage collection in the city of Casper is provided primarily by the City of Casper. Private hauling services are provided in Natrona County by BFI, as well as other smaller garbage haulers. The county has three landfills: in Casper, Alcova, and Midwest. (Dundas, 1993)

3.8 Waste Management

3.8.1 Hazardous Waste

The Resource Conservation and Recovery Act (RCRA) (42 USC 9601-9675 et. seq.) regulates the treatment, storage, and disposal of solid waste (both hazardous and non-hazardous). Much of the waste generated at the site is exempt under 40 CFR 261.4 (b)(5), which defines the following solid wastes as exempt from the designation of hazardous: "drilling fluids, produced waters, and other wastes associated with the exploration, development, or production of crude oil, natural gas, or geothermal energy". Crude oil, natural gas, and associated liquid petroleum gasses (LPG) are produced at NPR-3. (Lawrence Allison, 1987)

NPR-3 is listed as a conditionally exempt small quantity generator under RCRA. As such, NPR-3 could generate no more than 100 kg (220 lb) of hazardous waste per month and total on-site accumulation could not exceed 1,000 kg (2,205 lb) of hazardous waste, or 1 kg (2.2 lb) of acutely hazardous waste, at one time. During Calendar Year 1993, NPR-3 generated 88 kg of hazardous waste, which was mostly off-spec. PVC pipe cement. A Treatment, Storage and Disposal (TSD) permit is not currently required for NPR-3 under RCRA.

Drilling and production wastes at NPR-3 include oil, water, drilling mud, cuttings, well cement, produced waters, and sediments and sludges from produced water pits. Oil from wells is routed to test satellites and tank batteries, and water from the tank batteries is discharged into pits or injected into a UIC-permitted well. This water contains residual oil. Other RCRA-exempt wastes generated at NPR-3 include sediment and tank bottoms from pits and storage tanks, pigging wastes, soil contaminated with crude oil, and spent filters (DOE, 1992b).

In accordance with the Superfund Amendment Reauthorization Act (SARA) Title III, chemicals are evaluated to determine if any are listed as extremely hazardous substances, and if any of these are utilized at NPR-3 in reportable threshold planning quantities (TPQ). NPR-3 submits annual Tier II reports for items such as treating chemicals, hydrochloric acid, gasoline, diesel fuel, ethylene glycol, propane, and butane-gasoline mixture. The current maximum quantity of all chemicals stored at NPR-3 at any given time is 25,000 gallons (95 m^3) (DOE, 1990). Table 3-6 lists substances currently used at NPR-3 and the approximate annual usage.

There are three Underground Storage Tanks (USTs) at NPR-3: one 4,000 gallon (15.1 m^3) diesel tank, one 4,000 gallon (15.1 m^3) gasoline tank, and one 2,000 gallon (7.6 m^3) gasoline tank. Two other USTs were on-site: one 1,260 gallon (4.8 m^3) used oil tank and one 2,000 gallon (7.6 m^3) methanol tank, but these have since been removed. (Fosdick, 1990; FD Services, 1993)

The Comprehensive Environmental Response, Compensation, and Liability Act (CERCLA) (42 USC 9601-9675 et. seq.), establishes liability, compensation, clean-up, and emergency response by the Federal Government for hazardous substances released into the environment and for the clean-up of inactive hazardous waste disposal sites. A Phase I study of the site was completed in 1987 (Lawrence Allison, 1987). A Phase I study is designed to evaluate site history and records to locate and identify hazardous waste disposal sites. Historically, a variety of CERCLA-regulated substances have been used at NPR-3 (Table 3-7).

Other substances used in the past on NPR-3 include additives to drilling mud (crude oil, quebracho, phosphate), dehydrators (sulfonated oleic acid), aromatic solvents, emulsion breakers, polymers, oxyalkyl phenols, glycol, and isopropyl alcohol.

3.8.2 Pesticides

An inventory of pesticides conducted in 1990 identified the following substances at NPR-3: NalKil, Fenavar, and Ferti-Lome (Herbicides); Mouse-pruf (Rodenticide); and Bioguard (Insecticide) (DOE, 1990).

Until the summer of 1994, herbicide application was contracted to third parties. Company personnel began using the general-use herbicides Roundup, Banvil and Karmex for clearing parking lots, fence lines and areas around production equipment and buildings. Herbicides would be stored in a shed at the chemical dock. Herbicides would be purchased in small quantities and return agreements made with vendors whenever possible to limit the amount stored onsite. Rodenticides, such as D-Con and spray indoor insecticides, are kept at the warehouse.

3.8.3 Radioactive Waste

NPR-3 generates radioactive waste which is classified as "Naturally Occurring Radioactive Material" (NORM). These wastes are the by-products of oil and gas production in an area with naturally high radioactivity in the subsurface (UNC Remediation, 1990). Tests done to detect NORM have indicated a NORM level below proposed State limits.

The project also uses logging tools, which contain sealed radioactive sources, to measure the properties of the rock formations. In the event of an accident involving a sealed radioactive source, emergency procedures have been coordinated between the DOE, Contractor, and owner of the tools. These procedures would be used to minimize the potential exposure to radiation, and ensure that the source is properly contained. Small amounts of liquid radioactive tracers are also occasionally used. These isotopes are specially selected for their short half-life and quickly decay.

3.8.4 Waste Disposal

Disposal sites at NPR-3 include an industrial solid waste landfill, reserve pits, injection wells and the Bad Oil facility (DOE, 1992). Past disposal practices are fully covered in the Phase I study (Lawrence Allison, 1987) and are only repeated here when clarification is needed.

Thirteen solid waste disposal areas have been identified on the property. Eleven of these sites were used for non-hazardous waste. Two sites were used for the disposal of drilling mud (Lawrence Allison, 1987). Presently, NPR-3 has one industrial solid waste landfill which is 7.55 acres (1.9 ha) in size and would be developed in three phases. The landfill is currently in Phase I, which consists of the eastern third of the landfill (FD Services, 1992c).

Table 3-6 Substances Presently Used at NPR-3

Substance[a]	Monthly Amount[b] (gal)	Use
Nalco 19	0 (20 Pounds in inventory)	Removal of excess oxygen
Nalco 962	164	Scale inhibitor
Nalco 3211	113	Scale inhibitor
Nalco 3390	120	Oil/Water separator
Nalco 3403	0 used[c]; 19 gallons in storage at Gas Plant	Corrosion inhibitor
Nalco 3540	0	Acid pipeline cleaner
Nalco 3554	2	Glycol inhibitor
Nalco 3656	83	Corrosion inhibitor
Nalco 3900	0	Uni-treat packer fluid
Nalco 3903	220	Scale inhibitor
Nalco 3919	0 (discontinued 1993)	Scale inhibitor

Nalco 3940	4.5		Surfactant
Nalco 3999	0 (discontinued	1993)	Bactericide
Nalco 4453	38		Slugging compound
Nalco 4483	13		Emulsion breaker for Tensleep
Nalco 4493	117		Emulsion breaker
Nalco 4725	0		Batched into high paraffin oil
Nalco 4764	90		Solid dispersant
Nalco 4818	86		Reverse emulsion breaker
Nalco 4919	47		Corrosion inhibitor
Nalco 7290	5		Resin rinse
Nalco 8735	198		pH control
Nalco 89VE130	352		Paraffin dispersant
Methyl Mercaptan	1		Provides odor for leak detection in propane
Solvent 140	29		Solvent

Notes: [a] These substances are noted by the Manufacturers name.

[b] Usage is based on the month of August 1994 and does not include chemicals stored at the Chemical Dock. However, annual usage may be calculated by multiplying by 12 since monthly usage does not vary greatly.

[c] None used in August 1994, but the substance is used in other months.

Table 3-7 Hazardous Substances Historically Used at NPR-3

Substance	Approximate Dates of Usage	Use
Caustic Soda (Anhydrous sodium hydroxide)	1940-1950, 1970's- 1980's, 1993-Present	Treatment of native mud, drilling additive, water treatment plant
Chrome lignosulfonate	1960's	Corrosion inhibitor
Hydrochloric Acid	1950's - Present	Cleaning of wells and flowlines
Sodium chromate	Late 1970's	Drilling additive
Sodium bichromate	Late 1970's	Drilling additive
Hydrofluoric Acid	Unknown	Unknown
Xylene	Unknown to present	Well production

Ethylene glycol	Unknown to present	Gas processing
Methanol	Unknown to present	Gas processing
n-butyl alcohol	Unknown to present	Well production

Sources: Lawrence Allison, 1987; 40 CFR 302.

The NPR-3 industrial landfill is operated in a trench-and-fill method. The total landfill capacity is 15,500 cubic yards (11,900 cubic meters) (DOE, 1992; FD Services, 1992c). Industrial waste entering the landfill includes office waste, shipping boxes, oil-absorbent pads and booms, water filters, and other non-hazardous RCRA-exempt wastes. Special wastes entering the landfill include gas plant glycol filters and an occasional bag of unused non-hazardous chemicals such as potassium chloride or polyacrilimide (FD Services, 1992c). Spent iron sponge was disposed of three times in the last seven years with WYDEQ approval. However, iron sponge is no longer used at the gas plant, and has been replaced by Sulfatreat (FD Services, 1992c). Recycling of scrap metal, office paper, and aluminum cans is part of the Waste Minimization Program. In addition to the landfill, there is a landfarm which is used for the treatment of oil-contaminated soil. (FD Services, 1992c)

At the present time, NPR-3 contracts for solid waste collection and disposal. One 30-yard roll-off container is stationed in the field and is picked up and hauled to Casper as needed. On-going labor costs for operation and maintenance of the facility makes daily operation of the landfill impractical. Even though FD subcontracts with a private hauler to haul most of the solid waste, the landfill/landfarm remains in operation to maintain the WYDEQ permit, for treating oil-contaminated soils and for disposing of large quantity waste such as tank bottoms and empty sacks from drilling and workover operations.

Reserve pits handle wastes generated during well drilling, completion and workover (DOE, 1992). There are also four injection (disposal) wells on-site, used for backwash water from the water softener, produced water from oil reservoirs, and for disposal of other exploration and production (E&P) exempt wastes. Finally, there is a Bad Oil Facility which is used to hold oil for recycling, and sludge recovered from drilling pits, well servicing, tank and treater cleaning. Sludge from the Bad Oil Facility is collected in aboveground storage tanks and then applied to roads on-site in accordance with permits issued by WYDEQ (DOE, 1992).

3.9 Summary of the Affected Environment

The affected environment at NPR-3 considered by this Sitewide EA is summarized in Table 3-8.

Table 3-8 Summary of Affected Environment

LAND RESOURCES	3.1	
Land Use	3.1.1	Intensive development in central third, scattered development in northern third, little or no development in southern third and on bluffs.
Aesthetics	3.1.2	Typical of oilfields. Cleaner than most.

Recreation	3.1.3	No recreational facilities within or adjoining NPR-3.
AIR QUALITY AND ACOUSTICS	**3.2**	
Meteorology and Climate	3.2.1	Semi-arid with approximately 9 to 12 inches (23-30 cm) of precipitation annually; average low temperature in winter about 0°F (-18°C); average max temperature in summer 80 to 85°F (27-30 °C).
Air Quality	3.2.2	H_2S emissions from EOR activities.
Acoustics	3.2.3	Typical of oilfields.
WATER RESOURCES	**3.3**	
Surface Water Quantity	3.3.1	Ephemeral and intermittent streams in draws, small man-made ponds.
Ground Water Quantity	3.3.2	No high quality freshwater aquifers under NPR-3.
Surface Water Quality	3.3.3	Oil well production water discharged to draws under NPDES permits from WYDEQ.
Ground Water Quality	3.3.4	Water injection under UIC permits from WYOGCC.
Potable Water	3.3.5	Purchased from town of Midwest.
GEOLOGY AND SOILS	**3.4**	
Geology	3.4.1	Series of oil-bearing strata (reservoirs), several faults evidenced by the draws, seismically inactive.
Soils	3.4.2	Highly alkaline and saline soils derived from alkaline parent materials.
Prime Farmlands	3.4.3	None present within NPR-3 according to USDA Soil Conservation Service.
BIOLOGICAL RESOURCES	**3.5**	
Aquatic Biology	3.5.1	No fish reported in ephemeral and intermittent streams. One stocked pond exists at NPR-3.

Terrestrial Vegetation	3.5.2	Primarily rangeland, small areas of riparian vegetation (in draws) and pine-juniper forest (on bluffs). No forest management.
Terrestrial Wildlife	3.5.3	Typical of eastern Wyoming; No hunting or active wildlife management.
Threatened and Endangered Species	3.5.4	Federally-listed species possible: Blackfooted ferret, bald eagle (sighted, but no known nest within NPR-3), peregrine falcon. Previous blackfooted ferret survey negative.
Floodplains and Wetlands	3.5.5	Narrow zones within draws.
CULTURAL RESOURCES	**3.6**	Evidence of previous habitation by Native American tribes (likely Shoshoni and Sioux); Historical value of site due to Teapot Dome scandal in 1920's.
SOCIOECONOMICS	**3.7**	
Population and Employment	3.7.1	Natrona County characterized by slow population growth and unemployment rates similar to the state average.
Housing	3.7.2	No housing at NPR-3; housing availability abundant in Natrona County.
Transportation	3.7.3	All public highways servicing NPR-3 are free-flowing with no congestion.
Community Amenities	3.7.4	No shortages in Natrona County.
WASTE MANAGEMENT	**3.8**	
Hazardous Waste	3.8.1	Small quantities present at NPR-3. Off-site disposal if required.
Pesticides	3.8.2	Small quantities used and stored onsite at chemical dock.
Radioactive Waste	3.8.3	Only concern is low level of naturally occurring radioactive material (NORM's) generated by oil and gas production operations. Past tests show that the site is below proposed State limits for NORM.

Waste Disposal	3.8.4	Small quantities of waste disposal at the following on-site facilities: industrial solid waste landfill, reserve pits, injection wells, and bad oil facility.

4.0 ENVIRONMENTAL CONSEQUENCES AND MITIGATION MEASURES

Section 4.0 discusses environmental consequences (impacts) that could result from implementation of the Proposed Action and each alternative. The potential impacts of the Proposed Action are presented first. For each potential impact identified, specific mitigation measures have been proposed that would render the impact inconsequential. No potential impacts to any resource area from the Proposed Action have been identified for which practicable mitigation measures could not be developed.

Resource areas are addressed in the same order as the affected environment discussions in Section 3.0: land resources (Section 4.1), air quality and acoustics (Section 4.2), water resources (Section 4.3), geology and soils (Section 4.4), biological resources (Section 4.5), cultural resources (Section 4.6), socioeconomics (Section 4.7), and waste management (Section 4.8). The discussion under each resource area includes environmental consequences (impacts) and mitigation measures. Section 4.9 includes a summary of the impacts and mitigation measures under the Proposed Action and Section 4.10 covers a brief discussion of cumulative impacts.

4.1 Land Resources

4.1.1 Land Use

Environmental Consequences of the Proposed Action: Under the Proposed Action, approximately 250 to 300 acres of land would be directly disturbed. Most land disturbance would be confined to an approximately 2500-acre region located in the central third of NPR-3. Other parts of NPR-3 would experience limited disturbance, at most. Land use under the Proposed Action would remain compatible with existing land uses at NPR-3 and surrounding lands. Because mineral extraction activities are exempt from the county zoning resolutions, there are no zoning conflicts associated with the Proposed Action.

Scattered, minor land disturbances would result from general operations and support activities across the Reserve. Within the 2500-acre region where impacts would be concentrated, disturbance of approximately 30 acres (12 ha) of land would be required to support relocation and operation of each of the five steamflooding systems. Future development would occur on relatively gentle upland slopes rather than on steep side slopes, and no construction (other than road, pipeline, and utility crossings) is planned for riparian areas.

Minor land disturbances would also result from livestock grazing. The Department of Energy intends to lease the acreage for sheep grazing between January 1 and April 30. Increased wind and water erosion could occur in isolated areas as a result of over-grazing and the spread of noxious weeds could degrade range quality. Reduced range quality would also translate to decreased quality of wildlife habitat. Impacts of soil compaction would be concentrated primarily in bedding areas and along the trails which the sheep use. There would be some impact to riparian areas, although this will be limited because grazing would be in the spring when water demands are not as high.

Mitigation Measures: Temporarily disturbed areas would be mitigated in accordance with recommended reclamation procedures included in the plan cooperatively developed for NPR-3 by DOE and the Soil Conservation Service (SCS, 1992). Remaining areas used for ongoing development and oil production would be revegetated upon completion of those activities. Livestock numbers would be maintained at a low enough level to prevent major impacts. Additionally, grazing activities would be closely monitored and areas receiving use beyond their capacities would be isolated until the areas recover. A site-specific grazing management plan discussing the impacts and mitigation measures in detail, would be prepared.

Environmental Consequences of the EOR Technology Alternative: This alternative requires a similar level of drilling, road, pipeline and facility construction as the Proposed Action. The environmental consequences and mitigation measures are, therefore, generally the same.

Mitigation Measures: Mitigation measures would be generally the same as in Proposed Action.

Environmental Consequences of the Divestiture Alternative: A Finding of No Significant Impact (FONSI) was declared for the proposal to sell NPR-3. Subsequent to the publication and distribution of that EA, no further consideration has been given to the proposal to sell NPR-3. DOE ownership and management of NPR-3 are expected to continue into the foreseeable future. Therefore, sale to the private sector is not considered a likely alternative to the Proposed Action presented in this EA. If NPR-3 were sold to a private interest, it would likely be managed as an oilfield in a manner similar to that used by the DOE under the Proposed Action. However, an independent operator may choose to operate NPR-3 as a stripper oilfield and minimize new investment. The potential environmental impacts would basically be similar to, or less than, those under the Proposed Action. On the other hand, an independent operator may be less attentive to environmental protection than DOE, therefore the net impact is difficult to quantify.

Mitigation Measures: Mitigation measures might include provisions for oversight of operations, such as regulation by the Bureau of Land Management. Covenants in the sale contract might also be used to ensure that long-term environmental protection continues after the sale.

Environmental Consequences of the No-Action Alternative: Under the No-Action Alternative, existing wells and facilities would continue to be operated until the costs to lift a barrel of oil exceed the revenue gained on a well-by-well basis. There would be no additional new disturbed acreage, resulting in slightly lower levels of fugitive dust and less disturbance of natural habitat. Roads and facilities would be reclaimed to natural habitat as wells became uneconomical to continue production.

Mitigation Measures: There are no mitigation measures required under this alternative.

Environmental Consequences of the Decommissioning Alternative: Under this alternative, NPR-3 would cease production and begin environmental restoration. The level of activity would remain relatively high for several years while restoration and decommissioning occurs, but would cease at the completion of remedial action.

Mitigation Measures: There are no mitigation measures required for this resource.

4.1.2 Aesthetics

Environmental Consequences of the Proposed Action: Because of the existing state of disturbance throughout most of NPR-3 and the presence of other privately owned oilfields in the surrounding area, activities under the Proposed Action would have a negligible visual impact. Continued development would result in construction of additional roads, well locations, and support facilities in those parts of NPR-3 that already contain similar facilities. Any activities associated with oil extraction would be consistent with existing visual characteristics of the region. Because of the rim of bluffs surrounding much of NPR-3, activities performed at NPR-3 would not have an impact on any regional viewsheds. Development activities would not be visible to the general public or from the Wyoming Highway 259 corridor.

Mitigation Measures: No mitigation measures to offset minor visual changes resulting from the Proposed Action are necessary.

Environmental Consequences of the other Alternatives: None of the alternatives would generate any visual impacts, for the same reasons as discussed in the Proposed Action.

Mitigation Measures: No mitigation measures to offset minor visual changes resulting from the alternatives are necessary.

4.1.3 Ration

Environmental Consequences of the Proposed Action: There would be no impacts to recreational facilities as a result of continued development at NPR-3. No major recreational facilities exist at or in the immediate vicinity of NPR-3. The anticipated demand for regional recreational facilities would not be affected, since work force requirements associated with continued development would not change (Section 4.7.1).

Mitigation Measures: Because there are no major existing recreational facilities that could be adversely impacted by the Proposed Action and because the Proposed Action would not increase the demand for regional recreational facilities, no mitigation measures are necessary.

Environmental Consequences of the other Alternatives: None of the alternatives would generate any impacts to recreational resources, for the same reasons as discussed in the Proposed Action.

Mitigation Measures: No mitigation measures to offset resulting from the alternatives are necessary.

4.2 Air Quality and Acoustics

4.2.1 Meteorology and Climate

Environmental Consequences of the Proposed Action: No impacts to the meteorology and climate of the region containing NPR-3 would result from continued development at NPR-3.

Mitigation Measures: Because the Proposed Action would not adversely affect the regional climate, no mitigation measures are necessary.

Environmental Consequences of the other Alternatives: No impacts to the meteorology and climate of the region containing NPR-3 would result from adoption of any of the alternatives.

Mitigation Measures: Because the alternatives would not adversely affect the regional climate, no mitigation measures are necessary.

4.2.2 Air Quality

Environmental Consequences of the Proposed Action: Potential impacts to air quality from the Proposed Action would be limited. Emissions of air pollutants, including particulates, sulfur dioxide, carbon monoxide, hydrogen sulfide, nitrogen oxides and hydrocarbons would be maintained within permitted levels. Construction and operation of facilities and the drilling of additional injection and production wells under the Proposed Action could also cause limited increases in fugitive dust levels.

Continued EOR operations would stimulate the growth of anaerobic sulfur reducing bacteria, resulting in continued production of hydrogen sulfide (H_2S). An H_2S flare pilot facility to test disposal of excess H_2S emissions was put into operation in 1992. Installation of a permanent H_2S disposal system would likely require an air quality permit for construction (Raffelson, 1992). Improvements in the development of chemical, microbial and biocide treatment technologies, as a part of the Proposed Action, could potentially further reduce H_2S emissions.

Mitigation Measures: All air emissions associated with continued development would be maintained within permitted levels. A consultant firm has been hired to prepare a Title V Permit. Operating scenarios of the proposed Action would be included in the permit. H_2S levels would be controlled through microbial or biocide well treatment, chemical oxidation of the gas, and/or additional flaring. Fugitive dust emissions would be in direct proportion to disturbed acreage, but with reclamation should not exceed the WYDEQ standard within the project area or at the boundary. During project construction, fugitive dust would be reduced by wetting problem areas (perhaps using water obtained from the Madison formation), and by restricting vehicle travel wherever practicable. Crude oil sludge application to the roads reduces dust emissions. The application of sludge to the roads is permitted by WYDEQ for dust control.

Environmental Consequences of the EOR Technology Alternative: Depending on the choice of EOR method, substantial emissions of carbon dioxide, nitrogen, and flue gasses would be released. Since continued development of the steamflood would be halted, H_2S and nitrogen oxides emissions would decrease. Other impacts to air quality, such as fugitive dust, would be similar to the Proposed Action because the level of activity would be approximately the same.

Mitigation Measures: Mitigation measures for increased carbon dioxide emissions would include the construction of a gas sweetening module at the gas plant to remove and recycle carbon dioxide.

Environmental Consequences of the Divestiture Alternative: It is believed that private owners would manage the project in a manner similar to current operations. Impacts would be similar to those of the Proposed Action.

Mitigation Measures: Mitigation measures would be similar to those of the Proposed Action.

Environmental Consequences of the No-Action Alternative: Air emissions would start at the same level as the Proposed Action, and then slowly decrease in all criteria as the project sinks into non-profitability.

<u>Mitigation Measures</u>: Mitigation measures would be similar to those of the Proposed Action, except that only those measures that make sense in the context of a short remaining project life would be executed.

<u>Environmental Consequences of the Decommissioning Alternative</u>: Most major emissions sources would stop immediately. Other sources, such as fugitive dust and hydrocarbon emissions, would cease upon completion of restoration activities.

<u>Mitigation Measures</u>: No mitigation measures for the Decommissioning Alternative would be required.

4.2.3 Acoustics

<u>Environmental Consequences of the Proposed Action</u>: Noise emissions from construction activities and onsite operation of continued development activities would not be anticipated to result in any increases to ambient noise levels outside of the boundaries of NPR-3. During construction and operation activities, limited increases to ambient noise levels could potentially occur at NPR-3 and would primarily be associated with steam generator and water treatment facilities, drilling rigs, and vehicle traffic.

<u>Mitigation Measures</u>: Although no increase in noise levels are expected to occur outside the boundaries of NPR-3 as a result of the Proposed Action, ongoing measures for the protection of workers' hearing would continue to be implemented. These measures would include the use of standard silencing packages on construction equipment, and the use of OSHA-approved earmuffs or earplugs in designated areas or buildings which experience elevated noise levels.

<u>Environmental Consequences of the EOR Technology Alternative</u>: Noise levels from a similar level of industrial activity would generate environmental consequences similar to those in the Proposed Action.

<u>Mitigation Measures</u>: Noise levels from a similar level of industrial activity would require mitigation measures similar to those in the Proposed Action.

<u>Environmental Consequences of the Divestiture Alternative</u>: Noise levels from a similar level of industrial activity in the private sector would generate environmental consequences similar to those in the Proposed Action.

<u>Mitigation Measures</u>: Mitigation measures would be similar to those in the Proposed Action.

<u>Environmental Consequences of the No-Action Alternative</u>: Noise levels from the No-Action Alternative would generate environmental consequences similar to those in the Proposed Action. A generally reduced level of activity would not reduce high noise levels at specific sites. However, fieldwide noise levels would decrease over time as wells were shut in and activities reduced.

<u>Mitigation Measures</u>: Mitigation measures would be similar to those in the Proposed Action.

<u>Environmental Consequences of the Decommissioning Alternative</u>: Noise levels from the Decommissioning Alternative would generate environmental consequences similar to those in the

Proposed Action while industrial activity continued. A generally reduced level of activity would not reduce high noise levels at specific sites. However, fieldwide noise levels would decrease over time as wells were shut in and activities reduced.

Mitigation Measures: Mitigation measures would be similar to those in the Proposed Action.

4.3 Water Resources

4.3.1 Surface Water Quantity

Environmental Consequences of the Proposed Action: No water would be withdrawn from any surface water bodies under the Proposed Action.

The present discharges to surface water bodies (Little Teapot Creek and its tributaries) would remain the same or increase under the Proposed Action. The degree of increase would depend upon the success of the proposed biological treatment area. As indicated in Section 4.3.1, the principal source of discharged effluent is produced formation water generated by conventional oil recovery and EOR techniques. Under the Proposed Action, produced water would continue to be generated by both continued conventional in-fill development and by EOR activities. Small amounts of excess water not injected into UIC permitted wells or used as make-up water for EOR activities would continue to be discharged to the draws under the existing NPDES Permits (Table 3-2).

One research project proposed for RMOTC involves the creation of a biological treatment area designed to use halophytic (salt-loving) plant species to bind chlorides and remove oil and grease from produced water and lower its toxicity. If successful, the majority of produced water which is currently injected underground, or that does not meet present discharge limits, would be discharged through a new NPDES permit at the treatment area. This would result in a substantial increase in the amount of surface water discharge at NPR-3. The quality of this water would be equal to or better than that of current discharges.

Several production wells would be expected to be shut down in future years as their oil production rate diminishes. Closure of production wells could result in a decrease in the production of produced water. However, overall discharges are expected to remain the same or increase due to the biological treatment area.

Most activities considered under the Proposed Action would not generate large new areas of impervious surfaces which could increase storm water runoff discharges following rainfall events. New access roads and well pads servicing both the proposed conventional in-fill development and EOR activities would primarily comprise dirt, gravel, or other pervious surfaces. Small quantities of additional surface runoff could be generated due to soil compaction by heavy equipment.

Construction of some of the support facilities under the Proposed Action (such as a waste collection and treatment facility) could require the construction of small paved areas. The total increase in the area of impervious surfaces on NPR-3 (pavements and rooftops) would be less than 1 acre (0.4 ha), and incapable of generating measurable increases in flow in any stream channel.

A few large projects that could possibly be conducted at NPR-3 in the future could involve the construction of larger areas of impervious surfaces. If the facility were larger than 5 acres, a general stormwater discharge permit would be required.

Mitigation Measures: No mitigation is necessary to address small quantities of storm water runoff that could be generated by the Proposed Action. Surface water discharges are expected to remain the same or increase slightly which is not considered to be an adverse effect.

Environmental Consequences of the EOR Technology Alternative: Requirements for support facilities under the EOR Technology Alternative would be similar to that of the Proposed Action. Surface water flow rates would also be largely unchanged.

Mitigation Measures: Mitigation measures would be similar to those in the Proposed Action.

Environmental Consequences of the Divestiture Alternative: Surface water impacts from a similar level of industrial activity would generate environmental consequences similar to those in the Proposed Action. There would be no increase in produced water discharge from the biological treatment area, therefore discharges would be expected to remain the same or decrease slightly over time, as the amount of produced water decreases.

Mitigation Measures: Mitigation measures would be similar to those in the Proposed Action.

Environmental Consequences of the No-Action Alternative: Surface water flow would return to pre-development levels after the project reaches its economic limit and decommissioning begins.

Mitigation Measures: Mitigation measures would include continued use of Madison water supply wells to compensate for lost oilfield discharges.

Environmental Consequences of the Decommissioning Alternative: Surface water flow would be quickly returned to pre-development levels.

Mitigation Measures: Mitigation measures would include continued use of Madison water supply wells to compensate for lost oilfield discharges.

4.3.2 Ground Water Quantity

Environmental Consequences of the Proposed Action: Because no aquifers bearing high quality fresh water exist in the immediate vicinity of NPR-3, no such aquifers could be potentially depleted by the Proposed Action. Continuation of oil extraction by either conventional or EOR technologies would inevitably involve the simultaneous withdrawal of water from the oil bearing formations. However, this water is too high in total dissolved solids (TDS) and hydrocarbons to be suitable for use as potable water (Section 3.3.2). Water would continue to be withdrawn from the Madison formation to be used as make-up water for EOR activities under the Proposed Action. Because the salinity of the Madison formation water renders it unsuitable as potable water (Section 3.3.2), no adverse competition with regional demands for potable water would be possible. Since the Madison formation is deep and overlain by competent (rigid) strata not susceptible to compression, there is no potential for land subsidence due to groundwater withdrawals (Doyle, 1993).

Mitigation Measures: As there are no potentially competing uses for Madison formation water or other groundwater resources present at NPR-3, and because there is no potential for land subsidence, there is no need to mitigate for any potential overdraft of groundwater at NPR-3.

Environmental Consequences of the other Alternatives: No impacts to groundwater quantity at NPR-3 would result from adoption of any of the Alternatives.

Mitigation Measures: Because the Alternatives would not adversely affect groundwater quantity, no mitigation measures are necessary.

4.3.3 Surface Water Quality

Environmental Consequences of the Proposed Action: All discharges under the Proposed Action would comply with the terms of NPDES Permits. Existing NPDES permits (Table 4-5) would be renewed as required, but no modifications to the effluent limits would be sought. In cases where discharge requirements could not be met, the water would be injected into UIC injection wells permitted by the Wyoming Oil and Gas Conservation Commission.

One research project proposed for RMOTC involves the creation of a biological treatment area designed to use halophytic (salt-loving) plant species to bind chlorides in produced water and lower its toxicity. If successful, the majority of produced water which is currently injected underground, or that does not meet present discharge limits, would be treated to meet discharge limits and discharged through a new NPDES permit at the treatment area. This would result in a substantial increase in the amount of surface water discharge at NPR-3. The quality of this water would be equal to or better than that of current discharges.

The process water effluent originating from the deep Tensleep and Madison formations would continue to be hot. Although the temperature at the points-of-discharge would continue to fluctuate as it does presently (Section 3.3.3), the average temperature would not increase. As presently, the in-stream temperature would be expected to rapidly cool to ambient temperatures through atmospheric exchange.

As indicated in Section 4.3.1, minor quantities of surface runoff could reach the streams at NPR-3. Both the quantity and quality of this runoff would be similar to that runoff presently reaching the streams, for which NPR-3 is exempt from industrial source NPDES permitting requirements for surface runoff. Coverage under the general stormwater NPDES permit will be obtained for any facilities constructed which disturb more than 5 acres of ground.

Surface disturbance could result in the sedimentation of the intermittent and ephemeral streams if adequate erosion control is not practiced.

Spills of oil, produced water or hazardous chemicals could also affect surface water quality.

Mitigation Measures: Corrective action would be taken if any exceedances were noted in the course of monitoring each NPDES-permitted outfall in accordance with permit requirements. Standard erosion control practices selected in consultation with the Casper field office of the SCS would be used to prevent sedimentation of the stream channels. No mitigation measures are necessary to offset minor increases in surface runoff. The existing Spill Prevention Control and Countermeasure Plan would be revised as needed to ensure information is current. Existing spill response procedures would be maintained to ensure that spills are remedied in a timely manner. Finally, field inspections would continue to be performed regularly by Environmental staff to verify clean-up and to check for undetected leaks.

Environmental Consequences of the EOR Technology Alternative: It is not likely that wastewaters generated by EOR activities would be dischargeable. Surface disturbance and sedimentation would also continue to be a potential issue. Therefore, surface water quality impacts would not change from the proposed action, except that impacts from the biological treatment area would not be present under this alternative.

Mitigation Measures: Mitigation measures would be similar to those in the Proposed Action.

Environmental Consequences of the Divestiture Alternative: Operation by private industry would continue largely unchanged from current practices. Therefore, surface water quality impacts would not change from the proposed action.

Mitigation Measures: Mitigation measures would be similar to those in the Proposed Action.

Environmental Consequences of the No-Action Alternative: The economic life of the project would be shortened considerably. Tensleep formation discharges would cease earlier than expected in the Proposed Action. All NPDES permits would be deactivated as part of the early decommissioning process.

Mitigation Measures: Mitigation measures would include continued use of Madison water supply wells to make up for lost oilfield discharges.

Environmental Consequences of the Decommissioning Alternative: Tensleep formation discharges would cease immediately. All NPDES permits would be deactivated as part of the immediate decommissioning process.

Mitigation Measures: Mitigation measures would include continued use of Madison water supply wells to make up for lost oilfield discharges.

4.3.4 Ground Water Quality

Environmental Consequences of the Proposed Action: Steamflooding and waterflooding EOR activities using water from the Madison and Tensleep formations would locally dilute the formation water present with the hydrocarbons in the various oil producing formations at NPR-3. This dilution is not expected to render water in the shallow formations at NPR-3 suitable for potable purposes. As indicated in Section 3.3.4, the TDS level of Madison formation water, the highest quality groundwater present at NPR-3, renders it of marginal quality as potable water. Due to its depth, the quality of Madison formation water could not potentially be affected by activities at the surface or by the UIC-permitted injection wells.

Surface facilities such as reserve pits and disposal ponds could contaminate soil and local shallow groundwater. Spills of crude oil and other chemicals may have the same effect.

Mitigation Measures: Reserve pits and other production facilities would be designed, sited, constructed and operated according to WOGCC standards for critical areas (Rule 401) as applicable. The NPR-3 SPCC Plan would be regularly revised to ensure that it remains current. Existing spill response procedures would be maintained in order to ensure that spills are remedied in a timely manner. Finally, field inspections by Environmental staff would continue to be conducted regularly in order to verify clean-up and to check for undetected leaks.

Underground Injection Control wells for water injection, water disposal and steam injection would be tested for casing integrity in accordance with WOGCC regulations for UIC injection.

Finally, routine groundwater monitoring would continue around the NPR-3 landfill.

Environmental Consequences of the other Alternatives: Consequences of the other alternatives are similar to those of the Proposed Action.

Mitigation Measures: Mitigation measures would be similar to those in the Proposed Action.

4.3.5 Potable Water

Environmental Consequences of the Proposed Action: The potable water demands of NPR-3 would not increase due to the Proposed Action. Water would continue to be provided from the Casper and Midwest municipal systems and monitored as it is presently.

Mitigation Measures: Mitigation measures are not necessary to offset the limited use of potable water attributable to the Proposed Action.

Environmental Consequences of the other Alternatives: Potable water requirements at NPR-3 would change slightly as a result of adoption of any of the alternatives, but operation and monitoring of the potable water system would continue unchanged until decommissioning.

Mitigation Measures: Mitigation measures are not necessary to offset the limited use of potable water attributable to any of the alternatives.

4.4 Geology, Soils, and Prime and Unique Farmlands

4.4.1 Geology

Environmental Consequences of the Proposed Action: Other than the removal of oil from oil-bearing strata, no part of the Proposed Action would alter the geology of NPR-3 or the surrounding area. Because the oil-bearing strata are consolidated and not susceptible to consolidation (Doyle, 1993), there is no potential for sinkholes or land subsidence resulting from oil and water extraction.

Mitigation Measures: Since oil and water extraction at NPR-3 has no potential to result in adverse impacts, no mitigation measures are necessary.

Environmental Consequences of the other Alternatives: No impacts to the geology of NPR-3 would result from adoption of any of the alternatives.

Mitigation Measures: Because the alternatives would not adversely affect the local geology, no mitigation measures are necessary.

4.4.2 Soils

Environmental Consequences of the Proposed Action: Many activities under the Proposed Action would involve limited areas of surface soil disturbance. Surface soil disturbance would result in the removal of the generally thin topsoil and expose the highly alkaline and saline subsoils found throughout most of NPR-3. The exposed subsoils would be highly prone to erosion during the infrequent but intense downpours typical of eastern Wyoming. The SCS has determined that the majority of the soils mapped on the upland plains at NPR-3 present a severe hazard of water erosion (Figure 3-4 and Table 3-3).

Livestock grazing would increase soil compaction and erosion. The impacts would be most evident in bedding areas and along the trails the sheep would use. Some impacts would be noted in riparian areas, however these would be limited since grazing would be limited to spring when water demands are lower.

Since no part of the Proposed Action would involve more than scattered areas of surface disturbance and because most surface disturbing activities would not be performed simultaneously, large-scale soil erosion would not be anticipated. Furthermore, the surface disturbance would be very shallow and would not involve the redistribution or removal of large quantities of soil. The greatest potential for erosion could result from the implementation of new and expanded EOR activities. EOR activities (including steamflooding and waterflooding) could require the grading of contiguous or nearly contiguous areas of as much as 50 acres (20 ha).

Mitigation Measures: Standard erosion control practices selected in consultation with the Casper field office of the SCS would be used to prevent the sedimentation of the draws and other areas down slope of exposed soils. Exposed soils would be vegetatively reclaimed following a reclamation plan developed cooperatively for NPR-3 by the DOE and SCS. The SCS issued a list of recommended reclamation procedures for NPR-3 in 1992 (SCS, 1992), which would be followed until a sitewide reclamation plan is developed. To mitigate the impacts of livestock grazing, the number of AUMs would be maintained at a low enough level to prevent major impacts. Additionally, grazing activities would be strictly monitored and livestock would be impounded if necessary. A human herder would be required, during daylight hours, to ensure that animals do not congregate too long in any single spot. Areas receiving use beyond their capacities would be isolated until the areas recover. A site-specific grazing management plan would be prepared to discuss the impacts and mitigation measures in detail.

Environmental Consequences of the EOR Technology Alternative: Construction activities in support of EOR activities would result in surface disturbance to an extent and depth approximately equal to that of the Proposed Action. Potential for soil erosion would likewise be similar. No impacts would be felt from livestock grazing.

Mitigation Measures: Mitigation measures would be similar to those in the Proposed Action.

Environmental Consequences of the Divestiture Alternative: Operation by private industry would continue largely unchanged from current practices. Therefore, soil impacts would not change from the proposed action. Increased soil erosion due to overgrazing and erosion due to unrestricted damage to riparian areas could increase.

Mitigation Measures: Mitigation measures would be similar to those in the Proposed Action.

Environmental Consequences of the No-Action Alternative: New construction would be minimal and additional surface disturbance would be negligible. Soil erosion from new construction would not occur. Reclamation of abandoned wells and facilities would accelerate.

Mitigation Measures: To the limited extent necessary, mitigation measures would be similar to those in the Proposed Action.

Environmental Consequences of the Decommissioning Alternative: The project site would immediately begin decommissioning and restoration. Most surface occupancy would end. Leasing of the property for livestock grazing is possible.

Mitigation Measures: Mitigation of grazing impacts would be similar to those under the Proposed Action.

4.4.3 Prime and Unique Farmlands

Environmental Consequences: Because no prime or unique farmlands are present within NPR-3 (Davis, 1993b), no part of the Proposed Action has any potential for impact.

Mitigation Measures: As there are no prime farmlands present on or in the vicinity of NPR-3, no mitigation measures are necessary.

Environmental Consequences of the other Alternatives: None of the proposed alternatives has any potential for impact because no prime or unique farmlands are present within NPR-3.

Mitigation Measures: As there are no prime farmlands present on or in the vicinity of NPR-3, no mitigation measures are necessary.

4.5 Biological Resources

4.5.1 Aquatic Biology

Environmental Consequences of the Proposed Action: Ground disturbance under the Proposed Action could result in increased sedimentation of streams at NPR-3. While NPDES discharges have remained within the permitted effluent limits, a hydrocarbon sheen has been observed on containment ponds. The hydrocarbon sheen on ponds poses a toxicity threat to water birds (e.g. ducks, shorebirds) and other species attracted to these water sources.

Although the WGFD does not anticipate any direct impacts to any fisheries, the agency has expressed concern that the potential level of activity under the Proposed Action could magnify water quality impacts in the Powder River system. The Powder River is already impacted by poor water quality from other sources, and the river provides important habitat for sturgeon chubs and shovelnose sturgeon. However, as discussed in Section 4.3.3, the use of a biological treatment area for the treatment of produced water may actually improve the quality of water discharged, thereby offsetting impacts to the Powder River system.

Mitigation Measures: A soil erosion and sediment control plan would be implemented to prevent increased sedimentation of steams on NPR-3. If containment pond sediments are found to be contaminated, the ponds would be drained and the sediments removed. DOE/FD have contemplated elimination of all containment ponds as a possible mitigative measure. Discharge waters may also be recycled into the steam flooding/water flooding operations or diverted to a UIC disposal well, thus minimizing discharge into surface waters.

Mitigation measures would be developed in consultation with the WGFD. To ensure that impacts to fisheries in the Powder River basin are minimized, WGFD has recommended that special precautions be taken to prevent the release of pollutants from work areas at NPR-3. Where effluent must be discharged under existing NPDES permits, WGFD recommends that the creation of appropriately sized wetlands be considered as a means of improving water quality. The DOE is already investigating the use of a biological treatment area as a means of improving the aquatic habitat, and has included this activity in the Proposed Action. Alternately, WGFD would prefer that the effluent be stored where it could not enter surface waters.

Environmental Consequences of the EOR Alternative: Impacts of the EOR Alternative on surface water quality and quantity have been discussed previously. Kinds of impacts that may be expected as a result of implementing the alternative is similar to the Proposed Action, although the magnitude would vary slightly.

Mitigation Measures: Mitigation measures employed to protect aquatic biological resources would be similar to those of the Proposed Action.

Environmental Consequences of the Divestiture Action Alternative: Impacts of the Divestiture Alternative on surface water quality and quantity have been discussed previously. Kinds of impacts that may be expected as a result of implementing the alternative are similar to the Proposed Action, although the magnitude would vary slightly.

Mitigation Measures: Mitigation measures employed to protect aquatic biological resources would be similar to those of the Proposed Action.

Environmental Consequences of the No-Action Alternative: As facilities and wells are shut in the amount of produced water discharged would gradually decrease. This would have an effect on the streams and wetlands at NPR-3, and may also have a negative effect on the aquatic organisms.

Mitigation Measures: Mitigation measures employed to protect aquatic biological resources would be similar to those of the Proposed Action.

Environmental Consequences of the Decommissioning Alternative: As facilities and wells are shut in the discharge of produced water would cease. This would have a profound effect on the streams and wetlands at NPR-3 and their associated aquatic organisms.

Mitigation Measures: Mitigation measures employed to protect aquatic biological resources would be similar to those of the Proposed Action.

4.5.2 Terrestrial Vegetation

Environmental Consequences of the Proposed Action: Where surface disturbance is necessary to implement activities under the Proposed Action, it would result in the removal of existing vegetation. Surface disturbance would be largely limited to the rangeland associations at NPR-3, avoiding areas of riparian and pine-juniper vegetation (Figure 3-5). Small areas of riparian vegetation would be disturbed by certain activities such as road crossings and pipeline and utility installation. Expansion or establishment of EOR activities could potentially require the disturbance of small areas of riparian vegetation. Because developed areas at NPR-3 are highly scattered, incidental encroachment of machinery on areas of rangeland vegetation is inevitable.

Leasing of NPR-3 rangeland for grazing would also have a minor impact on the quantity and quality of vegetation.

Mitigation Measures: Exposed soils would be vegetatively reclaimed following a reclamation plan developed cooperatively by DOE and SCS. The SCS issued a list of recommended reclamation procedures for NPR-3 in 1992 (SCS, 1992), which would be followed until a sitewide reclamation plan could be developed. The number of AUMs would be maintained at a low enough level to prevent major impacts. Additionally, grazing activities would be closely monitored and areas receiving use beyond their capacities would be isolated to allow the areas to recover. A human herder would be required during daylight hours to prevent animals from congregating for too long in a single spot. A Grazing Management Plan would be developed to address impacts, monitoring and mitigation of vegetation quantity.

Environmental Consequences of the EOR Technology Alternative: Surface encroachment and displacement of vegetation would be of a similar magnitude as that of the Proposed Action, although the reasons for construction activity would differ.

Mitigation Measures: Mitigation measures employed to protect vegetation would be similar to those of the Proposed Action.

Environmental Consequences of the Divestiture Alternative: Operation by private industry would continue largely unchanged from current practices. Therefore, soil impacts would not change from the proposed action.

Mitigation Measures: Mitigation measures employed to protect vegetation would be similar to those of the Proposed Action.

Environmental Consequences of the No-Action Alternative: New construction would be minimal and additional surface disturbance would be negligible. Displacement of vegetation from new construction would not occur. Reclamation of abandoned wells and facilities would accelerate.

Mitigation Measures: To the limited extent necessary, mitigation measures would be similar to those in the Proposed Action.

Environmental Consequences of the Decommissioning Alternative: Surface disturbance would cease and the project would proceed to restoration of the original prairie. Non-indigenous plant species may be brought into the site in large quantities through reseeding efforts.

Mitigation Measures: Plant species would need to be carefully selected to ensure that indigenous species are used to the largest extent possible.

4.5.3 Terrestrial Wildlife

Environmental Consequences of the Proposed Action: The small areas of vegetation that would be disturbed under the Proposed Action represent an negligible loss of habitat for terrestrial wildlife. Livestock grazing would have a small impact on the amount of vegetation available for wildlife. The greatest impact would be to native grasses. Increased activity in localized parts of NPR-3 would not impact the pronghorn antelope and mule deer populations, whose natural mobility allows for movement throughout NPR-3 and adjoining undisturbed lands. The less mobile wildlife species (amphibian, reptiles and small mammals) could be killed by land disturbances.

Noise generated by activities under the Proposed Action would be generally consistent with noise generated by existing activities at NPR-3. Workers at NPR-3 have noticed that antelope and deer at NPR-3 have become conditioned to the noise (Halliburton NUS, 1993). Noise levels associated with oil drilling activities, such as those already present at NPR-3 and those proposed under the Proposed Action are not unusually high for industrial activities. Noise generated by construction under the Proposed Action would be minimal. Ambient drilling noise 50 feet (15 m) from a drill rig has been recorded at 75 dbA (DOE, 1990).

Produced water discharged to the draws under existing NPDES permits meets the Water Quality Standards established by the WYDEQ. However, the oil films which form on oil pits and the hydrocarbon sheen observed on containment ponds could be hazardous to wildlife, especially birds. Wildlife could be attracted to these pits and containment ponds as a source of drinking water or to retrieve insects trapped and struggling on the oily surface (Esmoil, 1991). Hydrocarbons could adhere to the feathers of parent birds and poison the bird while preening or contaminate and kill eggs during breeding. Complete net covering is a reliable deterrent to contamination of wildlife (Esmoil, 1991).

There is a potential for hydrogen sulfide (H_2S) generated during steamflooding and waterflooding operations to cause localized wildlife mortality. In one study, 237 animal deaths were attributed to H_2S gas (Esmoil, 1991); however, no wildlife mortality has been attributed to H_2S gas at NPR-3. In many oilfields this gas is vented through flare stacks (Esmoil, 1991) and most of the H_2S produced at NPR-3 is flared.

Mitigation Measures: No mitigation measures are necessary to compensate for the minor losses of wildlife habitat or increases in noise that would result from the Proposed Action. Impacts from livestock grazing would be mitigated by maintaining the number of AUMs at a low enough level to prevent major impacts, and closely monitoring grazing activities in accordance with the attached Grazing Management Plan.

Mitigation measures for hydrocarbon exposure would be developed in consultation with the FWS and the WGFD. The FWS has recommended that protective netting be placed over all containment ponds and any receiving wetlands displaying a hydrocarbon sheen, to prevent exposure of wildlife to oil. Most of the containment ponds would be closed, and the remaining few would be netted. One pond, located at the Tensleep battery was netted during the Summer of 1994.

It would be impossible to prevent dissipation of H_2S gas in areas where flare stacks are not used. When flare stacks are used, H_2S-caused wildlife mortality could be reduced by ensuring that igniters are operating efficiently so that the gas is properly flared and not vented directly into the environment. Also devices may be installed to inhibit raptors and other birds from perching on flares (Esmoil, 1991).

Environmental Consequences of the EOR Technology Alternative: The potential impacts due to noise and hydrocarbon emissions would be similar to those of the Proposed Action. The generation of hydrogen sulfide gas would decrease with time as the existing steam injection patterns became uneconomic to operate. Encroachment on habitat by construction is also of a similar magnitude to that of the Proposed Action, although the reasons for construction would differ.

Mitigation Measures: Mitigation measures would be largely similar to those in the Proposed Action. The decrease in production of hydrogen sulfide gas would require no mitigation.

Environmental Consequences of the Divestiture Alternative: Operation by private industry would continue largely unchanged from current practices. Therefore, impacts to wildlife would be similar to the Proposed Action.

Mitigation Measures: Mitigation measures employed to protect wildlife would be similar to those of the Proposed Action.

Environmental Consequences of the No-Action Alternative: The potential impacts due to noise and hydrocarbon emissions would start as being similar to those of the Proposed Action, but they would decrease over time as operations cease to be profitable. The generation of hydrogen sulfide gas would decrease with time as the existing steam injection patterns became uneconomic to operate. Encroachment on habitat by construction is also of a similar magnitude to that of the Proposed Action, although the reasons for construction would differ.

Mitigation Measures: Mitigation measures would be largely similar to those in the Proposed Action. The decrease in production of hydrogen sulfide gas would require no mitigation. Accelerated reclamation of oilfield pits and other facilities hazardous to wildlife would require no mitigation.

Environmental Consequences of the Decommissioning Alternative: Since current operations would be curtailed immediately, oilfield facilities that are hazardous to wildlife would immediately shut down and be promptly reclaimed.

Mitigation Measures: Mitigation of impacts under the Decommissioning Alternative would not be required, since the impacts would not be adverse to wildlife.

4.5.4 Threatened and Endangered Species

Environmental Consequences of the Proposed Action: There are no Federally-listed threatened or endangered species known to consistently inhabit NPR-3. Since the bald eagle and peregrine falcon (both endangered) are rare migrants, and the black-footed ferret is believed to be absent from the area (endangered), none of these species would be impacted by the Proposed Action.

Most of the Federal candidate species, although they occur in the region of NPR-3, are not known to exist at the NPR-3 site and thus are not expected to be affected. Since NPR-3 lies within the breeding range and contains suitable habitat for both the mountain plover (Category 1) and ferruginous hawk (Category 2), a field verification for nests of these species may be necessary prior to any disturbance of previously undisturbed land.

The loggerhead shrike (Category 2) is the only special status species known to occur regularly at NPR-3. Loggerhead shrikes, especially the young, have been shown to be vulnerable to oil contamination from oil pits in Wyoming (Esmoil, 1991).

The sturgeon chub (Category 2) and shovelnose sturgeon (Site Sensitive) are not known to occur at NPR-3. The distribution of the narrow-foot hygrotus diving beetle (Category 2) is unknown at NPR-3. The environmental consequences and mitigation measures applicable to these species are discussed in Section 4.5.1.

<u>Mitigation Measures</u>: Prior to the disturbance of any previously undisturbed land within mostly undeveloped areas of NPR-3, field surveillance would be conducted to determine whether Barr's milkvetch or nests of the ferruginous hawk, mountain plover or loggerhead shrike are present. A survey for the black-footed ferret would be performed before any prairie dog colonies are disturbed. Prior to the disturbance of any lands within the draws on NPR-3, a survey would be performed to determine whether the narrowfooted hygrotus diving beetle is present. In such a case guidance would be sought from FWS. Oilfield pits would be netted as funding becomes available in order to protect all migratory birds from harm.

<u>Environmental Consequences of the other Alternatives</u>: Continued operations under any of the proposed alternatives would result in impacts similar to those of the Proposed Action. The difference would be in the remaining life of the project, and the time until the project site would be returned to its former condition.

<u>Mitigation Measures</u>: Mitigation measures would be similar to those in the Proposed Action while operations continued. The restoration of the project after termination of operations would require no mitigation.

4.5.5 Floodplains and Wetlands

<u>Environmental Consequences of the Proposed Action</u>: Surface disturbance within the draws at NPR-3 would be largely limited to road crossings and to pipeline and utility installation. Additional surface disturbance would be caused by the increased use that wetlands would receive by livestock. These activities would not result in any permanent modification to the flood carrying properties of the affected draws. Small increases in the area of impervious surfaces within NPR-3 would not generate enough increased surface runoff during the 100-year or 500-year flood events to alter the area of the floodplains. Expansion or establishment of EOR activities could potentially require the filling of small swales. The impact on the floodflow capacity of the down slope draws would be minor.

Because encroachment into the draws would be minimal, the loss of wetlands from activities under the Proposed Action would be minimal. The only water bodies at NPR-3 that could possibly be filled under the Proposed Action would be process water containment basins, which do not meet the EPA definition of wetlands.

Wetlands in the draws could experience some change in the discharges of produced water released in accordance with NPDES permits. These wetlands already experience such discharges, and many areas of wetlands within the draws owe their existence to the NPDES discharges. Closure of existing or future wells by DOE, when they become uneconomic, would result in a decrease in water discharges under the NPDES permits and could result in the shrinkage or elimination of some wetlands. Development of a biological treatment area, which would treat and discharge most of the produced water at NPR-3, could create additional wetlands. Alternatively, the drilling of additional Tensleep wells could increase water discharges.

Sheep could damage riparian areas where they congregate to graze and cross creeks. Concentrated livestock grazing could beat down stream banks, foul surface waters, and damage riparian vegetation. However, the impacts to riparian areas would be less in the springtime, when grazing is planned, due to lesser water demands than the livestock would have during the summer months.

Mitigation Measures: In compliance with Executive Orders 11988 and 11990, DOE would investigate all practicable alternatives meeting the objectives of its mission at NPR-3 prior to even minor modifications to wetlands or floodplains. If an activity under the Proposed Action required permanent changes to the grade within a draw, the boundaries of the affected floodplain would be delineated, and the impact to the 100-year flood flow would be calculated using standard hydrological procedures.

If an activity under the Proposed Action required any temporary or permanent surface disturbance within a draw, DOE would delineate the boundaries of any affected wetlands using the Corps of Engineers Wetlands Delineation Manual (COE, 1987). It is expected that required wetland fill would qualify for Nationwide General permits under Section 404 of the Clean Water Act. The US Army Corps of Engineers (COE) would be notified in writing prior to any discharge of fill material to wetlands at NPR-3. If required, mitigation measures would be developed in consultation with the COE.

The number of AUMs would be maintained at a low enough level to prevent major impacts from livestock grazing. Additionally, grazing activities would be closely monitored, and areas receiving use beyond their capacities would be isolated to allow the areas to recover. Livestock would be closely supervised near riparian areas in order to prevent overutilization of these sensitive ecosystems. A site-specific grazing management plan would be prepared that discusses the impacts and mitigation measures in detail.

Environmental Consequences of the other Alternatives: Construction activities under the other alternatives would also be conducted in manner similar to that of the Proposed Action, in that wetlands would be generally avoided. Discharges of produced water would generally decrease with time, as production became uneconomic. None of the alternatives proposes drilling of additional wells in the Tensleep formation, which would increase water discharge volumes. None of the alternatives offers a project life as long as the Proposed Action, therefore wetlands would be adversely affected earlier.

Mitigation Measures: During operation of the project, mitigation would be similar to that of the Proposed Action. After reclamation, mitigation of lost wetlands would include the construction of nearby wetlands as compensation. Alternatively, the Madison water supply wells could continue to produce water and feed the existing wetlands at NPR-3.

4.6 Cultural Resources

Environmental Consequences of the Proposed Action: Impacts to cultural resources from the Proposed Action would be largely limited to the effects of ground disturbing activities. Under Section 106 of the National Historic Preservation Act (16 USC 470 and 36 CFR 800), Federal agencies (including the DOE) must consider the effects that actions would have on historic properties. As part of the Section 106 process, Federal agencies must consult with the State Historic Preservation Officer (SHPO) and the Advisory Council on Historic Preservation.

The Wyoming SHPO was consulted during the preparation of this Sitewide EA. The SHPO records indicate that much of NPR-3 was surveyed prior to 1980. However, surveys conducted prior to 1980 are considered unreliable by the SHPO. The SHPO has recommended that, prior to disturbance of previously undisturbed ground, a new archaeological survey be completed and the results sent to the SHPO. (Keck, 1993)

Because the specific locations of ground disturbing activities under the Proposed Action have not yet been identified, the Advisory Council on Historic Preservation and the Wyoming SHPO were not consulted concerning specific sites.

A Class III cultural resources inventory was completed for NPR-3 in June 1995. The inventory revealed two prehistoric sites which are eligible for listing on the National Register of Historic Places. Both sites are located in the southern end of the field in areas which are undesireable for drilling. Therefore, no impacts to the sites are expected. The oilfield itself is also eligible for listing as an historic site. Operation of NPR-3 as an oilfield is consistent with the history and setting of Teapot Dome, therefore there would be no effect to the site.

Mitigation Measures: The two eligible prehistoric sites would be avoided as potential development areas. A Programmatic Agreement (PA) would be developed to cover mitigation of potential impacts to the cultural resources found in the survey and to address the potential for ground disturbing activities to uncover cultural resources.

Should additional resources be found, avoidance of these resource areas would be the preferred mitigation. If cultural resources are identified, measures outlined in the Cultural Resources Management Plan (JBEC, 1991) would be followed. If avoidance of these resources is not feasible, techniques to preserve these natural resources could include data recovery and documentation. The Wyoming SHPO and/or Tribal Council would be notified if resources were identified. The discovery of cultural resources would result in work stopping immediately until an experienced individual could determine the eligibility of the site under the NHPA and associated laws.

Environmental Consequences of the EOR Technology Alternative: Construction under the EOR Technology Alternative would be of the same order of magnitude as the Proposed Action. Therefore, the potential impacts to cultural resources would be similar.

Mitigation Measures: Mitigation measures would be similar to those in the Proposed Action.

Environmental Consequences of the Divestiture Alternative: Operation by private industry would continue largely unchanged from current practices. Therefore, impacts to cultural resources would not change from the proposed action. However, because the operation would presumably utilize private funding on private land, the level of compliance required by Section 106 would be much less, if required at all.

Mitigation Measures: Mitigation measures employed to protect cultural resources might include covenants in the sale contract requiring protection of cultural resources.

Environmental Consequences of the No-Action Alternative: New construction under the No-Action Alternative would be halted. Only minor surface disturbance would occur until decommissioning of the field.

Mitigation Measures: A Section 106 consultation would be performed to address mitigation of potential impacts to the cultural resources that might still occur as a result of the limited operations.

Environmental Consequences of the Decommissioning Alternative: All activities at NPR-3 would be halted and the property would be promptly restored to is former state. No further disturbance of the surface would occur. Possible sale and dismantling of equipment and buildings could impact the integrity of the Teapot Dome site.

Mitigation Measures: A Section 106 consultation would be performed to address mitigation of potential impacts to the cultural resources that might still occur as a result of the limited operations.

4.7 Socioeconomics

4.7.1 Population and Employment

Environmental Consequences of the Proposed Action: Under the Proposed Action, employment at NPR-3 would remain at or close to the present levels. Minor fluctuations would be expected in response to project scheduling and political and economic shifts. FD (DOE's prime contractor for NPR-3) presently employs a work force of about 65 at NPR-3. As many as 20 additional persons hired by subcontractors to FD are working at NPR-3 at any given time. FD bases an additional staff of about 45 at the Casper office, which is responsible for the oversight of field operations both at NPR-3 and at the Naval Oil Shale Reserves (NOSRs) in Colorado and Utah. DOE maintains a staff of 15 in Casper for the oversight of NPR-3 and the NOSRs.

Mitigation Measures: Because the Proposed Action would not substantially change regional population or employment levels, no mitigation measures are necessary.

Environmental Consequences of the EOR Technology Alternative: Employment levels under this Alternative would not change substantially from that of the Proposed Action.

Mitigation Measures: Because this Alternative would not substantially change regional population or employment levels, no mitigation measures are necessary.

Environmental Consequences of the Divestiture Alternative: Private ownership of NPR-3 would result in a level of activity substantially unchanged from that of the proposed action, but a private operator would not likely use as large a work force to accomplish its goals. Unemployment would increase in Natrona County and adverse impact to the towns of Midwest and Edgerton would likely result.

Mitigation Measures: Although an adverse impact on employment levels might result, no mitigation of this Alternative would be possible because the new operator would not be under any obligation to mitigate staff reductions. However, it might be possible to incorporate such provisions into the sale contract.

Environmental Consequences of the No-Action Alternative: Employment levels would generally decline since oil production rates would begin to decline almost immediately.

Mitigation Measures: Job retraining and severance benefits would be awarded to those employees who are displaced as a result of declining activity at NPR-3

Environmental Consequences of the Decommissioning Alternative: Adverse impact to the towns of Midwest and Edgerton would be immediate. A substantial portion of these towns small employment pools are provided by NPR-3.

Mitigation Measures: Qualified employees would be offered positions for the decommissioning and reclamation work. Job retraining and severance benefits would be awarded to those employees who are displaced as a result of declining activity at NPR-3, and for the remainder of the work force after reclamation is complete.

4.7.2 Housing

Environmental Consequences of the Proposed Action: Because the Proposed Action would not substantially change employment levels at NPR-3, the supply of housing units in Natrona County would not be affected. Considering the high vacancy rate for housing units in Natrona County, any short-term increases in the demand for housing could easily be accommodated.

Mitigation Measures: Because of the adequacy of regional housing, no mitigation measures are necessary.

Environmental Consequences of the EOR Technology Alternative: No impacts to the local availability of housing would result from adoption of any of the Alternatives.

Mitigation Measures: Because of the adequacy of regional housing, no mitigation measures are necessary.

Environmental Consequences of the Divestiture Alternative: Private ownership of NPR-3 would likely reduce the size of the workforce and could in turn result in a decline in the housing values in Midwest, Edgerton and Casper.

Mitigation Measures: Although this would be an adverse impact no mitigation of this alternative would be possible because the new operator would not be under any obligation to maintain staffing levels.

Environmental Consequences of the No-Action Alternative: As employment levels decline with the oil production a slight effect might be seen in local housing values.

Mitigation Measures: This effect could not be mitigated.

Environmental Consequences of the Decommissioning Alternative: Because a significant portion of the positions at NPR-3 would be eliminated immediately this alternative would have an immediate effect on housing values in the area.

Mitigation Measures: This effect could not be mitigated.

4.7.3 Transportation

Environmental Consequences of the Proposed Action: Transportation of heavy machinery and materials to and from NPR-3 using Interstate 25 and Wyoming Route 259 would be necessary under the Proposed Action. Because the current level of service on these roads is substantially below capacity, no disruption of traffic flow would occur as a result.

Mitigation Measures: Because of the adequacy of regional transportation facilities, no mitigation measures are necessary.

Environmental Consequences of the other Alternatives: Highway traffic resulting from the adoption of any of the alternatives would be less than or approximately equal to that resulting from the Proposed Action.

Mitigation Measures: Because of the adequacy of regional transportation facilities, no mitigation measures are necessary.

4.7.4 Community Services

Environmental Consequences of the Proposed Action: Because employment and population levels are expected to remain generally constant under the Proposed Action, community services in Natrona County would not be affected.

Mitigation Measures: Because of the adequacy of regional community services, no mitigation measures are necessary.

Environmental Consequences of the other Alternatives: Employment and population levels resulting from the adoption of any of the alternatives would be less than or approximately equal to that resulting from the Proposed Action.

Mitigation Measures: Because of the adequacy of regional community services, no mitigation measures are necessary.

4.8 Waste Management

Environmental Consequences of the Proposed Action: Generation of hazardous waste is expected to decline. Product substitution and process changes have been successful in reducing hazardous waste generation despite relatively constant levels of activity.

Starting in Fiscal Year 1994, the bulk of the solid waste began to be hauled offside by a commercial hauler. With that arrangement and the NPR-3 landfill permit still active, disposal capacity for solid wastes would be adequate for the foreseeable future.

High level radioactive waste is not expected, but might be generated by an accident involving sealed radioactive sources. Naturally Occurring Radioactive Materials (NORM) would be present in production equipment in extremely low levels that would be below proposed state and Federal regulations.

Pesticides are not expected to be intentionally dispose, but may also be spilled or accidently released into the environment.

Mitigation Measures: Mitigation measures for hazardous substances would include waste minimization, product substitution and the monitoring of usage to ensure compliance with applicable laws and regulations. Proper disposal of all hazardous and non-hazardous materials would be ensured by training and environmental compliance audits.

Mitigation for high level radioactive and pesticide wastes would include training and operational procedures intended to prevent accidental releases. Prompt and effective spill response would minimize the quantity of waste generated in the event of a release.

NORM would be mitigated by continuing to assess the extent of its occurrence at NPR-3. If it is found to be at regulated levels, a scale prevention program would be investigated as a means to prevent the deposition of NORM-containing carbonate/sulfate scale. Inspection procedures would ensure that contaminated equipment is discovered, decontaminated, and that disposal of the NORM debris is properly administered.

Environmental Consequences of the EOR Technology Alternative: Work levels and waste generation rates under this Alternative would be similar to that of the Proposed Action.

Mitigation Measures: Mitigation measures would also be similar to those in the Proposed Action.

Environmental Consequences of the Divestiture Alternative: Operation by private industry would continue largely unchanged from current practices. Therefore, volumes of waste generated would not change from the proposed action.

Mitigation Measures: Private industry would be required to meet the same local regulations, therefore no mitigation is necessary.

Environmental Consequences of the No-Action Alternative: Waste generation rates would slowly decrease from current levels as the project becomes uneconomic due to declining oil production rates. At the point of decommissioning, generation rates for all types of wastes would dramatically increase as facilities are dismantled.

Mitigation Measures: Mitigation measures would also be similar to those in the Proposed Action during the operating phase. At decommissioning, a priority would be placed on salvaging and auctioning the decommissioned equipment. Other materials would be recycled as market conditions permitted.

Environmental Consequences of the Decommissioning Alternative: At the point of de-commissioning, generation rates for all types of wastes would dramatically increase as facilities are dismantled. Current pesticide inventory would be disposed of if it could not be sold, donated, or returned to the vendor.

Mitigation Measures: At decommissioning, a priority would be placed on salvaging and auctioning the decommissioned equipment. Other materials would be recycled as market conditions permitted.

4.9 Summary of Environmental Consequences and Mitigation Measures of Proposed Action

Impacts to each resource area potentially resulting from implementation of the Proposed Action are summarized in Table 4-1. Where mitigation measures are necessary to ensure that no potentially adverse impacts to any resource area result from the Proposed Action, they are listed in the adjacent column in Table 4-1.

Table 4-1
Summary of Environmental Consequences
of the Proposed Action
and Proposed Mitigation Measures

Resource	Section	Environmental Consequences	Proposed Mitigation Measures
LAND RESOURCES	4.1	See below	See below
Land Use	4.1.1	Minor land disturbances would result from many activities under the Proposed Action. As much as 30 acres (12 ha) of disturbance could be required for each new steamflood system site. Total land disturbance would total less than 250-300 acres (100-120 ha), mostly concentrated near the center of NPR-3. Minor land disturbances would also result from livestock grazing.	Areas of surface disturbance would be reclaimed using procedures recommended by the U.S. Soil Conservation Service. A Grazing Management Plan would be developed detailing impacts and mitigation.
Aesthetics	4.1.2	No potential for profound impacts.	No mitigation necessary.
Recreation	4.1.3	No potential for profound impacts.	No mitigation necessary.
AIR QUALITY AND ACOUSTICS	4.2	See below	See below
Meteorology and Climate	4.2.1	No potential for profound impacts.	No mitigation necessary.
Air Quality	4.2.2	Minor increases in emissions (primarily H_2S from steamflooding and waterflooding operations) would result. Temporary, minor increases in fugitive dust would result from construction activities. Stack emissions from steam generators and other sources would be maintained within permitted	Although all air emissions would be maintained within permitted levels, continued development would control H_2S production through use of flare facilities and microbial and/or chemical treatment of wells and/or produced gas. Obtain a Clean Air Act Title

		levels.	V permit to regulate air emissions.
Acoustics	4.2.3	Minor increases in noise levels from construction and drilling operations would be temporary and would not noticeably alter the existing acoustic levels at NPR-3.	No mitigation necessary.
WATER RESOURCES	4.3	See below	See below
Surface Water Quantity	4.3.1	The quantity of effluent discharged to surface water on NPR-3 may increase with the operation of a biological treatment area. The quality of the water should improve with the biotreatment area.	No mitigation necessary.
Ground Water Quantity	4.3.2	Water would continue to be withdrawn from oil-bearing strata and from Madison formation. Water from all formations underlying NPR-3 is unsuitable for potable use.	No mitigation necessary, except for UIC casing-integrity tests.
Surface Water Quality	4.3.3	No changes are proposed to the effluent limits in the existing NPDES permits. Use of a biological treatment area should improve discharge water quality. There is a potential for sedimentation of streams caused by erosion from soils exposed by construction activities.	Standard sediment and erosion control practices developed in consultation with the U.S. Soil Conservation Service would be followed.
Ground Water Quality	4.3.4	Aquifers exist that may be used for livestock and wildlife. These aquifers require protection to the level of their use, regardless of water quality.	UIC casing-integrity tests. Facility siting and construction standards, SPCC Plan, and spill response procedures

Potable Water	4.3.5	No potential for profound impacts.	No mitigation necessary.
GEOLOGY AND SOILS	4.4	See below	See below
Geology	4.4.1	No potential for land subsidence or other profound impacts.	No mitigation necessary.
Soils	4.4.2	Ground disturbing activities would expose small areas of soils to erosion and would result in loss of topsoil and mixing of subsurface soil horizons. Minor land disturbances would also result from livestock grazing.	Areas of surface disturbance would be reclaimed using procedures recommended by the U.S. Soil Conservation Service. A Grazing Management Plan would be developed detailing impacts and mitigation.
Prime Farmlands	4.4.3	No potential for profound impacts since no prime farmlands are present.	No mitigation necessary.
BIOLOGICAL RESOURCES	4.5	See below	See below
Aquatic Biology	4.5.1	Macroinvertebrates, minnows, and other small aquatic biota present in the streams could be affected by the elevated water temperatures at the NPDES discharge points. No profound impacts to aquatic biota in streams downstream of NPR-3 are expected.	A soil erosion and sediment control plan would be implemented to prevent increased sedimentation of streams.
Terrestrial Vegetation	4.5.2	Rangeland vegetation would be disturbed wherever ground disturbance under the Proposed Action occurs. No forested areas would be disturbed. Minor impacts to vegetation would also result from livestock grazing.	Areas of surface disturbance would be reclaimed using procedures recommended by the U.S. Soil Conservation Service. A Grazing Management Plan would be developed detailing impacts and mitigation.

Terrestrial Wildlife	4.5.3	Oil films and sheens in oil pits and containment ponds could adversely affect wildlife, especially birds. Grazing would reduce the amount of forage present for wildlife.	Oil pits and containment ponds would be netted or eliminated within 3 years. The number of animals allowed to graze at NPR-3 would be closely monitored and severely impacted areas isolated from use. A Grazing Management Plan would be developed detailing impacts and mitigation.
Threatened and Endangered Species	4.5.4	There is no potential for profound impacts due to the infrequent occurrence of listed species.	No mitigation necessary.
Floodplains and Wetlands	4.5.5	Construction in floodplains and wetlands would be limited to road, utility, and pipeline crossings and related structures. Surface topography within floodplains would not be notably modified. Minor land disturbances would also result from livestock grazing, especially near riparian areas.	The number of animals allowed to graze at NPR-3 would be closely monitored and severely effected areas isolated from use. A site specific grazing management plan would be developed.
CULTURAL RESOURCES	4.6	Ground disturbing activities could result in disturbance of surface and subsurface cultural resources.	Mitigation measures, if required, would be developed in consultation with the Wyoming SHPO.
SOCIO-ECONOMICS	4.7	See below	See below
Population and Employment	4.7.1	Employment levels are not expected to increase under the Proposed Action.	No mitigation necessary.
Housing	4.7.2	Existing housing supply in Natrona County is adequate to meet any increase associated with the Proposed Action.	No mitigation necessary.

Transportation	4.7.3	Existing transportation facilities serving the area around NPR-3 are under-utilized.	No mitigation necessary.
Community Amenities	4.7.4	Existing community amenities in Natrona County are adequate to meet any increased demand associated with the Proposed Action.	No mitigation necessary.
HAZARDOUS MATERIALS AND HAZARDOUS WASTE	4.8	See below	See below
Hazardous Waste	4.8	NPR-3 would remain a conditionally exempt small quantity generator (CESQG) under the proposed Action.	No mitigation necessary.
Pesticides	4.8	Pesticide usage would remain limited to occasional use of small quantities of properly labeled insecticides and herbicides as necessary. Potential for spills and other accidental releases.	Procurement and inventory control to minimize quantities. Training and operational procedures to minimize spills. Emergency Response Plan.
Radioactive Waste	4.8	Small quantities of NORM (naturally occurring radio-active materials). Potential for accidental release of sources.	The program to evaluate NORM issues would continue with added parameters.
Waste Disposal	4.8	The Proposed Action would increase the quantity of waste entering the solid waste landfill.	Most solid waste would be hauled offsite. However, the NPR-3 landfill would be kept open to allow continued use of the landfarm for oil-contaminated soil.

4.10 Cumulative Impacts of the Proposed Action and Alternatives

The cumulative impacts of continued development under the Proposed Action are expected to be minimal. Most areas within NPR-3 have previously been used for petroleum development and extraction, and activities would be concentrated within an already existing area of intensive oilfield development. Additionally, although 250 wells are proposed to be drilled over the next five years, approximately the same number of existing wells would be plugged and abandoned, so that the overall number of operating wells would not increase. By employing environmentally sound design, engineering, and mitigation practices, adverse impacts associated with continued development of NPR-3 would be reduced and made relatively short-term. The relative remoteness of NPR-3 from population centers and other sensitive environmental resources lessens the likelihood of cumulative impacts occurring to either the human or natural environment.

The cumulative impacts of the EOR Technology Alternative would be similar to those of the Proposed Action. The EOR techniques would affect those areas of NPR-3 where production efforts are currently underway. The various well stimulation techniques proposed under this alternative would change the types of air emissions, namely increasing the amount of carbon dioxide released. By using environmentally sound engineering and mitigation practices, the impacts associated with this alternative would also be reduced.

Divestiture of NPR-3 would produce individual impacts similar to those of the Proposed Action in regard to environmental concerns, however, the socioeconomic impacts would be greater. The methods that would be used by a private operator to manage NPR-3 would be similar to those proposed under the Proposed Action, but the number of employees required would be less. The resultant impacts from a reduction in force would be felt by all of the surrounding communities.

The greatest cumulative impact from the Divestiture Alternative, however, would be the difficulty in ensuring mitigation of the impacts of routine oilfield operation. Effects that would be detrimental to the environment, but that are not regulated by Federal, state or local laws, would be difficult, if not impossible, to mitigate even through covenants attached to the sale of the property.

The environmental impacts of the No-Action Alternative would slowly decrease as wells and facilities were shut in and abandoned. Coinciding with the decrease in environmental impacts would be a rise in socioeconomic impacts from the resultant reduction in force. Again, the reduction of staffing levels at NPR-3 would have a negative effect on the economy of the surrounding communities, especially Midwest and Edgerton. Although most of these impacts could be mitigated through career placement programs and other methods, the impacts to local housing values could not be mitigated. Additionally, the No-Action Alternative would not be consistent with the Congressional mandate to operate NPR-3 at the MER.

The cumulative impacts of the Decommissioning Alternative would be similar to those of the No-Action Alternative, except that the rates of all impacts would be increased. Under this alternative, operations at NPR-3 would cease immediately. Therefore, negative impacts to the socioeconomics of the region would also be immediate. Although most of these impacts could be mitigated through career placement programs and other methods, the impacts to local housing values could not be mitigated.

5.0 LIST OF PREPARERS

The following DOE employees contributed toward the preparation of this EA:

Miles, David A. Environmental Manager, Quality Assurance

Currently enrolled in Masters program in Environmental Management
Years of Experience: 23 (Government)

Rochelle, Joe Director of Engineering, Technical Review
B.S., Petroleum Engineering, 1979, University of Wyoming
Years of Experience: 14

The following employees of FD contributed toward the preparation of the Sitewide EA:

Doyle, David H. Environmental Engineer, Technical Reviewer & Editor
B.S., Petroleum & Natural Gas Engineering, 1979, Pennsylvania State University
Years of Experience: 15

Soehn, Lynette R. Environmental Specialist, Technical Reviewer & Editor
B.S., Biological Science, 1990, Colorado State University
Years of Experience: 5

The following employees of Halliburton NUS contributed toward the preparation of the Sitewide EA:

Barr, Ralph P. Technical Reviewer
B.A., Biology, 1972, Slippery Rock University
Years of Experience: 20

Doub, J. Peyton EA Coordinator, Soils and Geology, Biological Resources, Water Resources
M.S., Botany, 1984, University of California at Davis
B.S., Plant Sciences, 1982, Cornell University
Years of Experience: 10

Hoffman, Robert G. EA Coordinator, Land Resources, Air Quality and Acoustics
B.S., Environmental Resource Management, 1986, Pennsylvania State University
Years of Experience: 9

MacConnell, James M. Biological Resources
B.S., Zoology, 1974, University of Maryland
Years of Experience: 18

Ruff, J. Gregory Socioeconomics, Technical Review
M.P., Urban and Environmental Planning, 1992, University of Virginia
B.A., Political Science, 1987, James Madison University
Years of Experience: 6

Summerhill, Robin A. Cultural Resources, Waste Management
B.S., Environmental Studies, 1989, State University of New York and Syracuse University
Years of Experience: 5

Wunsch, David M. Project Manager
B.S., Animal Science, 1971, Rutgers University
M.S.C., Applied Physiology, 1973, University of Guelph
M.S., Pharmacology, 1976, New York Medical College

M.B.A., Management 1986, Loyola College
Years of Experience: 15

6.0 LIST OF AGENCIES AND PERSONS CONSULTED

Natrona County Public Library
307 East Second Street
Casper, WY 82601

Wyoming Department of Environmental Quality
Dennis Hemmer, Director
Herschler Building
122 West 25th Street
Cheyenne, WY 82002

Wyoming Department of Environmental Quality
Phil Ogle, Water Quality Division
Herschler Building
122 West 25th Street
Cheyenne, WY 82002

State Historic Preservation Officer
John Keck
2301 Central
4th Floor, Barrett Building
Cheyenne, WY 82002
ATTN: Karen Kempton

Wyoming Outdoor Council
201 Main Street
Lander, WY 82520
ATTN: Bonnie

7.0 BIBLIOGRAPHY

Brockmann, S. 1993. Personal communication dated January 13, 1993 between S. Brockmann of the US Fish and Wildlife Service and representatives of the Halliburton NUS.

Cadwell, M. 1993. Personal communication dated January 14, 1993 between M. Cadwell, Natrona County School District No. 1 and G. Ruff of Halliburton NUS.

Calder, D. 1993. Personal communication dated January 26, 1993 between Denise Calder, Natrona County Sheriff's Office, and G. Ruff of Halliburton NUS.

Casper Area Economic Development Alliance Inc. (CAEDA) 1992. Casper, Wyoming Economic Profile.

Casper Area Economic Development Alliance Inc. (CAEDA) 1993. Untitled information tables provided to G. Ruff of Halliburton NUS. January, 1993.

Catellier, L. 1993. Personal communication dated January 14, 1993 between L. Catellier, Natrona County School District No. 1 and G. Ruff of Halliburton NUS.

COE (US Army Corps of Engineers). 1987. Corps of Engineers Wetlands Delineation Manual. Environmental Laboratory, Waterways Experiment Station, US Army Corps of Engineers. Technical Report Y-87-1.

Collins, T. C. 1993. Letter dated January 22, 1993 from T. Collins, Environmental Coordinator, Office of Director, Environmental Services, Wyoming Game and Fish Department, to P. Doub of Halliburton NUS. Subject: Threatened and endangered species on NPR-3.

Cowardin, L. M., V. Carter, F. C. Golet, and E. T. LaRoe. 1979. Classification of Wetlands and Deepwater Habitats of the United States. US Fish and Wildlife Service, Office of Biological Services, Washington, DC. FWS/OBS-79/31.

Davis, C. P. 1993. Letter dated January 14, 1993 from C. Davis, Site Supervisor, Wyoming State Office, U.S. Fish and Wildlife Services, to P. Doub of Halliburton NUS. Subject: threatened and endangered species on NPR-3.

Davis, G. 1993a. Soil survey map and descriptive information for NPR-3. Provided by G. Davis of the USDA Soil Conservation Service to P. Doub of Halliburton NUS.

Davis, G. 1993b. Letter dated January 4, 1993 from G. Davis of the USDA Soil Conservation Service to P. Doub of Halliburton NUS. Subject: Prime Farmlands.

Davis, G. 1993c. Natrona County Hydric Soils. List of hydric soil mapping units provided by G. Davis of the USDA Soil Conservation Service to P. Doub of Halliburton NUS on January 12, 1993.

Doyle, D. 1993. Personal communication dated February 9, 1993 between D. Doyle of FD Services, Inc. and P. Doub of Halliburton NUS. Subject: Water quality.

Dundas, R. 1993. Personal communication dated January 26, 1993 between Bob Dundas, State of Wyoming Department of Environmental Quality and G. Ruff of Halliburton NUS.

Dunn. 1993. Letter dated January 12, 1993 from C. S. Dunn of FD Services, Inc. to C. R. Williams of US Department of Energy. Subject: NPDES Semi-Annual Report.

Esmoil, B. J. 1991. Wildlife Mortality Associated with Oil Pits in Wyoming. Final Reports, Wyoming Cooperative Fish and Wildlife Research Unit. February 1991.

FD Services, Inc. 1992a. 1991 Site Environmental Report. Naval Petroleum Reserve No. 3, Naval Oil Shale Reserves Nos. 1, 2 and 3. 37 pp.

FD Services, Inc. 1992b. NPOSR-CUW Subcontractor/Vendor/Visitor Environmental/Safety & Health Orientation (w/ Hydrogen Sulfide Data Sheet), revised 10/15/92

FD Services, Inc. 1992c. Application to Renew Industrial Solid Waste Disposal Facility Permit. FD

Services, Inc. December 9, 1992.

FD Services, Inc. 1993. Fax dated January 26, 1993 from A. Khatib of FD Services, Inc. to Robin Summerhill of Halliburton NUS. Subject: Underground Storage Tanks and the attachments for the Internal Memorandum regarding Chemical Usage dated December 3.

FD Services, Inc. 1994. Internal FD Services Memorandum dated September 1, 1994 from the Water Facilities Supervisor to the Environmental Manager. Subject: Chemical Usage - NPR-3.

Fosdick, M. R. 1990. Letter dated November 21, 1990 from M. Fosdick of John Brown E & C, Inc. to C. R. Williams of the US Department of Energy. Subject: Underground Storage Tank Management Plan.

Fosdick, M. R. 1992b. Letter dated March 18, 1992 from M. R. Fosdick of John Brown E & C, Inc. to J. Strohman of the Wyoming Department of Environmental Quality. Subject: Discussion of Well 17 WX 21, Kill Procedure. Letter No. 92-191:MRF.

Halliburton NUS. 1993. Data collection trip meeting notes of P. Doub and R. Hoffman of Halliburton NUS, during visit to NPR-3 the week of January 11, 1993.

Hatcher, J. and J. Goss. 1995. Class III Cultural Resource Inventory of the naval Petroleum Reserve No. 3 (Teapot Dome Oil Field). 203 pp. plus appendices.

Hill, D. 1993. Personal communication dated January 27, 1993 between David Hill, Director of Public Utilities, City of Casper, Wyoming and G. Ruff of Halliburton NUS.

Honeycutt, V. 1993. Personal communication dated January 26, 1993 between Vicki Honeycutt, City of Casper Police Department and G. Ruff of Halliburton NUS.

JBEC. 1990. Environmental Department Pesticide Management Plan. October 1, 1990. 3 pp.

JBEC. 1991. Naval Petroleum Reserve No. 3, Naval Oil Shale Reserves 1, 2 and 3, Cultural Resource Management Plan. John Brown E & C, Inc. US Department of Energy, Naval Petroleum and Oil Shale Reserves, Casper, Wyoming. 7 pp.

Keck, J. T. 1993. Letter dated January 12, 1993 from J. Keck of the Wyoming Department of Commerce, Division of Parks and Cultural Resources, State Historic Preservation Office to P. Doub of Halliburton NUS. Subject: Naval Petroleum Reserve No. 3 (NPR-3) Development, SHPO #0193JKW012.

Keller, T. 1993a. Personal communication dated January 13, 1992 between T. Keller of the Natrona County Planning Commission and representatives of the Halliburton NUS.

Keller, T. 1993b. Personal communication dated January 27, 1993 between Tom Keller, Natrona County Planning Department and G. Ruff of Halliburton NUS.

Khatib, A. 1993a. Personal communication dated February 9, 1993 between A. Khatib of FD Services, Inc. and R. Hoffman of Halliburton NUS.

Khatib, A. 1993. FAX dated January 27, 1993 from A. Khatib of FD Services, Inc. to P. Doub and R. Hoffman of Halliburton NUS.

Khatib, A. 1993. Personal communication dated January 26, 1993 between A. Khatib of FD Services, Inc. and R. Summerhill of Halliburton NUS. Subject: Underground Storage Tanks.

Lane, T. 1992. Internal memorandum from the Environmental Manager of John Brown E & C, Inc. Subject: placement of bird netting on NPR-3 ponds.

Lawrence Allison. 1987. CERCLA Phase I Assessment for Naval Petroleum Reserve No. 3. May 1987. Lawrence Allison and Associates, West. 32 pp. plus appendices.

Leech, H. B. 1966. The *Pedalis*-Group of *Hygrotus*, With Descriptions of Two New Species and a Key to the Species (Coleoptera: Dytiscidae). Proceedings of the California Academy of Sciences (Fourth Series) 23(15): 481 - 498.

Leek, D. 1993. Personal communication dated January 12, 1993 between Dennis Leek of the Wyoming Transportation Department and P. Doub of Halliburton NUS.

Loomis, D. 1993. Personal communication dated January 26, 1993 between Donna Loomis, City of Casper Fire Department and G. Ruff of Halliburton NUS.

Morris, L. 1993. Personal communication dated January 19, 1993 between Linda Morris of the US Bureau of the Census and G. Ruff of Halliburton NUS.

NTCHS. 1991. Hydric Soils of the United States. In cooperation with the National Technical Committee for Hydric Soils. USDA Soil Conservation Service. Miscellaneous Publication Number 1491.

Natrona County. 1978. Land Use Plan, Natrona County, Wyoming. Prepared for the Natrona County Board of Commissioners. 143 pp.

Neighbours, M. L. 1993. Letter dated January 11, 1993 from M. Neighbours of the Nature Conservancy, Wyoming Natural Diversity Data Base to P. Doub of Halliburton NUS. Subject: Review of Wyoming Natural Diversity Database for threatened and endangered species recorded on NPR-3.

Page, L. M. and B. M. Burr. 1991. A Field Guide to Freshwater Fishes, North America North of Mexico. Houghton Mifflin Company, Boston.

Raffelson, C. N. 1992. Letter dated November 24, 1992 from C. Raffelson of the Wyoming Department of Environmental Quality to A. Khatib of FD Services, Inc. Subject: Air Quality Permitting for pilot flaring system to control H_2S at NPR-3.

Rand McNally. 1992. Commercial Atlas and Marketing Guide.

SCS. 1992. Recommended Reclamation Procedures for Naval Petroleum Reserve No. 3. Prepared for L. LaFreniere of FD Services by G. Davis of the USDA Soil Conservation Service.

Soehn, L. 1993. Personal communication dated February 8, 1993 between L. Soehn of FD Services, Inc. and R. Hoffman of Halliburton NUS. Subject: Bald eagle observation on NPR-3.

Soehn, L. 1993a. Letter dated July 14, 1994 from L. Soehn of Fluor Daniel (NPOSR), Inc., to C. Williams of the US Department of Energy. Subject: 1993 Wyoming Department of Environmental Quality Emission Inventory Report.

Stark, C. 1991a. Internal memorandum from the Environmental Specialist of John Brown E&C Inc. regarding dead golden eagle found under power pole near 46-A-21.

Stark, C. 1991b. Internal memorandum from the Environmental Specialist of John Brown E&C Inc. regarding injured shrike at NPR-3.

Stark, C. 1993. Personal communication dated February 4, 1993 between C. Stark, Environmental Specialist, FD Services and J. MacConnell of Halliburton NUS. Subject: threatened and endangered species and wildlife.

State of Wyoming. 1992a. Wyoming Income and Employment Report, Department of Administration and Information, Division of Economic Analysis. April, 1992.

State of Wyoming. 1992b. Wyoming Economic Forecast Report, Department of Administration and Information, Division of Economic Analysis. February, 1992.

Stebbins, R.C. 1985. A Field Guide to Western Reptiles and Amphibians. Houghton Mifflin Company, Boston.

Sullivan, J. 1993. Personal communication dated January 26, 1993 between Jim Sullivan, Natrona County Fire Department, and G. Ruff of Halliburton NUS.

Thiele, D. 1993. Personal communication dated January 14, 1993 between D. Thiele of the Wyoming Game and Fish Department and representatives of the Halliburton NUS.

Thiele, D. and Condor, A. 1993. Personal communication dated February 3, 1993 between D. Thiele and A. Condor of the Wyoming Game and Fish Department and J. MacConnell of Halliburton NUS. Subject: threatened and endangered species, water quality, and aquatic biota.

UNC Remediation. 1990. Naval Petroleum Reserve No. 3, Waste Petroleum Program Plan. Prepared for John Brown E & C, Inc. April 1, 1990. P.O. No. 00309.

DOE. 1990. Environmental Assessment for Continued Development of Naval Petroleum Reserve No. 3. U.S. Department of Energy, Naval Petroleum Reserves in Colorado, Utah and Wyoming. EA-0442.

DOE. 1992a. Fossil Energy NEPA Guidance Manual. December 2, 1992.

DOE. 1992b. Draft Tiger Team Assessment of the Naval Petroleum and Oil Shale Reserves, Colorado, Utah and Wyoming. US Department of Energy, Office of Environment, Safety and Health. Volume 1 of 2.

USGS. 1974. Arminto, Wyoming 1:250,000 scale topographic quadrangle.

US National Park Service. 1991. National Register of Historic Places, 1966 -1991. United States National Park Service in Cooperation with American Association of State and Local History and the National Conference of State Historic Preservation Officers. Washington, DC.

US Navy. 1976. Final Environmental Impact Statement, Development of Naval Petroleum Reserve No. 3. U.S. Department of the Navy, Naval Petroleum and Oil Shale Reserves. 152 pp. plus appendices.

Watson, J. P. 1987. Letter dated April 15, 1987 from J. Watson of Lawrence-Allison & Associates West, Inc. to C. R. Williams of the US Department of Energy. Subject: Contract No. DE-AC01-86FE60896. Letter No. 87-187:JPW.

Western Environmental Services and Testing, Inc. 1988. Subsurface Soil Study for 2,4 Dichlorophenoxyacetic Acid (2,4-D), Naval Petroleum Reserve #3, Midwest, Wyoming. November 1988. File Number 8840-425.

WGFD. 1991. Draft Distribution and Status of Wyoming Birds and Mammals. Wyoming Game and Fish Department, Game Division. Edited by Bob Oakleaf, Bob Luce, Sharon Ritter and Andrea Cerovski. 152 pp. plus attachment.

WGFD. 1993. Output dated January 7, 1993 from the Wildlife Observation System for Township T-38-39N, Range 78W. Wyoming Game and Fish Department.

Wyoming State Data Center. 1992. WSDC Bulletin.

Young, J. F. 1986. Letter dated September 9, 1986 from J. Young of the USDA Soil Conservation Service, Casper Field Office to David R. McCallister of Lawrence-Allison & Associates West, Inc. Subject:

APPENDIX A - RESPONSE TO COMMENTS

The following concerns and comments were noted during the public comment phase. Each issue is listed below and is immediately followed by a response, in bold. Copies of all letters received appear at the end of this section.

Issues 1 through 4 were raised by the Wyoming Department of Environmental Quality.

1a. The first issue pertains to the potential for contamination of soil, groundwater, and surface water from reserve pits and disposal activities during drilling, and from waste water treatment and disposal ponds during production activities. This concern arises from WQD's responsibility to protect surface and ground water and to resolve violations of the standards when they occur. The Wyoming Oil and Gas Conservation Commission has regulatory authority over the construction, location, operation and reclamation of oil field pits within a lease, unit or communitized area which are used solely for the storage, treatment, and disposal of drilling, production and treater unit wastes; and WQD encourages close coordination with the Commission.

The only reference to groundwater quality is contained in section 3.3.4 entitled "Ground Water Quantity" (page 3-8). The EA indicates that Steele shale occupies the interval from the surface to an approximate depth of 2,000 feet. If there are no shale-isolated marine sand body aquifers with this interval, the EA presents an accurate assessment of groundwater quality and potential use.

There are two porous and permeable sandstone formations within the Steele Shale. The Sussex sandstone outcrops in a ring near the center of the Teapot Dome structure, and does not appear to be an aquifer. The second sandstone body is the Shannon sandstone which is an oil reservoir in much of the field. A fault separates the oil reservoir from the Shannon outcrop at Salt Creek to the north. Groundwater is encountered in the Shannon in some areas north of the fault, but the concentration of Total Dissolved Solids exceeds 10,000 mg/l.

1b. In section 3.3.2 (page 3-7), the discussion of "USDWs" and "or other fresh water aquifers: is confusing. If the groundwater is less than 10,000 mg/l of TDS, it is considered a USDW and a fresh water since both USDW and fresh water must meet the criteria of being less than 10,000 mg/l of TDS.

The statement that "If the groundwater is less than 10,000 mg/l of TDS, it is considered a USDW" is not entirely correct. Exempted aquifers are not USDW's under the Safe Drinking Water Act, which permits aquifer exemptions for fresh water aquifers being used for Class II injection. Several such aquifer exemptions exist at NPR-3. In addition, aquifers that contain crude oil, natural gas, or other contaminants that make it undesirable for a water supply can also be exempted. Several other aquifers at NPR-3 qualify for exemption under this criteria, although the actual exemption has not been pursued with the Wyoming Oil & Gas Conservation Commission.

As a result, there is a strong distinction at NPR-3 between "fresh water aquifers" and "USDWs". Produced water from oil and gas production is put to beneficial use for livestock and wildlife at NPR-3, but there would be no intention to protect it as a source of municipal water supply. The Madison formation, at the bottom of the geologic column, could be considered a USDW, but activities at NPR-3 are not likely to impact this aquifer.

1c. The permeability of the Steele shale below the surface coupled with the weathering of shale at the surface provides a pathway for shallow groundwater to recharge surface water drainages. Recharge water could come from precipitation or seepage from waste water facilities. Existing problems have required groundwater pollution remediation systems in the Salt Creek area to prevent oil seeps from entering surface waters via groundwater recharge. This indicates consideration for adequate design, construction and operation of reserve pits and disposal facilities is necessary to protect water resources.

Wyoming Groundwater Standards contained in Chapter VIII require the protection of all groundwaters of the state for existing and potential uses. The standards prohibit the discharge of biological, hazardous, or toxic materials or substances into shallow groundwater. The main purpose of the standards is to protect all uses for which the groundwater is suitable and to protect against migration of contaminants to useable groundwaters or surface waters.

The Department of Energy recognizes its responsibility to protect shallow groundwater and to comply with DEQ regulations. Reserve pits and other production facilities will be sited and constructed according to WOGCC standards for critical areas (Rule 401) when they are applicable. Further, the Department of Energy is not aware of any situations at NPR-3, where surface waters have been threatened by oil being transported by groundwater recharge, such as the comments describe.

1d. The statement in section 4.3.4 Ground Water Quality Mitigating Measures (page 4-11), which indicates protection of groundwater is not necessary because there are no fresh water aquifers yielding potentially useable potable water, is incorrect. Appropriate design, construction and operation of reserve pits, earthen waste water treatment facilities and disposal ponds need to be included as mitigation measures to prevent soil, groundwater and surface water contamination. Spill prevention and clean up procedures should be addressed in section 4.3.4 along with other groundwater quality protection methods.

The design, construction and operation of oilfield facilities have added as a mitigation measure. Comments addressing spill prevention, cleanup procedures and other groundwater quality protection methods have also be added to the final EA.

2. The second issue concerns WQD's policy on coverage under the general NPDES storm water discharge permit. Runoff from construction activities has been defined as a point source by EPA and a permit is required. In September 1992, the WQD issued its general permit for storm water discharges form construction activities. WQD's policy for construction activities associated with oil and gas development is that a pollution prevention plan must be prepared and notification for coverage under the Wyoming general permit must be given for each well and associated facilities (roads, pipelines, tank batteries, etc.) that disturb five acres or more.

Additionally, construction of other facilities would require coverage under the general permit if they disturb five acres or more. An indication that coverage under the general permit is required should be given in section 1.5 and should be included as a mitigation measure.

The need for a general stormwater discharge permit has been identified in the EA for facilities that disturb five acres or more.

3. The third issue pertains to livestock grazing on the Naval Petroleum Reserve. This action is not fully analyzed in the EA. Although listed in Table 2-4 (page 2-14) and mentioned in section 2.1.3 (page 2-13), oversight and management of livestock grazing are not adequately presented as part of the proposed action. Discussions provided in section 4.1.1, 4.4.2, 4.5.2 and 4.5.5 regarding the impacts of grazing are inadequate, and a site specific grazing management plan should be prepared. The plan should specify current range and riparian plant community condition; planned season of use; utilization standards; planned improvements; grazing Best Management Practices (BMPs) that will be used to protect rangeland and riparian resources and water quality; and any other factors relevant to the management of livestock grazing on the Reserve. Preparation of the grazing management plan could be considered a mitigation measure for impacts of livestock and wildlife utilization. If a grazing management plan is prepared as a separate document and not incorporated in the final EA or Decision Document, WQD requests that the plan be made available for review.

A site specific grazing management plan has been prepared and included as an attachment to the EA. Discussion of impacts and mitigation has been expanded upon in the EA to include the topics of concern described above.

4. Review of the Environmental Assessment for the Naval Petroleum Reserve No 3. indicates that there are several proposed projects such as pumping facilities, H_2S treatment facilities, natural gas compressors, flares to burn H_2S contaminated natural gas, expansion or modification of the natural

gas processing plant, and enhanced oil recovery technologies which are potential sources of air emissions. Wyoming Air Quality Standards and Regulations, Section 21, requires that any person who plans to construct any new facility or source, modify any existing facility or source, or to engage in the use of which many cause the issuance of or an increase in the issuance of air contaminates into the air shall obtain a construction permit from the State of Wyoming, Department of Environmental Quality before any actual work is begun on the facility. Section 21 (b) further explains the permit application procedures.

Also note that page 3-5 of the Environmental Assessment states that predicted emissions of Nitrogen Oxides (NOx) increased to greater than 100 tons per year in 1993. This level exceed the Title V Operating Permit Program threshold and will subject NPR No. 3 to the permitting requirements of Section 30 of the Wyoming Air Quality Standards and Regulations.

The Department of Energy is aware of the need for a Title V Operating Permit. A consulting firm is in the process of preparing the permit application, which will address all regulations applicable to NPR-3's current and planned operations.

Issue 5 was raised by the Wyoming State Office of Historic Preservation (SHPO).

5. The last cultural resource inventory performed at NPR-3 was in 1976. Cultural resource surveys conducted prior to 1980 usually do not meet current inventory or evaluation standards. Additionally, "it is stated on page 4-21 that, 'Impacts to cultural resources from the Proposed Action would be limited to the effects of ground disturbing activities'. This statement is inconsistent with standard practices for determining effect, as outlined in *National Register Bulletin 15, How to Apply the National Register Criteria for Evaluation*. Oil field activity has the potential to impact integrity of setting, design, workmanship, feeling, and association for sites significant under National Register criteria A, B, or C. Both direct and indirect effects to historic properties must be considered on a project by project basis."

A Class III Cultural Resource Inventory was conducted of NPR-3 from February to June of 1995. This information will be sent to the SHPO. As stated in the text, only two sites were found to be eligible for listing on the National Register. These sites will be avoided. To address the possibility of effects to cultural resources from groundbreaking activities, and to meet requirements of the National Historic Preservation Act, a Programmatic Agreement will be developed with the Wyoming SHPO and the National Advisory Council.

A Class III cultural resources inventory has been contracted for NPR-3. Work is expected to start in February, 1995 and will be completed within a 60-day project schedule (depending on snow cover). Copies of the final report will be made available to the SHPO when it is published. At that time, the Department of Energy will also submit a proposed Memorandum of Agreement to cover mitigation of potential impacts to the cultural resources that were found in the survey. The MOA will be signed before a decision document is completed by the Department of Energy. If appropriate, the decision document would be a Finding of No Significant Impact (FONSI).

Department of Energy Continued Development - Naval Petroleum Reserve No. 3 Finding Of No Significant Impact

Proposed Actions: The Proposed Action is the continued development of the Naval Petroleum

Reserve No. 3 (NPR-3) for five years. Continued development includes all activities typically required to profitably manage a mature oilfield such as NPR-3.

In addition to the continued development of oil and gas resources, it is proposed to fully develop the Rocky Mountain Oilfield Testing Center (RMOTC). The mission of RMOTC would be to provide facilities and necessary support to government and private industry, for testing and evaluating new oilfield and environmental technologies, and to transfer these results to the petroleum industry through seminars and publications.

Type of Statement: Final Environmental Assessment (EA)

Lead Agency: The United States Department of Energy

Cooperating Agencies: None

For Further Information:
David A. Miles
NEPA Compliance Officer
United States Department of Energy
Naval Petroleum and Oil Shale Reserves in Colorado,
Utah and Wyoming
907 North Poplar, Suite 150
Casper, Wyoming 82601
(307) 261-5161, ext. 5071

Abstract: Continued development activities under the Proposed Action (the preferred alternative) would include the drilling of approximately 250 oil production and injection (gas, water, and steam) wells, the construction of between 25 and 30 miles of associated gas, water, and steam pipelines, the installation of several production and support facilities, and the construction of between 15 and 20 miles of access roads. This work would be performed over the next five years. These drilling and construction estimates include any necessary activities related to RMOTC operations.

Continued development activities either have no potential to result in adverse environmental impacts or would only result in adverse impacts that could be readily mitigated. Resource types discussed in detail include land resources, air quality and acoustics, water resources, geology and soils, biological resources, cultural resources, socioeconomics, and waste management. Continued development is not expected to result in substantial changes in the types and quantities of air emissions and wastewater discharges already generated by existing operations at NPR-3. Continued development, especially where it involves expansion of EOR activities, would result in small areas of new land disturbance at several locations on NPR-3, especially in the already intensively developed central area.

AGENCY: Naval Petroleum and Oil Shale Reserves

U.S. Department of Energy (DOE)

ACTION: Finding of No Significant Impact (FONSI) for the Continued Development of Naval Petroleum Reserve No. 3 (DOE/EA-1008)

SUMMARY: The Secretary of Energy is required by law to "explore, prospect, conserve, develop, use, and operate" the Naval Petroleum and Oil Shale Reserves. The Naval Petroleum Reserves Production Act of 1976 (Public Law 94¡258), requires that the Naval Petroleum Reserves be produced at their maximum efficient rate (MER), consistent with sound engineering practices, for a period of six years. The President has authorized five 3-year extensions to the six year period since 1982. The United States Department of Energy (DOE) has managed NPR-3 for oil recovery at the "Maximum Efficient Rate" (MER) since 1976.

To fulfill this mission, DOE is proposing continued development activities which would include the drilling of approximately 250 oil production and injection (gas, water, and steam) wells, the construction of between 25 and 30 miles of associated gas, water, and steam pipelines, the installation of several production and support facilities, and the construction of between 15 and 20 miles of access roads. This work would be performed over the next five years and will mainly utilize practices standard to the industry. These drilling and construction estimates include any necessary activities related to the operation of the Rocky Mountain Oilfield Testing Center (RMOTC). The development of the RMOTC at NPR-3 is included as part of continued development activities. The purpose of RMOTC will be to provide facilities and necessary support to government and private industry for testing and evaluating new oilfield and environmental technologies, and to transfer these results to the petroleum industry through seminars and publications.

Continued development activities either have no potential to result in adverse environmental impacts or would only result in adverse impacts that could be readily mitigated. Continued development is not expected to result in substantial changes in the types and quantities of air emissions and wastewater discharges already generated by existing operations at NPR-3. Continued development, especially where it involves expansion of EOR activities, will result in small areas of new land disturbance at several locations on NPR-3, especially in the already intensively developed central area. The small amounts of disturbed surface area will be reclaimed to its original natural state when production operations terminate.

DOE prepared an environmental assessment (DOE/EA-1008) that analyzes the proposed projects involved with continued development of NPR-3. Based on the analyses in the EA, the DOE finds that the proposed action is not a major Federal action significantly affecting the quality of the human environment within the meaning of the National Environmental Policy Act of 1969 (NEPA). The preparation of an environmental impact statement is not required, and the DOE is issuing this Finding of No Significant Impact (FONSI).

PUBLIC AVAILABILITY: Copies of the EA and FONSI will be distributed to persons and agencies known to be interested in or affected by the proposed action and will be made available for public inspection at the Natrona County Public Library, Kelly Walsh High School, Natrona County High School and the U.S. Department of Energy Reading Room. Anyone wishing to receive copies of either document, or further information on the proposal, should contact:

Clarke D. Turner
Director
Naval Petroleum and Oil Shale Reserves in Colorado, Utah and Wyoming
U.S. Department of Energy
907 North Poplar, Suite 150
Casper, WY 82601
Phone: (307) 261-5161

For further information on the NEPA compliance process, contact:

David A. Miles
NEPA Compliance Manager
Naval Petroleum and Oil Shale Reserves in Colorado, Utah and Wyoming
U.S. Department of Energy
907 North Poplar, Suite 150
Casper, WY 82601
Phone: (307) 261-5161

SUPPLEMENTAL INFORMATION: Section 7422 of Title 10, United States Code, charges the Secretary of Energy with the authority and responsibility to "explore, prospect, conserve, develop, use, and operate the naval petroleum reserves." This section further provides that the "...naval petroleum reserves shall be used and operated for their protection, conservation, maintenance and testing," and production when authorized.

NPR-3, or Teapot Dome, is a 9,481-acre (3,837 ha) oilfield located in Natrona County, Wyoming, approximately 35 miles (56 km) north of the City of Casper. Production at the Naval Petroleum Reserve No. 3 in Natrona County, Wyoming, began in the 1920s during a time of substantial exploration and production, when leases were issued by the Interior Department under the Mineral Leasing Act. Production was discontinued after 1927 and renewed between 1959 and 1976 in a limited program to prevent the loss of U.S. Government oil to privately-owned wells on adjacent land. In 1976, Congress passed the Naval Petroleum Reserves Production Act (Public Law 94¡258), which requires that the Naval Petroleum Reserves be produced at their maximum efficient rate (MER), consistent with sound engineering practices, for a period of six years. The law also provides that at the conclusion of the initial 6¡year production period, the President (with the approval of Congress) could extend production in increments of up to three years each, if continued production was found to be in the national interest. The President has authorized five 3-year extensions since 1982, extending production continuously through April 5, 1997.

The Proposed Action is the continued development of NPR-3 for the next five years. Continued development includes all activities typically required to profitably manage a mature stripper oilfield, such as NPR-3, at the MER. Continued development comprises four general categories of activity: continued development drilling utilizing conventional oil recovery technologies; continued and expanded use of Enhanced Oil Recovery (EOR) techniques that are necessary for continued oil production from reservoirs after primary or secondary recovery; continuation of general operations and support activities; and full implementation of the Rocky Mountain Oilfield Testing Center.

Continued development activities either have no potential to result in adverse environmental impacts or would only result in adverse impacts that could be readily mitigated. The Sitewide EA summarizes the potentially affected environment at NPR¡3 as of 1994, discusses all potentially adverse environmental impacts, and proposes specific mitigation measures that offset each identified adverse impact. Resource types discussed in detail include land resources, air quality and acoustics, water resources, geology and soils, biological resources, cultural resources, socioeconomics, and waste management.

Continued development of NPR-3, as outlined in the Proposed Action, would not substantially alter the character of existing operations and would be consistent with NPR-3's historic role as an oilfield.

Continued development is not expected to result in major changes in the types and quantities of air emissions and wastewater discharges already generated by existing operations at NPR-3. Continued development, especially where it involves expansion of EOR activities, would result in small areas of new land disturbance at several locations on NPR-3, especially in the already intensively developed central area.

Alternatives to the Proposed Action that were reviewed include: other chemical and thermal EOR technology alternatives to maintain oil and gas production, divestiture of NPR-3 by the Federal government, a no-action alternative of continuing operation of NPR-3, but without further development, and the immediate decommissioning of the project.

DETERMINATIONS: Based on the findings of the EA, DOE has determined that the proposal does not constitute a major Federal action significantly affecting the quality of the human environment within the meaning of NEPA. Therefore, an environmental impact statement is not required, and DOE is issuing this FONSI.

Issued in Casper, WY, _____, 1995

_____

Clarke D. Turner
Director
Naval Petroleum and Oil Shale Reserves in Colorado, Utah and Wyoming

Appendix F.1

Example 3: Environmental Assessment for the Transfer of the Department of Energy Grand Junction Office to Non-DOE Ownership (Preliminary Final)

Agency: U.S. Department of Energy
Preparer: Tetra Tech, Inc., Falls Church, Virginia and Gaithersburg, Maryland
Date: February 2000
Action: Transfer of real property out of Federal ownership

Overview

The subject document is one of several EAs that have been prepared since the early 1990s for the transfer of government property out of federal ownership. Such transfers have become frequent since the end of the Cold War and have become quite numerous in recent years as a result of an emphasis upon smaller, leaner government. Large numbers of small and/or functionally redundant military bases have been closed under the Base Closure and Realignment Act of 1988 (BRAC). Land and/or buildings on many other military bases and other government installations have been determined to be excess, i.e. not required to carry out current and reasonably foreseeable mission-related activities. Excess real property and property on closed bases is typically transferred to local governments or commissions interested in reuse and redevelopment of the property, or it is sold directly to development interests. Although the likelihood of future rounds of base closures under BRAC are uncertain, it is highly probable that the excessing of federal property will continue to be frequent as long as current political trends continue.

The National Environmental Policy Act (NEPA) document for the disposal and reuse of entire bases closed under BRAC has typically been an Environmental Impact Statement (EIS). Most EISs for disposal and reuse of closed military bases have focused on issues such as environmental contamination from past military activities; human health concerns such as asbestos and lead-based paint; traffic, noise, and other issues related to the regional compatibility of proposed reuses; the potential for air and water discharges from reuse activities; disturbance of historical and archaeological (cultural) resources; and disturbance of ecological resources, including wetlands and threatened and endangered species. The latter two issues (cultural resources and ecological resources) are frequently of critical importance when assessing reuse alternatives involving new development of previously undeveloped buffer lands. NEPA documentation for BRAC closures is technically limited to a consideration of potential impacts associated with reuse of the closed bases, using the closed base as the baseline (technically, the actual decision to close a base under BRAC is not subject to review under NEPA). However, the documents usually also consider the economic effects of the closure decision as well.

Transfer of smaller bases or government installations, or portions of larger installations, out of federal ownership is sometimes addressed in an EA and FONSI instead of an EIS. The use of EAs rather than EISs for the transfer of property will likely become increasingly frequent as federal agencies continue to downsize real property holdings in less dramatic, less controversial increments than the large-scale base closures under BRAC.

The following example is an EA prepared by the U.S. Department of Energy (DOE) for the transfer to private ownership of a small, specialized industrial installation of decreased importance in the post-Cold War era. Although the DOE is not completely vacating the facility, it plans to privatize the entire facility and then lease those facilities that continue to be needed. The installation is a 62-acre tract with 35 mostly industrial and administrative structures. The tract is almost completely paved except for approximately 25 acres of open space bordering an adjoining river.

The decision to prepare an EA rather than an EIS was not clear-cut. The transfer involves an entire installation, the installation has a long history of industrial activity, and the installation has experienced extensive environmental cleanup and restoration efforts, including the cleanup and restoration of a series of wetlands adjoining the river. The preparers of the EA had to clearly demonstrate that the range of possible privately-sponsored development activities anticipated for the property would not result in significant environmental impacts.

NEPA documents for the transfer of property out of federal ownership typically consider a series of reuse scenarios (usually three to five), each differing with respect to the distribution of industrial, commercial, residential, and recreational land uses across the property. This EA did likewise, but with one key difference: rather than addressing each reuse scenario as a separate alternative, it considered four reuse scenarios as four potential outcomes resulting from a single action, which is transferring the property out of federal ownership. The only other alternative considered in the EA is a no-action alternative, which is to keep the property in DOE ownership.

SPECIFIC COMMENTS

The following specific comments are targeted to specific sections and figures in the EA.

1. Page vi, Definitions: Although not mandatory, a glossary such as this helps make an EA more understandable to the public. Although EAs, and other NEPA documents, are technical analyses written by technical experts, they must be prepared for a non-technical audience. A good NEPA document must serve to educate readers so that each reader can draw meaningful conclusions on his/her own from the information presented. Although the public involvement process for most EAs is substantially less than for typical EISs, an EA is as much a public relations document as is an EIS.

2. Executive Summary, Page ES-1, measurements: Throughout the document measurements are presented in metric form with U.S. Customary equivalents in parentheses. Although the DOE uses dual measurement values in most of its NEPA documents, this practice detracts from the readability of the text. A better approach would be to use only one measurement system in the text and provide a one-page conversion table as a appendix. For readability and public relations purposes, use of the most recognizable measurement form (usually the U.S. Customary form) is preferable.

3. Page 1-3, Location Map: This rather ingenious approach provides tiers of increasingly detailed maps on one page, thereby avoiding multiple pages of maps that interfere with the flow of the text.

4. Page 2-1, Purpose and Need for Agency Action: Although brevity is desirable in an EA, this is too brief. Additional sentences explaining why the installation is not needed for foreseeable future DOE activities would help convince the reader that the action is necessary. Although defending the purpose and need for an action is more important in an EIS (which addresses actions with a potential for significant impacts), a solid defense of actions documented in an EA is a good public relations practice.

5. Pages 3-1 to 3-3, Description of Proposed Action and No Action Alternatives: Readers must approach this section with care as it describes a Proposed Action with four land use scenarios, followed by a No Action Alternative. One might expect each land use "scenario" to be treated as a separate alternative. The rationale is that the Proposed Action is merely the transfer of the property; the subsequent use by private recipients is out of DOE's control. Comparing the impacts from each scenario provides a way of addressing secondary impacts from with possible reuse scenarios without

formally adopting any reuse activities as part of the Proposed Action.

6. Page 4-1, Geology/Soils/Topography: The lumping of these topics into a single subsection is common in NEPA documents. Unless geological issues (e.g. seismic hazards, effects on mineral extraction) play a key role in the consideration of environmental consequences, the existing environment descriptions can often afford to be simplified.

7. Page 4-10, Floodplains and Wetlands: Although merely referencing flood insurance rate maps (FIRMs) is adequate, a better approach for EAs addressing small sites is to attach copies of the maps as an appendix. The reading and interpretation of FIRMs (and other flood map products) is surprisingly complex; some descriptive guidance in the text as to how the maps were interpreted is a good idea.

8. Page 6-1, References: This format for listing references (literature cited) is commonly used in EAs and EISs prepared by DOE. The format used in scientific literature (Author/Date/Title/Other Data) is prevalent among NEPA documents prepared by other agencies and is preferable.

9. Appendix B: This table lists plant and wildlife species actually observed on the affected site during a systematic survey. It is preferable to a list of species which could potentially occur on the site, which is often provided when site observation data is unavailable. However, the value of such lists to impact assessment is often questionable. A focus on habitats, listed species, food chains, and certain other species of special interest (e.g. landscape plants in urban settings) is usually preferable.

10. It is noted that the use of a table to summarize the information presented in Chapter 4 would augment the readability of this document.

Appendix F.2

**Environmental Assessment For The Transfer
of the Department of Energy Grand Junction Office
to Non-DOE Ownership**

DOE/EA-1338

FINAL

Environmental Assessment for the Transfer of the Department of Energy Grand Junction Office to Non-DOE Ownership

April 2000

U.S. Department of Energy
Grand Junction Office
2597 B ¾ Road
Grand Junction, CO 81503

TABLE OF CONTENTS

List of Figures

List of Tables

Abbreviations and Acronyms

CERCLA	*Comprehensive Environmental Response, Compensation, and Liability Act*
CFR	Code of Federal Regulations
CO	Colorado
DOE	U.S. Department of Energy
DOE-GJO	Department of Energy, Grand Junction Office
EA	environmental assessment
EPA	U.S. Environmental Protection Agency
GJO	Grand Junction Office
GJPO	Grand Junction Projects Office
GJPORAP	Grand Junction Projects Office Remedial Action Project
NEPA	*National Environmental Policy Act*
OSHA	Occupational Safety and Health Administration
PCBs	polychlorinated biphenyls
RCRA	*Resource Conservation and Recovery Act*
USC	United States Code
UMTRA	Uranium Mill Tailings Remedial Action

Definitions

Curie: A unit of activity of a radionuclide. A curie is equal to 3.7×10^{10} disintegrations (i.e., nuclear transformations) per second. (DOE Order 5400.5, Radiation Protection of the Public and Environment)

Decommissioning: Actions taken to reduce the potential health and safety impacts of inactivated DOE contaminated facilities, including activities to stabilize, reduce, or remove radioactive materials or to demolish the facilities. (DOE Order 430.1, Life Cycle Asset Management)

Decontamination: The removal of radioactive contamination from facilities, equipment, or soils by washing, heating, chemical or electrochemical action, mechanical cleaning, or other techniques. (DOE Order 430.1, Life Cycle Asset Management)

Hazardous waste: Those wastes that are listed by EPA in 40 CFR Part 261, Subpart D; exhibit any of the four hazardous waste characteristics (ignitability, corrosivity, reactivity, or toxicity) identified in 40 CFR Part 261, Subpart C; or are a mixture of non-hazardous and hazardous wastes. Listed wastes are divided into three groups according to their origin: nonspecific sources (e.g., spent solvents such as toluene), specific sources (e.g., bottom sediments from the treatment of waste waters from wood preserving), and discarded commercial chemical products, all off-specification species, containers, and spill residues thereof. The Federal statute concerning hazardous wastes is the *Resource Conservation and Recovery Act* (RCRA) (Title 42 U. S. Code [USC] 6901 et seq.). In Colorado, hazardous wastes are also defined by the Colorado Hazardous Waste Regulations at Part 6 Colorado Code of Regulations Sections 1007-3.

High-level waste: The highly radioactive waste material that results from the reprocessing of spent nuclear fuel, including liquid waste, produced directly in reprocessing and any solid waste derived from the liquid, that contains a combination of transuranic waste and fission products in concentrations requiring permanent isolation. (*Nuclear Waste Policy Act* of 1982 [42 USC 10101, et seq.]) and DOE Order 435.1, Radioactive Waste Management)

Institutional Controls: Restrictions on use of land or natural resources. Some examples of implementation methods are by deed restrictions, restrictive covenants, access controls (e.g., fencing), posting of notices, and information distribution. (Adapted from U.S. EPA guidance)

Low-level waste: Waste that contains radioactivity and is not classified as high-level waste, transuranic waste, spent nuclear fuel, or uranium mill tailings (UMT) waste. (*Nuclear Waste Policy Act* of 1982 [Title 42 USC 10101, et seq.] and DOE Order 435.1, Radioactive Waste Management)

Mixed waste: Waste containing both radioactive and hazardous components as defined by the RCRA (42 USC 6901 et seq.), as amended by the *Federal Facilities Compliance Act* (Public Law 102-386). Mixed waste can also include hazardous or radioactive waste containing asbestos or polychlorinated biphenyls (PCBs) which are regulated under the *Toxic Substances Control Act* (15 USC 2601 et seq.).

Naturally occurring radioactive material: Naturally occurring materials not regulated under the RCRA (42 USC 2011 et seq.), as amended, whose composition, radionuclide concentrations, availability, or proximity to man have been increased by or as a result of human practices. Naturally occurring radioactive material does not include the natural radioactivity of rocks or soils, or background radiation. (DOE Order 435.1)

Non-hazardous solid waste: Waste that is not subject to stringent storage, treatment, or disposal requirements and that can be disposed of in a municipal landfill or other *Resource Conservation and Recovery Act* (42 USC 6901 et seq.) Subtitle D facility.

Radiological contamination: Contamination of property, material, or equipment by radionuclides at levels above those specified in DOE Order 5400.5, Radiation Protection of the Public and the Environment, Table IV-1.

Regulated waste: Waste that is deemed to be hazardous, radioactive, mixed, or toxic under the RCRA (42 USC 6901 et seq.), *Atomic Energy Act* (42 USC 2011 et seq.), and *Toxic Substances Control Act* (15 USC 2601 et seq.). These wastes are subject to stringent storage, treatment, and disposal regulations.

Release: The exercising of DOE's authority to release property from its control after confirming that residual radioactive material has been determined to meet the guidelines for residual radioactive material in accordance with DOE 5400.5, Radiation Protection of the Public and the Environment and other applicable radiological requirements. (Adapted from DOE Order 5400.5)

Release CERCLA: Any discharging, dumping, emitting, emptying, escaping, injecting, leaching, leaking, pouring, pumping, spilling of radioactive substances into the environment including abandoning any type of receptacle containing radioactive substances, but does not include disposal in a permitted disposal facility. (42 USC 9601 et seq.) (Adapted from DOE Order M435.1-1)

Release Survey: A radiological inspection conducted in order to verify that buildings are free of radiological hazards. The objective of the release survey is to demonstrate that radiological conditions in buildings satisfy DOE guidelines for unrestricted use upon completion of remediation. (Adapted from DOE Order 5400.5)

Residual radioactive material: Waste in the form of tailings resulting from the processing of ores and other waste related to such processing; these wastes are regulated under the *Uranium Mill Tailings Radiation Control Act* (42 USC 7901 et seq.). It also refers to any radioactive material which is in or on soil, air, equipment, or structures as a consequence of past activities or operations. (DOE Order 5400.5)

Spent nuclear fuel: Fuel that has been withdrawn from a nuclear reactor following irradiation, but that has not been reprocessed to remove its constituent elements. (DOE Order 435.1, Radioactive Waste Management)

Transuranic waste: Without regard to source or form, waste that is contaminated with alpha-emitting transuranium radionuclides with half-lives greater than 20 years and concentrations greater than 100 nanocuries per gram at the time of assay. Applies to isotopes with an atomic number greater than 90 (*Waste Isolation Pilot Plant Land Withdrawal Act* of 1992 [Public Law 102-579, Section 2], as amended).

Uranium mill tailings waste: Uranium mill tailings and associated wastes derived from the processing of ores and related activity and controlled by the Grand Junction Projects Office Remedial Action Project. UMT wastes are defined as such to maintain a distinction with residual radioactive material, which is defined under the *Uranium Mill Tailings Radiation Control Act* (42 USC 7901 et seq).

EXECUTIVE SUMMARY

Pursuant to the *National Environmental Policy Act* (NEPA) (Title 42 U.S. Code [USC] Section 4321 et seq.), the Department of Energy (DOE) has prepared this environmental assessment (EA) of the proposed transfer of real and personal property at the Grand Junction Office (GJO) to non-DOE ownership. The GJO consists of 24.7 hectares (61.7 acres) and approximately 35 structures. The DOE has determined that it no longer needs to own the facility to perform its ongoing missions. To reduce costs, DOE proposes to transfer real and personal property at the site to non-DOE ownership using its authority under Section 161(g) of the *Atomic Energy Act* (42 USC 2011 et seq.). In addition, DOE will conduct a Federal screening of the property to determine other agency interest following guidance found in the Federal Property Management Regulations (Title 41 Code of Federal Regulations [CFR] Part 101).

The property is composed of the following three tracts or land uses:

(1)   3.2 hectares (8 acres) Army Tract: This tract is currently under lease to the U.S. Army Reserve. It will be transferred in 2000 with no changes in use (e.g., administrative offices and combat construction vehicle maintenance). The U.S. Army Reserve has prepared a separate Environmental Baseline Survey for this project. The DOE will conduct a separate NEPA review of this agency-to-agency transfer.

(2)   10 hectares (24.7 acres) Open Space with Ponds and Wetlands: Approximately 40% of the site consists of ponds, wetlands, upland and riparian vegetation. Most of this area has previously been disturbed and currently remains as the only undeveloped area of the site. The current land use is considered to be Open Space/Recreational.

(3)   11.7 hectares (29 acres) Developed Area: This remaining portion of the site contains most of the buildings, pavement, and grassy areas. Due to the remediation efforts that have been ongoing at the site for the past ten years, most of the non-paved areas have been disturbed and back-filled. There are approximately 35 structures in this area, most of which date to the late 1940s or early 1950s. Approximately 3.2 hectares (8 acres) of the 11.7 hectares (29 acres) of space is currently leased to the Western Colorado Business Development Center's Small Business Incubator. The property does not include the 0.5 hectares (1.3 acres) of land currently leased by the DOE's facility operations and support contractor from the Union Pacific Railroad.

The Proposed Action is to transfer real and personal property at the GJO Site to a non-DOE entity. The DOE plans to complete the transfer by September 30, 2000, in one transaction. The DOE and its contractors plan to remain at the site as a tenant. At this time, the DOE intends to lease office space at the GJO until its existing contracts with its prime contractors end on September 30, 2001, or longer. Potential future users of the site could be other Federal agencies, state/local quasi-governmental agencies, or the private sector.

The transaction itself does not have the potential for causing impacts to the human environment. The DOE will not control future land uses at the site, and future uses may cause impacts to the human environment. In this EA, the DOE will address reasonably foreseeable land use scenarios and the environmental effects that may result from the transfer of the facility. The land use scenarios considered under the Proposed Action are: (1) Commercial; (2) Industrial (Gravel Pit) ; (3) Mixed Used; and (4) Open Space. The No Action Alternative is also evaluated. In this case, DOE would retain ownership of the site and continue its present day activities as the owner.

The EA addresses the following environmental resource areas: geology/soils/topography, groundwater/surface water, floodplains and wetlands, land use and infrastructure, human health, ecological resources, cultural resources, air quality, noise, visual resources, solid and hazardous waste management, transportation, and socioeconomics/environmental justice. As part of this analysis, the DOE has performed a floodplains and wetlands assessment as required by 10 CFR Part 1022, *Compliance with Floodplain/Wetlands Environmental Review Requirements*. The GJO property lies within the 1000-year floodplain. In addition, two jurisdictional wetlands were created on the GJO Site in 1994. DOE has incorporated the results of the assessment in this EA.

Environmental consequences associated with most of the land use scenarios were found to be similar to those of the present day, No Action Alternative. A Memorandum of Agreement between the DOE and the State Historic Preservation Officer commits the DOE to providing the Historic American Engineering Record documentation for the potential historic district located on site. This would allow for building demolition to take place under all the land use scenarios, include clearing of the site under the Industrial and Open Space scenarios.

No impacts are anticipated in the geology and soils resource area, although the site topography would be altered under the gravel pit scenario. Upon completion of mining activities, the area would be left as a topographic depression that would fill with groundwater.

Excavation of the gravel layer, which lies below the water table, could increase the suspended solids load (i.e, turbidity) in the groundwater. Potential impacts to groundwater quality and also to the Gunnison River (because of the interconnection of the shallow aquifer with the river) during normal mine operations would be addressed during permitting of the mine by the Colorado Department of Public Health and Environment.

Environmental consequences associated with air quality and noise are expected to be minimal assuming dust suppression techniques are employed during demolition activities and gravel mining operations, and truck traffic is restricted to normal business hours. The Floodplains and Wetlands Assessment found that the proposed action would be consistent with Executive Orders 11988 (Floodplains Management) and 11990 (Protection of Wetlands) if the recipient of the property complies with applicable regulations before initiating wetland fills and fulfills mitigation responsibilities. The subject wetlands can be readily recreated through mitigation and certain offsite mitigation options could result in functionally superior wetlands.

Environmental impacts to land use and infrastructure would be minimal. The site would be considered industrial if subject to zoning. However, there is no assurance that the necessary permits or approvals could be obtained for certain types of industrial uses. Upgrades to the utility system may be necessary under the commercial and mixed use alternatives but this would not result in any environmental impacts.

Impacts to ecological resources would be minimal. Wetlands lack maturity and are likely to be easily replaced in more desirable locations upstream or downstream of the site. There would be no impacts to the dike that run adjacent to the Gunnison River and protect the site.

Volumes of hazardous materials and wastes generated and stored at the site would likely decrease under all the potential land uses. Large volumes of solid waste would be generated under two of the scenarios due to building demolition but demolition waste represents less than one-tenth of 1 percent of the capacity of the Mesa County Landfill.

Impacts to transportation are considered to be negligible under all alternatives. The industrial alternative would generate an increase in heavy truck volume but would have little effect on traffic volumes on local surface roads and state highways. Socioeconomic impacts are considered negligible as well. Any increases or decreases to population, employment, housing, or services are minor when compared to the entire Metropolitan Statistical Area which is the entire Mesa County. There would be no impacts to minority or low-income populations.

1.0   INTRODUCTION

The scope of this environmental assessment (EA) is to analyze the potential consequences of the Proposed Action on human health and the environment. Accordingly, this EA contains an introduction to the site and the history of the Grand Junction Office (Chapter One), a description of the Purpose and Need for Agency Action (Chapter Two), a description of the Proposed Action and Alternatives (Chapter Three), and the description of the Affected Environment and the Environmental Consequences (Chapter Four). Resource categories addressed in this EA include geology, soils and topography, groundwater and surface water, floodplains and wetlands, land use and infrastructure, human health, ecological resources, cultural resources, air quality, noise, visual resources, solid and hazardous waste management, transportation, and socioeconomics and environmental justice.

1.1   BACKGROUND

The U.S Department of Energy (DOE) Grand Junction Office (GJO) has prepared this EA to present the public with information on the potential impacts associated with transfer of real and personal property at the site to non-DOE ownership. DOE is required to assess the potential consequences of its activities on the human environment under the regulations of the *National Environmental Policy Act* (NEPA) (Title 42 U.S. Code [USC] 4321 et seq., codified at Title 40 Code of Federal Regulations [CFR] Parts 1500-1508). Currently, DOE's primary missions at the site are environmental restoration, environmental science, technology development, and long-term stewardship of inactive waste sites. The mission of the office has decreased over the past several years, and the DOE has determined that it no longer needs to own the GJO facility to perform its assigned missions. To lower its operating costs, the DOE intends to transfer the facility to non-DOE ownership in 2000. Non-DOE ownership could include other Federal entities, quasi-state/local government agencies, or the private sector.

The DOE completed a site-wide EA for the GJO in 1996, which resulted in a Finding of No Significant Impact (DOE 1996a). This EA considered the impacts due to ongoing operations and activities at the site, but did not consider transfer of the property to non-DOE ownership. As a result, the DOE has determined that a new EA must be prepared to consider the potential environmental consequences of a range of reasonably foreseeable future land use scenarios if the facility is transferred to non-DOE ownership.

If the impacts associated with the transfer of the property are identified as insignificant as a result of this EA, DOE shall issue a Finding of No Significant Impact and will authorize transfer of the property to a non-DOE entity. If impacts are identified as significant, an Environmental Impact Statement will be prepared.

As part of this analysis, the DOE has performed a floodplains and wetlands assessment as required by 10 CFR 1022, *Compliance with Floodplain/Wetlands Environmental Review Requirements*. The GJO property lies within the 1000-year floodplain. In addition, two jurisdictional wetlands were created on the GJO Site in 1994. DOE has incorporated the results of the assessment into this EA.

The GJO has developed the *"U.S. Department of Energy Grand Junction Office Site Transition Plan"* (DOE 1999b) which outlines the process that DOE will use to determine the best option for achieving cost savings, or mortgage reduction. This EA is part of that decision process and allows the DOE to consider the potential environmental consequences of its decisionmaking process. It is DOE policy to enable beneficial reuse of excess or underutilized property by making it available to other DOE programs, agencies, or communities. DOE recognizes the fact that transfer of entire parcels of land for local economic development purposes can also benefit the Federal Government by reducing or eliminating DOE's landlord costs and generating revenue from payroll taxes through job creation.

1.2 SITE DESCRIPTION AND HISTORY

The facility is located about 3 kilometers (2 miles) south of the main business district of the city of Grand Junction at an elevation of 1,390 meters (4,570 feet) above sea level (Figure 1.2-1). Situated on a bend of the Gunnison River, the facility is bounded on the north, south, and west by the Gunnison River. Bordering the east side of the facility is a 0.5 hectares (1.3 acres) tract of land leased from the Union Pacific Railroad and a city-owned cemetery (DOE 1996a).

The facility consists of 24.9 hectares (61.7 acres) of land and approximately 35 structures. The property is composed of the following three tracts or land uses:

(1) 3.2 hectares (8 acres) Army Tract: This tract is currently under lease to the U.S. Army Reserve. It will be transferred in late 2000 with no changes in use (e.g., administrative offices and combat construction vehicle maintenance). The U.S. Army Reserve has prepared a separate Environmental Baseline Survey for this project. The DOE will conduct a separate NEPA review of this agency-to-agency transfer.

(2) 10 hectares (24.7 acres) Open Space with Ponds and Wetlands: Approximately 40 percent of the site consists of ponds, wetlands, upland and riparian vegetation. Most of this area has previously been disturbed and currently remains as the only undeveloped area of the site. The current land use is considered to be Open Space/Recreational.

(3) 11.7 hectares (29 acres) Developed Area: This remaining portion of the site contains most of the buildings, pavement, and grassy areas. Due to the remediation efforts that have been ongoing at the site for the past ten years, most of the non-paved areas have been disturbed and back-filled. There are approximately 35 structures in this area, most of which date to the late 1940s or early 1950s. Approximately 3.2 hectare (8 acres) of the 11.7 hectares (29 acres) of space is currently leased to the Western Colorado Business Development Center's Small Business Incubator. The property does not include the 0.5 hectares (1.3 acres) of land currently leased by the DOE's facility operations and support contractor from the Union Pacific Railroad.

Formerly a gravel pit, the GJO facility lands were acquired by the U.S. War Department in August 1943 to procure uranium for the Manhattan Project. Under contract with the Federal Government, the U.S. Vanadium Corporation constructed and operated a refinery from 1943 to 1947 in which green sludges of uranium oxide were roasted and concentrated. A 20-percent uranium oxide sludge and a vanadium concentrate ("fused black flake") were produced in the refining process. Wastes from the refinery consisted of dust, several hundred tons of alumina cake, and liquid discharges.

In December 1947, the U.S. Atomic Energy Commission established the Colorado Raw Materials Office at the GJO facility to manage the domestic uranium exploration and procurement programs. Personnel at the office were responsible for the receipt, sampling, and analysis of uranium and vanadium concentrates purchased from ore-processing operations in the western United States. Between 1948 and 1971, a total of 157,500 metric tons (173,650 tons) of uranium oxide and 12,970 metric tons (14,300 tons) of vanadium oxide were received and stockpiled in steel drums at the facility. The last shipments of vanadium and uranium to the facility occurred in 1965 and 1975, respectively (DOE 1996a).

A pilot-plant program was initiated in 1953 with the construction of a small plant that was used for research into the development of a resin-in-pulp milling process. After 1954, the pilot-plant program was dedicated to amenability testing of uranium ores and to the development and testing of new uranium milling processes. A new larger pilot plant, consisting of two mill buildings, a crushing and sampling plant, office, laboratory, warehouse, and maintenance shop, was constructed in the south portion of the GJO facility. From 1954, until it was closed in

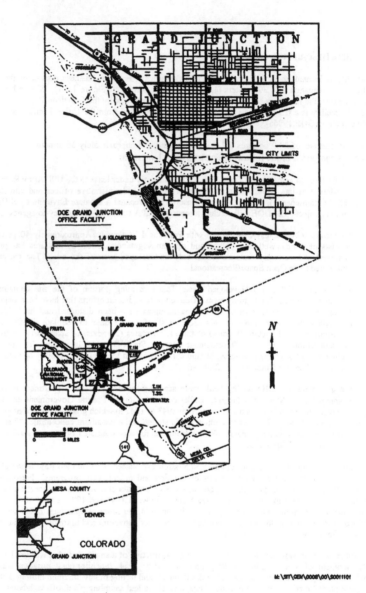

M:\SIT\GEN\0006\00\S0011101

Figure 1.2-1. *Location Map*

1958, the pilot plant operated three circuits on a 24-hour-a-day, 7-day-a-week basis. Uranium mill tailings from this plant, at first, were allowed to pond just west of what were once Buildings 31A and 33. A slurry line was later constructed to carry the tailings to a gravel pit located at the present-day site of the South Pond (DOE 1996a).

After the closure of the pilot plant in 1958, the GJO facility was used as a regional office for a variety of DOE programs directed toward uranium procurement, domestic uranium resource evaluation, and the advancement of geological and geophysical techniques. In recent years, the GJO has provided technical and administrative support personnel for various DOE, U.S. Department of Defense, and U.S. Environmental Protection Agency (EPA) programs, including laboratory and construction services that are required to support environmental restoration activities (DOE 1996a).

In the 1970s, the office conducted the National Uranium Resource Evaluation, a nationwide assessment of available uranium reserves. In the 1980s, the office mission shifted to environmental restoration and long-term surveillance and maintenance. As part of the Uranium Mill Tailings Remedial Action (UMTRA) Project, the cleanup of uranium mill tailings from over 4,000 properties in the Grand Junction area was managed at the GJO. In 1988, the Long Term Surveillance and Maintenance Program was assigned to GJO. The office was also assigned responsibility for the cleanup of the Monticello Millsite and Monticello Vicinity Properties *Comprehensive Environmental Response, Compensation, and Liability Act* (CERCLA) (42 USC 9601 et seq.) sites in Utah. The GJO also began cleanup of the GJO facility under a voluntary CERCLA cleanup. That effort continues to the present day. In 1996, GJO was assigned responsibility for the UMTRA Groundwater Program, which addresses groundwater contamination at 24 former uranium processing sites around the United States. In 1998, the GJO was assigned responsibility for the environmental restoration work at the former Pinellas Plant Site in Florida. GJO continues to manage the Uranium Lease Management Program, which administers uranium lease tracts in the western United States. GJO also performs Work-for-Others projects and technology development projects in the field of environmental restoration.

1.3   GRAND JUNCTION PROJECTS OFFICE REMEDIAL ACTION PROJECT

In 1984, site characterization and remedial action studies were initiated to assess the extent of radiological contamination on the facility from early GJO operations. The studies and subsequent cleanup were conducted under the Grand Junction Projects Office Remedial Action Project (GJPORAP). The studies showed that the site did not meet the criteria to be placed on the CERCLA National Priorities List. However, DOE decided to conduct a voluntary CERCLA cleanup. A Remedial Investigation/Feasibility Study-EA (DOE 1989) was prepared to determine clean-up strategies and to satisfy requirements of NEPA and the CERCLA of 1980 (42 USC 9601 et seq.). DOE issued a Finding of No Significant Impact for the cleanup in 1989, and the GJPORAP Record of Decision (DOE 1990a) was made final and approved by the DOE Idaho Operations Office in April 1990.

Removal of uranium mill tailings and associated radioactive materials (see Definitions section) began in 1989 and continues today. By July 1, 1994, all known exterior (i.e., open-land-area) uranium mill tailings waste had been removed from the facility and transported to the UMTRA Project Cheney Disposal Cell (DOE 1995a). This cell is located 18 miles southeast of Grand Junction and is designed to permanently contain residual radioactive materials. A description and analysis of the cell is in the *Final Environmental Impact Statement, Remedial Actions at the Former Climax Uranium Company Uranium Mill Site, Grand Junction, Mesa County, Colorado* (DOE 1986).

The GJPORAP Record of Decision (DOE 1990a) selected natural flushing as the remedy for contaminated groundwater and surface water. Contaminant levels are expected to reach acceptable levels in 50 to 80 years. The DOE will maintain restrictions on groundwater and surface water use after the transfer until acceptable levels are reached.

2.0   PURPOSE AND NEED FOR AGENCY ACTION

The historic missions that required ownership of the DOE-GJO Site have been completed, and appropriate regulatory response actions have been or will be taken to address contamination at the site. DOE does not need to retain ownership for GJO to perform its missions. To reduce mortgage liabilities, the DOE intends to transfer ownership of the site using Section 161(g) of the *Atomic Energy Act* (42 USC 2011 et seq.) and other statutory and regulatory authorities.

In addition, DOE will screen the property as surplus to determine other agency interest following guidance found in the Federal Property Management Regulations (41 CFR 101).

3.0 DESCRIPTION OF PROPOSED ACTION AND NO ACTION ALTERNATIVES

This chapter describes both the Proposed Action and the No Action Alternatives. The No Action Alternative is continued ownership of the GJO facility by the DOE.

3.1 DESCRIPTION OF THE PROPOSED ACTION–DISPOSAL OF REAL AND PERSONAL PROPERTY TO ANOTHER ENTITY

The Proposed Action is to transfer real and personal property at the GJO Site to a non-DOE entity. Potential future users of the site could be other Federal agencies, state/local quasi-governmental agencies, or the private sector. DOE will conduct a Federal property screening to determine other agency interest following guidance found in the Federal Property Management Regulations (41 CFR 101). The Proposed Action does not include the 3.2 hectare-tract (8 acres) that will be transferred to the U.S. Army Reserve. A separate NEPA review will be conducted for the transfer of this parcel to the U.S. Army Reserve.

The real property under consideration in this EA consists of approximately 24.9 hectares (61.7 acres) of land and 35 structures. The transfer is planned to be completed by September 30, 2000, and will be conducted in one transaction. The DOE and its contractors plan to remain at the site as a tenant. At this time, the DOE intends to lease office space at the GJO, at least until its existing contracts with its prime contractors end on September 30, 2001.

The transaction itself does not have the potential for causing impacts to the human environment. The DOE will not control future land uses at the site, and future uses may affect the human environment. In this EA, the DOE will address reasonably foreseeable land use scenarios and the environmental effects that may result from the transfer of the facility. Not all of the land uses may, in fact, be feasible. There is no assurance that any given land use or activity will receive the necessary approvals or permits.

The following subsections describe the potential land use scenarios under the Proposed Action Alternative.

3.1.1 All Commercial Land Use

Under the all commercial land use scenario, the site would be used for a mixture of office space, retail space, and wholesale space. It is possible that future occupants could be other Federal, state or local agencies in addition to the private sector. The types of activities occurring at the site would be very similar to the current land uses as a DOE Site. Under this scenario, it is estimated that the site would employ approximately 200-400 individuals with annual income levels ranging from $15,000 to $40,000. A mixture of businesses would occupy the current buildings. It is envisioned that some existing buildings on the site could be torn down and replaced with new office or commercial structures. Demolition debris would result from this activity. It is also possible that development could occur in the northern portion of the site that is now open space, including the wetland areas.

A variety of commercial ventures could inhabit the site. Typical uses could include: analytical laboratory, professional services firms (public accounting, consulting services), caterers, medical care, software manufacturing, furniture making, and book publishing. Some types of businesses may be required to obtain environmental or other types of permits.

3.1.2 All Industrial Land Use

Under the all industrial land use scenario, the site would be entirely used for heavy manufacturing or processing operations. To estimate the environmental effects of such use, the DOE will assess the effects of the property reverting back to its previous land use as a gravel pit prior to being acquired by the War Department in 1943.

Under this scenario, the site would be redeveloped into a gravel pit. All or almost all buildings located on the site would be demolished and the debris would be transported to an appropriate landfill. The site would become a full-scale gravel mining operation similar to the gravel pit located approximately 0.8 kilometers (0.5 miles) south of the site. Approximately ten individuals would be employed on-site with annual incomes ranging from $15,000 to $40,000. Under this scenario, the existing open space area of the site that includes wetland/upland environments would also be disturbed by the gravel pit operation. This all industrial land use is envisioned to be the bounding case scenario for potential impacts. Bounding case scenarios are used to provide an upper limit of anticipated impact levels. Bounding cases are not necessarily representative of what will actually take place. As such, this land use is being evaluated strictly to assess worst-case environmental effects; there is no assurance that a gravel pit would receive the necessary approvals to operate.

3.1.3 Mixed Use Land Use

Under the mixed use land use scenario, it is envisioned that the site would house a variety of uses including office space, commercial, light and heavy industrial, manufacturing, research and development, and high technology. It is possible that future occupants could be other Federal, state, or local agencies, in addition to the private sector. Land use would be similar to the current DOE land use. Estimated employment levels would be similar to the all commercial land use scenario with between 200-400 individuals being employed at the site. Annual incomes would range from $15,000 to $40,000. Typical businesses would include: analytical laboratory, professional services firms (public accounting, consulting services, etc), caterers, medical care, software manufacturing, furniture making, book publishing, auto repair, and light to medium manufacturing. It is anticipated that only minor quantities of hazardous materials and hazardous wastes would be handled by these businesses. Some of the businesses may be required to obtain environmental permits. The potential exists that the open space areas of the site could be filled in for possible expansion. Some of the existing structures may be renovated or replaced by new construction.

3.1.4 All Open Space Land Use

Under the all open space land use scenario, the site is envisioned as reverting back to open space. The buildings on the site would be demolished and the debris would be transported to an appropriate landfill. The site would attract local recreational users and school children on field trips/science class trips. Possible expansion of the existing wetlands might take place. Future uses could also include a state or Federal wildlife preserve.

3.2 NO ACTION ALTERNATIVE

The DOE is required to assess the potential environmental consequences of the No Action Alternative in addition to the Proposed Action and any other alternatives. The DOE plans to transfer ownership of the site to a non-DOE entity by October 2000. As a result, the likelihood of the No Action Alternative taking place is minimal. The NEPA, however, requires analysis of the No Action Alternative. In the case of this EA, the No Action Alternative represents continuation of present day activities on the site and continued ownership by the DOE. It serves as the baseline against which the other alternatives are compared.

Under the No Action Alternative, present day activities would continue at the site. DOE would retain ownership of the site. The 3.2 hectares (8 acres) of the U.S. Army Reserve tract would still be transferred on schedule (on December 31, 2000). Under the terms of the lease, the Western Colorado Business Development Center's Small Business Incubator would continue to occupy approximately 3.2 hectares (8 acres) at the southern end of the site. The 10 hectares (24.7 acres) of the open space area occupying the north end of the site would continue in its undeveloped state. The remainder of the site would contain a mixture of commercial and light industrial uses.

Mixed uses continuing at the site include an analytical laboratory, an environmental sciences research laboratory, and office space for DOE employees and contractors *WASTREN,* Inc., and Mactec-ERS. Other uses include instrument calibration facilities, copying services, telecommunication services, facility maintenance support, hazardous materials storage facilities, office space, and a small laboratory for a tenant from the Oak Ridge National Laboratory.

Current employment at the site is approximately 325 individuals with 23 DOE government employees and approximately 300 contractors. Annual salaries average between $30,000 and $50,000. Approximately 300 vehicle trips are generated daily to and from the site. Employment at the site will decrease over the course of the next two years leveling off at approximately 125 employees with 20 DOE employees and approximately 100 contractors.

Electrical power and natural gas would continue to be supplied by the Public Service Company of Colorado. Water supply would continue to be provided by the city of Grand Junction. GJO would continue to route its sanitary sewer effluent to the publicly owned treatment works operated by the city of Grand Junction. An outside firm would continue to perform facility ground maintenance and any fertilizers or herbicides would be applied by licensed applicators and materials would be stored off-site at the firm's facility.

Environmental monitoring activities would continue at the site. DOE would continue to maintain institutional controls preventing use of contaminated groundwater and surface water.

3.3   ALTERNATIVES AND LAND USE SCENARIOS CONSIDERED BUT ELIMINATED FROM FURTHER ANALYSIS

DOE considered but eliminated the alternative of leasing the facility to other entities. DOE has determined that this permanent transfer of the site to non-DOE ownership best meets its objectives for mortgage reduction and beneficial reuse.

Residential and educational use as a primary or secondary school were considered but eliminated for the following reasons:

(1)   Location adjacent to railroad with frequent freight traffic.

(2)   Nearby location of police firing range.

(3)   Heavy truck traffic from nearby gravel pit.

However, under commercial or mixed use, professional or vocational training may occur at the site.

4.0   AFFECTED ENVIRONMENT AND ENVIRONMENTAL CONSEQUENCES

4.1   GEOLOGY/SOILS/TOPOGRAPHY

Affected Environment

The GJO is located in the Canyonlands section of the Colorado River Plateau physiographic province at the interface of the Grand Valley (carved by the Colorado River) and the Uncompahgre Plateau. It is located within the floodplain of the Gunnison River. The site is protected from flooding by a levee or dike constructed along the west, south, and north side of the site. Because of this protection, the GJO is considered to be in the 1,000-year floodplain. The topography of the area is mainly flat, with elevations ranging from approximately 1,390 meters to 1,393 meters (4,560 feet to 4,570 feet) above mean sea level.

Rocks exposed in the Grand Junction area range in age from Precambrian to Cretaceous. The Precambrian basement complex consists of meta-sedimentary, meta-igneous, and igneous rocks. This basement is directly overlain by the Chinle Formation, composed of red siltstones and thin discontinuous conglomerate. Directly over the Chinle Formation is the Wingate Sandstone, a fine-grained, cross-bedded sandstone of Late Triassic age. Over the Chinle Formation, and also of Late Triassic age, is the Kayenta Formation, a white, buff, or gray, fine- to medium-grained fluvial sandstone containing minor lenses of siltstone, shale and conglomerate.

Overlying the Kayenta is the Entrada Sandstone, of Late Jurassic age. This is the main artesian aquifer of the Grand Junction area. Above the Entrada Sandstone lies about 15 meters (50 feet) of alternating sandstone, siltstone, and shale, formerly called the Summerville Formation, but recently classified as the basal part of the Salt Wash Member of the Morrison Formation.

The Upper Jurassic Morrison Formation comprises four members, but only two of them are present in the Grand Junction area. These are the Salt Wash and the Brushy Basin Members. A measured section of the Brushy Basin, about 4 kilometers (2.5 miles) southwest of the GJO Facility, indicates the total thickness of this member is approximately 90 meters (300 feet). The Brushy Basin consists of mudstone, siltstone, and shale (about 90 percent), and minor lenses of sandstone (approximately 10 percent), and is overlain by Quaternary alluvium at the GJO.

The alluvium consists of poorly sorted, unconsolidated basal gravel with a silt and sand matrix and an overlying unit of sand and silt. A cross section of the GJO area is presented in Figure 4.1-1. Geologic logs from the 1989 Remedial Investigation/Feasibility Study monitoring well-installations indicate that both the gravel unit and the overlying silts and sands are laterally continuous throughout the GJO area. The basal gravel unit was likely deposited as the Gunnison River migrated east to its present position. As migration occurred, older alluvial sediments to the west were eroded and a new layer of sediment was deposited, resulting in a laterally continuous layer of gravel, sand, and silt (DOE 1989).

According to well logs included in the 1989 Remedial Investigation/Feasibility Study, most borings encountered river gravel at depths approximately 7 to 10 feet below the surface. The river gravel deposit is described as a silty, sandy unit that varies in thickness from approximately 5 to 25 feet (DOE 1989), although in some areas the thickness is unknown because the borings were terminated within the river gravel deposit.

Groundwater typically occurs at approximately 8 to 10 feet below ground surface. Groundwater is discussed in Section 4.2.

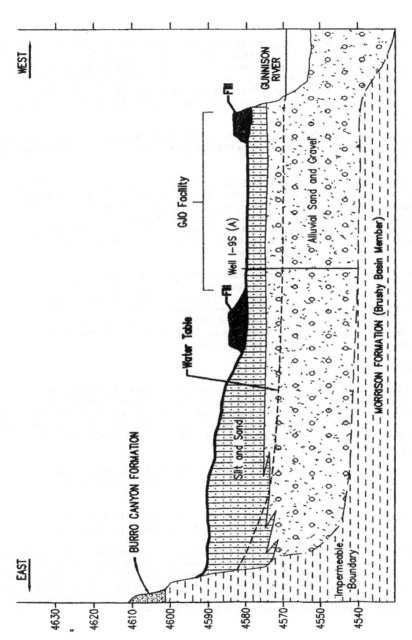

Figure 4.1-1. Cross Section of the GJO Area

Potentially active faults of Late Cenozoic age have been identified in the Grand Junction area. The closest to GJO are those of the Jacobs Ladder Fault complex, located less than 3.2 kilometers (2 miles) southwest of the facility. There are also two small faults with displacements of 9 meters (30 feet) and 1.5 meters (5 feet) located on the canyon wall adjacent to the facility, and beneath the north end of the GJO, respectively. The largest has displaced the Morrison Formation. However, it is not likely to affect the groundwater within the alluvial aquifer because of the small displacement and the presence of clay particles from the Morrison Fault that would tend to secure the fault plane (DOE 1989).

According to the Uniform Building Code (ICBO 1997), the GJO is located in seismic zone 1, which indicates a low damage risk. The facility could be used by other DOE offices for similar activities without additional seismic considerations (DOE 1998a).

The GJO facility lies above the Gunnison River, where approximately 11 meters (35 feet) of alluvial sand and gravel have been deposited. Soils formed in this material modulate from a thickness of a few inches to a few feet, and are classified as fine-loamy, mixed (calcareous), mesic Typic Torrifluvents, which are young, undeveloped soils formed in alluvium. Sediment accumulations at the site have been primarily fluvial and derived from igneous, metamorphic, and sedimentary parent material. Soil textures fluctuate from sandy loam to loam, and soil pH ranges from 7.9 to 8.4. There are minor quantities of colluvial debris and alluvial outwash from contiguous highland areas along the valley boundaries.

Before remediation began, soils contaminated by radium-226, thorium-230, and uranium covered approximately 9 hectares (23 acres) of the facility, essentially in areas of buried uranium mill tailings waste. These areas were remediated to meet the standards in 40 CFR 192, "Health and Environmental Standards for Uranium and Thorium Mill Tailings." By July 1, 1994, remediation of open-land areas had been completed, and those soils are now considered uncontaminated. Soils beneath paved areas and adjacent to buried utility lines have been characterized and are considered to be uncontaminated, as defined by the standards in 40 CFR 192. Characterization continues of soils adjacent to buried septic tanks.

Environmental Consequences

Commercial Use. Under this scenario, there would be a slight increase in the potential impacts to soils compared to the No Action Alternative. It is assumed that the Analytical Chemistry Laboratory would continue to operate at a similar level of effort, and would generate a similar amount of waste. An increase in hazardous waste generation could be expected if new light industrial small businesses are located at GJO. By increasing the amount of waste transported to hazardous waste storage areas, there would be a corresponding slight increase in the potential for soil contamination from spills.

Another potential source of soil contamination would be leaks in the buried sewer pipelines. Because there is no leak detection system, unknown leaks could occur. If a leak were discovered, contaminated soils would be treated and/or disposed of properly.

Construction of new buildings and potential demolition of existing buildings would cause some disturbance to soils. The effect would be similar to the disturbance caused by the environmental cleanup at the site.

Industrial Use. As discussed in Section 3.1.2, the bounding case for the all industrial scenario is use of the entire facility as a gravel mining operation. In this case, the gravel underlying the area would be excavated and used as a valuable geologic resource. Prior to removal of the gravel deposits, the overlying soils would be removed and most likely stockpiled for use as backfill when the gravel deposits are depleted. Soil disturbance would be similar to that caused by the environmental cleanup at the site. Upon depletion of the gravel deposits and cessation of mining activities, the topography of the area would be left as a depression that would fill with groundwater.

Because of the presence of the earthen dike between the possible excavation area and the Gunnison River, there is no potential for erosion and subsequent sediment deposition into the river.

Mixed Use. Under this scenario, the potential for impacts to soils would be slightly less than that described above for the all commercial scenario. It is assumed that the Analytical Chemistry Laboratory would continue to operate at a similar level of effort, and would generate a similar amount of waste. A small increase in hazardous waste generation could be expected if new light industrial small businesses joined the incubator facilities. By increasing the amount of waste transported to the hazardous waste storage areas, there would be a corresponding slight increase in the potential for soil contamination from spills.

Another potential source of soil contamination would be leaks in the buried sewer pipelines. Because there is no leak detection system, unknown leaks could occur. If a leak were discovered, contaminated soils would be treated and/or disposed of properly.

Construction of new buildings and potential demolition of existing buildings would cause some disturbance to soils. The effect would be similar to the disturbance caused by the environmental cleanup at the site.

Open Space. Some soil disturbance would occur during the demolition of the existing buildings. No other impacts to soils would be expected.

No Action Alternative. DOE would continue present day operations. Hazardous, low-level and mixed low-level radioactive, and/or polychlorinated biphenyls (PCB) wastes from replacing old light ballasts would continue to be transported across the facility from satellite accumulation areas to the current waste storage facilities. During transport, there is potential for soil contamination from waste spillage. However, spills would be unlikely because of the primary and secondary containment features of the packaging. If a spill did occur, the affected area would be small because of the relatively small volumes (less than 40 gallons) of waste typically transported. Procedures established in Chapter 12 of the *GJPO Emergency Preparedness and Response Plan* – Hazardous Materials Contingency Plan and Emergency Procedures – would be followed if a spill occurred. In order to minimize the affected area, contaminated soils would be immediately treated and/or contained.

Another potential source of soil contamination would be leaks in the buried sewer pipelines. Because there is no leak detection system, unknown leaks could occur. If a leak were discovered, contaminated soils would be treated and/or disposed of properly.

4.2   GROUNDWATER

Affected Environment

Three hydrogeologic units of interest underlie the GJO facility. In descending order, they are the shallow unconsolidated alluvial aquifer along the Gunnison River, the Morrison Formation aquitard, and the Entrada Sandstone aquifer. The alluvial aquifer occupies approximately 22.8 hectares (56.4 acres) of the Gunnison River floodplain; its thickness varies from 6 to 21 meters (20 to 70 feet), with an average of 6 to 8 meters (20 to 25 feet).

The aquifer is bounded on the west and north by the Gunnison River, and on the east by the shales and sandstones of the Morrison Formation. It is open to the south where the alluvium continues along the east margin of the river. Recharge is predominantly from river fluctuations, although it is also affected by precipitation to a lesser extent. Groundwater is discharged into the Gunnison River along the north and west boundaries of the GJO. The aquifer has a hydraulic conductivity of about 30 feet per day, and the specific yield is about 0.05. The water table averages 7 feet below the ground surface, but fluctuates with changing river levels. The aquifer has a tendency to be salty (almost brackish) and is rarely used for agricultural purposes. It is not used as a significant source of water regionally and is not used at all at GJO (DOE 1998a).

Underlying the alluvial aquifer is the Morrison Formation, comprising the Brushy Basin and Salt Wash Members in the Grand Junction area. These primarily shale formations are described in more detail in Section 4.1. The Morrison Formation serves as an aquitard beneath the facility, as it inhibits downward groundwater flow and prevents communication between the overlying alluvial aquifer and the underlying Entrada Sandstone aquifer.

As shown on Table 4.2-1, groundwater in the alluvial aquifer is contaminated with low levels of arsenic, radium, uranium, selenium, and molybdenum. This contamination is attributed to the past uranium milling and processing activities on the site. Of the components measured in 1998, concentrations of arsenic, molybdenum, nitrate, selenium, total dissolved solids, uranium-234+238, and net gross alpha exceeded standards (DOE 1999a). Groundwater monitoring data suggests that contamination levels may be declining over time and continued monitoring will verify this if it is the case. The remedy selected in the 1990 Record of Decision was natural attenuation (flushing). Groundwater modeling of the alluvial aquifer predicts that the groundwater will be cleaned to below standards within 50 to 80 years after the removal of the exterior uranium mill tailings waste (DOE 1989), which was completed July 1, 1994.

Environmental Consequences

Commercial Use. Impacts to groundwater quality could occur as a result of a fuel or waste spill, or a sewer line leak and subsequent migration of contaminants through the soil profile to the groundwater table. A spill directly into the surface water bodies onsite could also potentially affect the groundwater quality because the shallow aquifer and the surface water bodies are hydraulically connected. However, it is expected that under this scenario the quantities of fuel or waste transported or stored on the facility would be small and not significantly greater than those expected under the No Action Alternative described below.

Institutional controls would be in place to ensure that there continues to be no use of the shallow groundwater. Groundwater quality would be expected to improve over time via natural flushing.

Industrial Use. Under the gravel pit scenario, disturbance to the gravel layer below the water table during excavation could increase the suspended solids load (i.e, turbidity) in the groundwater. The magnitude of this increase would be dependant on the amount of organic matter and fine material such as silt and clay present in the void spaces of the gravel deposit (in other words, how "clean" the gravel is). There is currently no data on the grain size distribution of the gravel layer, although the unit is described as a silty, sandy gravel in the 1989 Remedial Investigation/Feasibility Study (DOE 1989).

Because of the shallow water table in the area, the mine would have to be dewatered to allow for gravel excavation. Dewatering of the mine would alter the hydraulic gradient in the area, causing groundwater to flow from the surrounding area towards the pit, or mine. This would limit transport of the more turbid groundwater away from the mined area. Computer modeling of the groundwater/surface water hydraulics of the area would be necessary to determine exact flowpaths and the radius of influence of the dewatering activities. Potential impacts to groundwater quality and to the Gunnison River (because of the interconnection of the shallow aquifer with the river) during normal mine operations would be addressed during permitting of the mine by the Colorado Department of Public Health and Environment.

Because the groundwater is contaminated, the Colorado Department of Public Health and Environment would require that the mine operators properly handle the effluent from mine dewatering. The ultimate treatment and disposition of the extracted groundwater would be regulated by the Colorado Department of Public Health and Environment.

Table 4.2-1. *Comparison of Federal and State Groundwater Quality Standards to 1998 and Historical Maximum Concentrations in the Alluvial Aquifer.*[a,b]

Constituent	Federal/ State Standard	1998 Maximum[c,d]			Historical Maximum[c,d]		
		Upgradient	On Site	Down Gradient	Upgradient	On Site	Down Gradient
Common Ions (mg/L)							
Nitrate (as N)[f]	10	<0.004	16.74	0.271	1.581	69.573	33.883
Total Dissolved Solids[g]	2,138	1,710	5,690	2,890	2,180	10,200	8,620
Metals (mg/L)							
Arsenic	0.05	–0.0018	0.35	–0.0058	0.0114	0.68	–0.031
Barium	1.0	–0.0177	–0.0464	0.037	–0.0187	0.4	0.038
Cadmium	0.01	<0.001	<0.0034	<0.0051	<0.002	0.055	<0.005
Chromium (total)	0.05	<0.002	–0.01	<0.0216	0.010	0.039	0.112
Lead	0.05	<0.001	0.0006	0.0007	–0.0026	0.0571	0.0046
Mercury	0.002	-	-	-	0.0002	0.00023	0.0002
Molybdenum	0.1	–0.0067	0.54	0.138	0.023	19	0.413
Selenium	0.01	<0.001	0.107	0.0092	–0.0025	0.685	0.073
Silver	0.05	-	<0.0076	<0.0114	<0.01	0.006	0.0056
Radionuclides (pCi/L)							
Gross Alpha (excluding radon and uranium)[h]	15	<12.37	113.59	-3.2912	71.02	1073.14	620.52
Radium 226+228	5.0	0.32	0.57	0.25	1	36	2.70
Thorium 230+232	60	-	-	-	0.2	18	4.3
Uranium 234+238[i]	30	5.50	1140.7	192.58	22.77[j]	6039	1006.5

[a]Standards from the Uranium Mill Tailings Radiation Control Act, revised in 1986.

[b]Colorado Department of Public Health and Environment Water Quality Control Division, *Basic Standards for Groundwater,* effective August 30, 1997. Standards in the "Potentially Usable Quality" classification were used for GJO groundwater.

[c]"-" indicates no data available; "<" indicates that the maximum concentration was below the detection limit (number shown is detection limit); "--" indicates an approximate value (the value was outside the limits for which the instrument was calibrated).

[d]The units are indicated in the "Federal/State Standard" column.

[e]Based on maximum concentrations observed from 1984 through 1997.

[f]Nitrate (as N) was derived for measured nitrate using the conversion N=NO_3 + 4.427.

[g]This is a site-specific standard calculated as background x 1.25. The background value is based on the June 1998 sampling event.

[h]Measured values represent total gross alpha minus uranium activity. Negative values indicate uranium concentrations exceeded gross alpha activity. Uranium concentrations that were measured in grams were converted to pCi/L. The conversion assumes equilibrium and an activity of 0.687 pCi/μg.

[i]Total uranium concentrations that were measured in grams were converted to uranium 234+238 in pCi/L for comparison. The conversion assumes equilibrium and an activity of 0.671 pCi/μg.

[j]Extreme-values testing of uranium results from samples collected in 1985 and 1989 indicated that two values (201 pCi/L and 84 pCi/L) were outliers; these values from upgradient wells were not included in this table.
Source: DOE 1999a.

Groundwater quality could also be affected by a fuel or hydraulic fluid spill or leak from the heavy equipment used in the mining operation. It is expected that the mine operators would have spill response procedures in place that would minimize potential impacts.

Since all known uranium mill tailings waste that historically contaminated the groundwater of the shallow aquifer was removed by July 1, 1994, groundwater quality should improve over time via natural flushing or attenuation. As mentioned above, concentrations of water quality parameters associated with the historic leaching of uranium mill tailings waste are expected to be below applicable standards within 50 to 80 years.

Institutional controls would be in place to ensure that there continues to be no use of the shallow groundwater. Groundwater quality would be expected to improve over time, via natural flushing. Dewatering during mining operations could accelerate the natural flushing of the aquifer.

Mixed Use. Impacts to groundwater quality from this scenario would be expected to be similar to those described for the No Action Alternative described below.

Institutional controls would be in place to ensure that there continues to be no use of the shallow groundwater. Groundwater quality would be expected to improve over time, via natural flushing.

Open Space. Barring a heavy equipment fuel spill or leak during building demolition, no adverse impacts to groundwater quality would be expected from this scenario. Institutional controls would be in place to ensure that there continues to be no use of the shallow groundwater. Groundwater quality would be expected to improve over time, via natural flushing.

In the case of a fuel spill or leak during building demolition, it is expected that the heavy equipment operators would have spill response procedures in place that would minimize potential impacts.

No Action Alternative. Impacts to groundwater quality could occur as a result of a fuel, hazardous materials or waste spill, or a sewer line leak. A spill could cause contaminants to move through the soil profile to the groundwater table. A spill directly into the surface water bodies onsite could also potentially affect the groundwater quality because the shallow aquifer and the surface water bodies are hydraulically connected. Spill response procedures that minimize the potential for any spilled materials to reach the groundwater table would be in place.

Because all known uranium mill tailings waste that historically contaminated the groundwater of the shallow aquifer was removed by July 1, 1994, groundwater quality should improve over time via natural flushing or attenuation. As mentioned above, concentrations of contaminants associated with uranium mill tailings waste are expected to drop below applicable standards within 50 to 80 years. Meanwhile, institutional controls would be in place to ensure that there continues to be no use of the shallow groundwater.

4.3   SURFACE WATER

Affected Environment

Surface water bodies at or near the GJO include the Gunnison River, the North Pond, and the South Pond, as well as some wetland areas adjacent to the North Pond (see Figure 1.2-2). All of these water bodies contain water perennially. The ponds and wetland areas are located within the GJO, and the Gunnison River is contiguous to the facility's western and northern boundaries. The state has designated four use classifications for the segment of the Gunnison River near the GJO facility: (1) Recreation-Class I; (2) Cold Water Aquatic Life-Class 1; (3) Domestic Water Supply; and (4) Agriculture (DOE 1999a). Other than wildlife habitat, there is no use of the North or South Ponds, and there is no known consumptive use of the Gunnison River between the facility and the confluence with the Colorado River (DOE 1996a).

The wetland area and South Pond (in its current configuration) were created in the spring of 1994 during remediation of the uranium mill tailings waste-related contamination. The North Pond is the remnant of a gravel pit mining operation that occurred on the site in the early 1920s (DOE 1989).

The Gunnison River, which is hydraulically connected to the shallow aquifer, is subject to the Colorado Water Quality Control Commission's general narrative water-quality standards and specific water-quality standards for radioactive materials and organic pollutants. These standards are also used to evaluate the water quality of the

North and South Ponds. Water in the North Pond, South Pond, and wetland areas is contaminated with the same constituents as the groundwater because these surface waters are recharged by the shallow alluvial aquifer. They contain comparable concentrations of substances associated with uranium mill tailings waste, including arsenic, manganese, uranium, selenium, molybdenum, vanadium and sulfate.

Tables 4.3-1 and 4.3-2 show the 1998 sampling results for the Gunnison River and the ponds/wetlands, respectively. The Water Quality Control Commission's standards are also shown for comparison, as are historical maximums for the Gunnison River.

In addition to the parameters shown on Table 4.3-2, the North Pond, South Pond, and wetland areas samples were also analyzed for gross alpha, gross beta, and radium-226 activity. Although gross alpha and beta activities were above instrument detection limits, no surface water quality standards exist for comparison. The state and Federal standard for radium 226+228 of 5 picocuries per liter was not exceeded in any on-site surface water sample.

Radionuclide concentrations in samples collected in 1998 from three locations in the Gunnison River (upstream, adjacent to, and downstream of the GJO) were below applicable Colorado Department of Public Health and Environment Water Quality Control Commission's standards. In addition, total uranium and radium-226 concentrations in 1998 were relatively constant in all samples, indicating the contaminated groundwater underlying the GJO is not impacting the Gunnison River.

Environmental Consequences

Commercial Use. The water quality of the North Pond, South Pond, and wetland areas would be expected to improve over time through passive remediation of the shallow aquifer. Under this scenario, the potential for spills or other releases that could affect surface water would be slightly greater, but similar to that described below for the No Action Alternative. Institutional controls would be in place to restrict use of these surface waters. Because the dike isolates the Gunnison River from the site, no impacts to the river would be expected.

Industrial Use. Under the bounding case for this scenario, a gravel mining operation, the existing surface water bodies at the facility would be destroyed during excavation of the gravel and dewatering of the excavated areas. As mentioned above in the groundwater section, under the gravel mining scenario there is a potential for excavation activities to cause an increase in the suspended solids load, or turbidity, of the groundwater. Because the shallow groundwater at the site and the Gunnison River are hydraulically interconnected, there is potential for the more turbid groundwater to reach the river and potentially increase its total suspended solids load. Because of the lack of data concerning the grain size distribution of the gravel layer below the water table, a quantitative analysis of the potential for impacts to the Gunnison River is not possible at this time.

As mentioned above in the groundwater section, dewatering of the mine would alter the hydraulic gradient in the area, thus limiting transport of the more turbid water away from the mined area. However, computer modeling of the groundwater/surface water hydraulics of the area would be necessary to determine if dewatering activities would affect the Gunnison River.

Mixed Use. Under this scenario, the water quality of the North Pond, South Pond, and wetland areas would be expected to improve over time through passive remediation of the shallow aquifer. The potential for spills or other releases that could affect surface water would be similar to that described below for the No Action Alternative. Institutional controls would be in place to restrict use of the surface waters. Because the dike isolates the Gunnison River from the site, no impacts to the river would be expected.

Open Space. Under this scenario, the water quality of the North Pond, South Pond, and wetland areas would be expected to improve over time through passive remediation of the shallow aquifer. Since there would be no use of hazardous materials on the property, there would be no potential for spills or other releases that could affect

Table 4.3-1. *Comparison of State Surface Water Quality Standards to 1998 Historical Maximum Concentrations in the Gunnison River.*[a,b]

Constituent	State Standard	1998 Maximum[d]			Historical Maximum[d]		
		Upgradient	Adjacent to Site	Down-Gradient	Upgradient	Adjacent to Site	Down-Gradient
Common Ions (mg/L)							
Chloride	250	4.83	4.93	5.03	12.4	12.6	80
Nitrate (as N)[e]	10	0.707	0.673	0.642	6	6	6
Nitrite (as N)[f]	0.05	-	-	-	<0.304	-	<0.304
Sulfate	480	215	215	215	513	512	584
Field Measurements							
Dissolved Oxygen[g]	7.0 mg/L	-	-	-	9.5	9.3	9.5
pH	6.5-9.0	8.44-8.44	8.45-8.50	8.58-8.58	7.20-9.04	7.29-9.19	7.33-9.01
Inorganics							
Fecal Coliform[h]	200	-	-	-	1500	300	1300
Metals (mg/L)[i]							
Arsenic	0.05	<0.001	<0.001	<0.001	0.011	~0.0086	0.011
Cadmium	0.001	<0.001	<0.001	<0.001	0.002	~0.00063	~0.00032
Chromium+6	0.011	<0.002	<0.002	<0.002	~0.0038	~0.0045	~0.0034
Copper	0.013	-	-	-	0.056	0.013	0.05
Iron	0.300	~0.0135	~0.0264	<0.0199	0.44	0.1	0.32
Lead	0.004	<0.001	<0.001	<0.001	0.059	0.0193	0.027
Manganese	0.050	~0.0065	~0.0068	~0.0065	0.2	0.0766	0.122
Mercury	0.0001	-	-	-	<0.002	<0.002	<0.002
Nickel	0.101	-	-	-	0.005	<0.025	0.021
Selenium	0.008	~0.0038	~0.0038	~0.0036	0.0096	0.014	0.0148
Silver	0.0001	-	-	-	<0.0005	<0.0005	0.0005
Zinc	0.113	-	-	-	1.07	0.86	1.72
Radiological (pCi/L)							
Radium 226+228	5	0.68	0.39	0.38	16.8	15.5	16.3
Uranium[j]	40	~2.7	~2.6	~2.7	10.42	14.39	23.35

[a]Colorado Department of Public Health and Environment Water Quality Control Commission surface water standards; Regulation No. 31 and 35, effective March 2, 1998 and May 30, 1998, respectively.

[b]"-" indicates no data available; "~" indicates an approximate value (the value was outside the limits for which the instrument was calibrated; "<" indicates that the maximum concentration was below the detection limit (number shown is detection limit).

[c]The units are indicated in the "State Standard" column.

[d]Based on maximum concentrations observed from 1980 through 1997.

[e]Nitrate (as N) was derived for measured nitrate using the conversion N=NO₃ + 4.427.

[f]Nitrite (as N) was derived for measured nitrite using the conversion N=NO₃ + 3.285.

[g]The standard value for dissolved oxygen represents a minimum concentration. Measured values must be greater than 7.0 mg/L to comply with this standard. Listed values represent the minimum measurements observed.

[h]Number of colonies per 100 mL.

[i]All values given are for dissolved constituents.

[j]Uranium concentrations that were measured in milligrams per liter were converted to picocuries per liter for comparison. The conversion assumes equilibrium and an activity of 0.687 pCi/μg.

Source: DOE 1999a.

Table 4.3-2. *Comparison of Onsite Surface Water Quality in 1998 with State Standards*

Constituent	State Standard	North Pond	South Pond	Wetland Area
Chloride	250 mg/l	334	116	651
Sulfate	480 mg/l	2,240	1,600	6,780
Uranium	40 pCi/l	102	269	111

Source: DOE 1999[a]

surface water quality. Institutional controls would be in place to restrict use of the surface waters. This land use scenario would not be expected to impact the Gunnison River.

No Action Alternative. Surface water quality of the North Pond, South Pond, and wetland area is expected to improve over time through passive remediation of the shallow aquifer (see Section 4.2). Impacts to surface water quality could occur as a result of a fuel or waste spill near or directly into a surface water body or into the storm drainage system. However, it would be unlikely for the spilled contaminant to adversely affect water quality because of the generally small quantities of waste and fuel stored or transported on the facility. In the event of a spill, procedures are in place for rapid containment and removal of potential contaminants before they migrate to water bodies. Institutional controls would be in place to restrict use of the surface waters. Because of the presence of the dike, no impacts to the Gunnison River would be expected.

4.4　FLOODPLAINS AND WETLANDS

Affected Environment

The U.S. Army Corps of Engineers has determined that those areas of the GJO property inside (north and east) of the riverside dike are not in the 100-year or 500-year floodplain of the Gunnison River (DOE 1999a; DOE 1996a). The dike follows the southern and eastern shores of the river and was constructed to protect the developed areas from the river (Figure 1.2-2). A Flood Insurance Rate Map for Mesa County dated July 1978 indicates that areas inside the dike still lie within the 1,000-year floodplain (DOE 1999a). To comply with Executive Order 11988, *Floodplain Management*, GJO obtained a permit in 1989 authorizing operations within the designated 1,000-year floodplain.

An updated Flood Insurance Rate Map for Mesa County, dated July 15, 1992, designates areas inside the dike as "Other Areas, Zone X" (FEMA 1992). Areas with that designation have been determined to be outside of the 500-year floodplain. The updated Flood Insurance Rate Map designates areas between the dike and the river as Zone A (100-year floodplain), but no base flood elevations are indicated.

Wetlands totaling approximately 0.6 hectares (1.45 acres) (see Figure 1.2-2) were restored in 1994 and 1995 on open land that had been disturbed by the removal of UTM (DOE 1996a; DOE 1999a). The wetlands were restored to comply with a *Clean Water Act* Section 404 permit issued by the U.S. Army Corps of Engineers (Number 10040, dated March 13, 1989) (COE 1989). The permit authorized GJO to discharge approximately 26,132 cubic meters (34,200 cubic yards) of permanent fill material and approximately 1,528 cubic meters (2,000 cubic yards) of temporary fill material into wetlands on the property. The authorized fill was necessary to remove uranium mill tailings waste from the property and to reconstruct the dike. Other non-delineated wetlands exist at the site; there are approximately 0.8 hectares (2 acres) of wetlands on GJO. There is an on-going effort to further delineate the presence of wetlands at GJO. This information will be available prior to the transfer.

The restored wetlands were revegetated using willow (*Salix* sp.) and Fremont cottonwood (*Populus fremontii*) seedlings, common cattail (*Typha latifolia*) plants, hardstem bulrush (*Scirpus acutus*) cuttings, and a seed mix of native and adapted grasses. GJO also restored 1.26 hectares (3.11 acres) of jurisdictional riparian vegetation between the dike and the Gunnison River using willows, cottonwoods (*Populus* ` sp.), silver buffaloberry (*Shepherdia argentea*), skunkbush sumac (*Rhus trilobata*), and a grass seed mix. Additionally, GJO seeded grasses and planted willow seedlings on 1.7 hectares (4.20 acres) of other (non-jurisdictional) riparian lands and on 4.35 hectares (10.69 acres) of uplands inside the dike. The entire planted area has been monitored annually since 1995. Much of the planted vegetation had to be replaced in 1996 due to flooding.

Environmental Consequences

Commercial Use. The existing dike and land between the dike and the river would not be disturbed by the recipient of the property, and the action would be in compliance with Executive Order 11988 (*Floodplain Management*). This scenario could result in the permanent filling of all or part of the approximately 0.8 hectares (2 acres) of wetlands inside the dike to accommodate commercial development. The wetlands subject to disturbance are isolated wetlands inside of the dike, which prevents surface connection with the river, even during most major floods. Loss of the wetlands would therefore affect neither regional flood patterns, nor increase erosion of the riverbanks. In addition, water quality and the availability of nutrients or biomass in the river also would not be affected. Because the majority of the wetlands were recently created on land disturbed by environmental remediation, they are not irreplaceable natural resources of exceptional significance. Loss of the wetlands would result in the loss of a small amount of riparian wetland habitat favored by species such as the snowy egret (*Egretta thula*). However, the value of the onsite wetlands as habitat for most wildlife has already been reduced due to the urban setting.

The recipient would be responsible for complying with Section 404 of the *Clean Water Act* and other applicable regulations prior to disturbing the wetlands. The recipient would also be responsible for implementing any mitigation required by regulation. As long as the recipient complies with any applicable wetland regulations, transfer of the property by DOE is consistent with Executive Order 11990 (*Protection of Wetlands*). Because the majority of the affected wetlands were created as a mitigation project within the last five years, new wetlands with similar properties could be rapidly established.

Industrial Use. The existing dike and land between the dike and the river would not be disturbed by the recipient of the property. There would be no changes to the 100-year floodplain and the action would be in compliance with Executive Order 11988 (*Floodplain Management*).

The recipient of the property could excavate all or part of the approximately 0.8 hectares (2 acres) of wetlands inside the dike to establish a gravel pit. As noted for the Commercial Use scenario, the wetlands are hydrologically isolated from other surface water features and not expected to contribute substantially to regional flood protection, water quality, or wildlife habitat. Transfer of the property would be in compliance with Executive Order 11990 (*Protection of Wetlands*) if the recipient complies with applicable regulations before excavating in wetlands and fulfills mitigation responsibilities. The subject wetlands can be readily recreated, and certain offsite mitigation options could result in functionally superior wetlands.

Mitigation options could include creating wetlands on the gravel pit site once extractive operations cease, or establishing wetlands elsewhere along nearby reaches of the Gunnison or Colorado Rivers. Such offsite wetland mitigation could result in the establishment of higher quality wetlands in a less urban setting that are of greater value to the regional river system and more closely resembling the wetlands formerly present on the GJO Site prior to its initial development.

Mixed Use. The existing dike and land between the dike and the river would not be disturbed by the recipient of the property, and the action would be in compliance with Executive Order 11988 (*Floodplain Management*). The Mixed Use scenario could result in the permanent filling of all or part of the approximately 0.8 hectares (2 acres) of wetlands inside the dike to accommodate various types of development. As described for the Commercial Use scenario, the restored wetlands are hydrologically isolated from other surface water features and not expected to contribute substantially to regional flood protection, water quality, or wildlife habitat. Transfer of the property would be in compliance with Executive Order 11990 (*Protection of Wetlands*) if the recipient complies with applicable regulations before filling wetlands and fulfills mitigation responsibilities. As described for the Commercial and Industrial Uses, the subject wetlands can be readily recreated, and certain offsite mitigation options could result in functionally superior wetlands.

Open Space. The wetlands and areas within the 100-year floodplain on the site would remain undisturbed under the Open Space scenario. The existing dike would remain unchanged and would continue to be maintained as necessary to prevent deterioration. The Open Space scenario would therefore be in compliance with Executive Orders 11988 (*Floodplain Management*) and 11990 (*Protection of Wetlands*). Conversion of adjoining developed areas to naturally vegetated lands could improve the value of the wetlands and riparian areas as wildlife habitat.

No Action Alternative. The wetlands and areas within the 100-year floodplain on the site would remain undisturbed under the No Action Alternative. The existing dike would remain unchanged and would continue to be maintained as necessary to prevent deterioration. The No Action Alternative would therefore be in compliance with Executive Orders 11988 (*Floodplain Management*) and 11990 (*Protection of Wetlands*).

4.5 LAND USE

Affected Environment

The DOE-GJO Site is a 24.9 hectare (61.7 acres) triangular tract of land wedged between the Union Pacific Railroad on the east and the Gunnison River on the north, west, and south. The main parcel of land was acquired from a private landowner in 1943 by a predecessor agency to the DOE. A more detailed history of the site is provided in Section 1.2 of this EA. The legal description of the site is available in the *Facility Condition Assessment* (DOE 1998a). A title search (DOE 1999c) of the property was ordered by the DOE in April 1999 and completed in late April. This title search shows the property to be in the possession of the DOE with the exception of a 200-foot right-of-way that runs the entire length of the eastern edge of their property. In addition, there is a road right-of-way that was granted in 1959 for purposes of extending 25 7/8 Road to connect it with the Black Bridge (subsequently demolished) into town. The Denver and Rio Grande Western Railway granted right-of-way back to DOE for the south gate entrance. Maps depict two more right-of-way grants back to DOE for the North Gate and the far northern end of the property, but the title search did not reveal these grants.

The nearest residence to the GJO is approximately 0.4 kilometers (0.25 miles) from the site. Land used for agricultural purposes lies across the Gunnison River to the north, west, and south of the site. Patented lands, subsequently acquired by the city of Grand Junction for use as a cemetery, lie to the east of the site across the railroad right-of-way. A police firing range and the main railroad track are adjacent land uses to the GJO Site. A quarry operation exists approximately one mile (1610 meters) southeast of the site.

The land is near the Orchard Mesa neighborhood, a community located to the south of both the Colorado River and the city of Grand Junction. The GJO Site is approximately 3.2 kilometers (2 miles) south of Grand Junction's main business district and less than 4.8 kilometers (3 miles) northeast of lands included in the Colorado National Monument administered by the National Park Service. It is approximately 1 kilometer (one-half mile) east of public lands administered by the Bureau of Land Management. The Redlands Dam on the Gunnison River is upstream from the site.

The facility is located within Mesa County, outside the city limits of Grand Junction. The site is located within the boundaries covered by the *Orchard Mesa Neighborhood Plan* (Mesa 1996), which was adopted jointly by the city of Grand Junction and Mesa County Planning Commissions in March 1995. If the land were privately owned it would be zoned "industrial." Under the existing Mesa County zoning ordinance, a range of manufacturing, commercial, and related uses could be allowed. Federal lands are not subject to zoning.

The *Orchard Mesa Neighborhood Plan* adopts the following goals that may impact future uses of the site:

• No additional areas on Orchard Mesa should be zoned industrial.

• Complete planning and design for a pedestrian/bicycle/emergency vehicle bridge across the Gunnison River at or near the Old Black Bridge Site from 1998 to 2001 and construction from 2002 to 2005.

The entrances to the GJO Site face east. The site entrances are accessible from U.S. Highway 50, through a large cemetery, by way of the narrow, two-lane, city-maintained streets of Canon Street (0.8 kilometers [0.5 miles]) and B 3/4 Road (0.8 kilometers [0.5 miles]). Until the mid 1980s, the site was also accessible from the west side of the Gunnison River across Black Bridge which was located just north of the facility. Black Bridge was removed due to disrepair.

In 1979, a 2.15 hectare (5.32 acres) parcel on the north side of the site adjoining the Gunnison River was deeded out of the DOE tract to Mesa County. The property came to be known as "Black Bridge Park". The deed stipulated that the property be used as a public park and public recreation area for its exclusive and perpetual use. The deed also provided that the property would revert to the Federal Government in the event of a breach or failure to maintain the property for the specified use. By 1993, Mesa County determined that it could not properly maintain Black Bridge Park as a public recreation area. There were numerous complaints of vagrancy, property destruction, disorderly conduct, and other disturbances. On May 27, 1994, DOE-GJO reimbursed the General Services Administration $5,000 to re-acquire the property. The GSA transferred the parcel back to DOE on June 10, 1994.

Immediately to the east of the enclosed site area is a parking lot, partially owned by the railroad and leased to the GJO facility contractor. Further to the east are railroad tracks, and another parking lot within land owned by the railroad (but leased to the GJO contractor for parking). The lease of land west of the tracks consists of three 18.15 meter (60 feet)-wide strips of land. The lease can be terminated by either party within thirty days written notice. The facility contractor holds a commercial lease from the railroad for a fourth 18.15 meter (60 feet)-wide strip of land for use as a parking lot on the east side of the tracks. Rent, which was $564 per year in 1992, is paid annually and is adjusted annually based on the Consumer Price Index. This lease is for a term of thirty days and continues on a month-to-month tenancy that is terminable by either party within thirty days written notice. In 1992, the previous facility contractor purchased a Private Way License from the railroad for two 8-foot concrete pedestrian walkway crossings at grade across the right-of-way and trackage. The private way license is "a strictly private one and is not intended for public use (DOE 1998c)." The leases and private way license are not transferable and are not part of the property transfer addressed in this EA.

There are two railroad crossings for vehicular traffic and two additional crossings for pedestrians. None of the crossings have warning guards or lights. The railroad line in front of the DOE facility is a spur line used from a few times to several times daily for the transport of coal. Because trains must maneuver a sharp curve near the DOE Site, they travel past the site at very slow rates of speed. A train occasionally stops on the tracks for brief periods to switch cars, but an informal arrangement with the DOE provides that while a train may block one entrance to the site during these stops, it will not block both crossings.

A gravel road, 26 3/8 Road, runs from B 3/4 Road south past a police firing range to a Gunnison River access area south of the GJO Site. Because of vegetation and the curve of the river, this access area is neither accessible nor visible from the GJO Site. The river bank across from the GJO Site to the west is steep and rises to form a mesa. Only a few houses are built on and near the edge of the mesa, in the Little Park Road area. These houses are accessible from Grand Junction only by crossing a bridge several miles from the GJO Site and traversing a road which winds up the west side of the mesa.

Two ponds, wetland areas and open space occupy approximately 8.1 hectares (20 acres) at the north end of the GJO Site. The North Pond, South Pond, and much of the wetland areas now located on the GJO Site were actually the location of a "gravel pit lake" and the "gravel development." These features appear on the tract map recorded with the warranty deed that conveyed the property to the United States in 1943.

The U.S. Army Corps of Engineers determined, in its Flood Hazards Study of 1976, that the GJO facility was not in either the 100-year or 500-year floodplain of the Gunnison River because of the protection afforded by the earthen dike, which is located between the facility and the river. The dike must be maintained in order to secure the facility from 100-year and 500-year flood events. The Mesa County Housing and Urban Design Flood Insurance Rate Map (FEMA 1992) places the GJO facility within the 1,000-year floodplain.

Environmental Consequences

Commercial Use. The site would be a mix of office space and retail space but no industrial uses would be present.

Industrial Use. This use would be similar to land uses approximately 1.6 kilometers (1 mile) south of the site where another gravel pit is located. It would not affect the firing range adjacent to the site.

Mixed Use. Under the mixed use scenario, it is envisioned land use would be nearly the same as the current use.

Open Space. Under the open space scenario, the land would revert to open space.

No Action Alternative. There would be no changes in land use under the No Action Alternative.

4.6 INFRASTRUCTURE

Building utilities include electricity, natural gas (heating), water, sewer, and telecommunications. Other than electricity, none of the utilities extend beyond the main grouping of buildings. Distribution for each of the utility systems can be seen in Figures 6 through 11 in Section 3.5 of the *Facility Condition Assessment* (DOE, 1998a). The following data is compiled from the *Facility Condition Assessment* (DOE 1998a).

Electrical Power. Electrical power is supplied to the facility by the Public Service Company of Colorado through a 13,000-volt main feeder to the main substation south of Building 810. GJO owns the primary and secondary electrical systems on the facility. Average monthly on-peak and off-peak electrical usages for the GJO are 929 and 787 kilowatts, respectively.

Forty electrical transformers are located on the facility. In 1988, all on-site electrical transformers owned by DOE that contained 50 or more parts per million PCBs were retrofitted with dielectric fluid containing less than 50 ppm of PCBs. An off-site contractor disposed of the removed dielectric fluid at a permitted EPA treatment facility. Several of the on-site transformers belong to the Public Service Company of Colorado; the PCB content in the dielectric fluid in these transformers is unknown.

Natural Gas. The Public Service Company of Colorado and the Western Natural Gas and Transmission Corporation supply natural gas to the facility through a feed-line located in Building 40. From Building 40, natural gas is distributed through polyethylene lines to all the gas-fired hot-water boilers on the facility. There are 18 water boilers located in 13 facilities. Heat is generated in these cast-iron sectional hot water boilers and distributed to individual rooms. Exhaust gases produced by the heating system include negligible amounts of carbon monoxide and nitrous oxide.

Water Supply. The GJO contracts with the city of Grand Junction for domestic water. The city generally obtains water from Kannah Creek and rarely draws water from the Colorado and Gunnison Rivers. Domestic water is used for drinking water, laboratory purposes, fire protection, and lawn irrigation. In 1989, drinking water from all water coolers on site was tested for lead. The analytical results indicated that the drinking water was in compliance with the Colorado drinking water standard for inorganic lead (0.005 milligram per liter). The DOE conducted lead testing and installed back flow preventers.

Fuel Storage. Fuel is stored in four areas on the GJO facility. An approximately 100-gallon above ground storage tank with secondary containment is located west of Building 3022. A small quantity (approximately 10 gallons) of unleaded gasoline for use in the maintenance shops is stored in approved 2.5- and 5-gallon gasoline containers inside Building 3022. About 10 gallons of diesel fuel is stored in a metal fuel tank in Building 20 for use in operating an emergency generator in the event of a power outage. The fourth source is a 500-gallon propane tank located south of Building 20. Propane fuel is piped into Building 20 and is used to operate fusion furnaces in the Analytical Laboratory. Protection of the tank is ensured by the placement of six steel posts around the tank.

Storm-Drain System. A series of drain lines underlie the GJO facility and collect storm-water runoff. During precipitation events, storm water is routed through the buried lines into a lift station near the southern terminus of the South Pond. It is discharged into the South Pond once the water level within the lift station reaches the elevation of the discharge line. Because the storm-water effluent consists of runoff from the facility parking lots, office buildings, and paved areas, EPA determined in 1992 that a National Pollutant Discharge Elimination System permit was not required for the facility. Current site activities and operations are continually evaluated for applicability to National Pollutant Discharge Elimination System regulations. To date, no activities that would require DOE to obtain a National Pollutant Discharges Elimination System storm-water permit have been identified.

Sanitary Sewer. Sewer effluent from the GJO facility is routed to the publicly owned treatment works operated by the city of Grand Junction. The effluent consists of domestic sewage, discharges from the Analytical Chemistry, Radon, and Environmental Sciences Laboratories, detergent wash water from the cafeteria, and water used for facility maintenance purposes.

In March 1989, the city issued an Industrial Pretreatment Permit (No. 0023) to the GJO in accordance with provisions in Article 10 of Chapter 25, Code of Ordinance for the city of Grand Junction. Article 10 sets forth uniform requirements for users of city and county publicly owned treatment works and enables the city to comply with the *Clean Water Act of 1977*, as amended, the General Pretreatment Regulations (40 CFR 403), and the *Colorado Water Quality Control Act*, as amended. The permit was revised by the city in February 1993. The revision required that the sewer effluent be sampled for biological oxygen demand, oil and grease, PCBs, pH, silver, total suspended solids, total dissolved solids, ammonia, and temperature. The revised permit established threshold limits for temperature, pH, oil and grease, PCBs, and silver. The permit expired in June 1999. The city did not require GJO to renew the permit because it is no longer a significant industrial user due to lower flow rates. GJO continues to sample its sewer effluent as part of its ongoing environmental monitoring program, although no longer required to do so by the city. The City of Grand Junction maintains its own NPDES permit for the city owned treatment plant.

Telephone Service. Telecommunications specialists install, program, and maintain telephones and telephone lines at the GJO facility. The U.S. West central office in Grand Junction, Colorado, provides phone service for off-site local calls; the Federal Telephone System furnishes long-distance phone service.

Environmental Consequences

Commercial Use. Under this scenario, some utility systems might need to be retrofitted or upgraded to accommodate individual users or tenants. Currently, most buildings are not individually metered. The system has the design capacity to handle in excess of 600 personnel.

Industrial Use. Under this scenario, the existing utility systems would be disconnected at or near the facility boundary and capped. The utility system would likely be demolished and removed or abandoned in place.

Mixed Use. Environmental effects would be similar to the all Commercial scenario and the No Action Alternative. Some upgrades might be required to the existing system and individual metering would likely be needed at individual buildings. The system has the design capacity to handle in excess of 600 personnel.

Open Space. Environmental effects would be similar to the Industrial scenario. The on-site utility system would be capped and abandoned in-place. Impacts to the local and regional utility system would be negligible.

No Action Alternative. Environmental consequences under the No Action Alternative would be similar to those under the Proposed Action. The site operated with over 600 personnel as late as 1996, so the utility system is properly sized to handle the projected future population at the site. Some upgrades/retrofits may be necessary in the near future due to the age of the systems.

4.7 HUMAN HEALTH

Affected Environment

Because of the GJO facility's history as a uranium milling operation, contamination in buildings, soils, groundwater and surface water posed risks to human health. Ongoing cleanup of the site and controls restricting use of the groundwater and surface water will minimize the risks to workers and the general public.

All buildings have been surveyed for radiological contamination and will be remediated if necessary, then released for unrestricted use or demolished. Building 2 was released with supplemental limits that were determined through a Public Dose Evaluation (DOE 1996b) to pose no unacceptable risk to the general public. Building 20 is also undergoing a Public Dose Evaluation.

Current on-site operations include chemical and radiological analytical laboratories. Situations in which an on-site worker potentially could be exposed to above-background levels of radiation would be during preparation and analysis of radiological samples in the analytical laboratory and during handling of radon sources. If exposure were to occur, the primary pathways would be inhalation and ingestion of airborne particulates; inhalation of radon and radon daughters; or, direct exposure to gamma radiation from samples.

The primary risks to human health under present conditions are from nonradiological hazards such as (1) falling, tripping, or slipping; (2) industrial accidents; or (3) exposure to chemicals. The risk of hazards in the first two categories is about the same as for workers in any office setting or on any construction site. Implementation of health and safety measures, such as job-site safety meetings helps to reduce these risks. The potential for laboratory workers to be exposed to chemicals is reduced by implementation of laboratory hygiene plans.

Currently, use of the groundwater and surface water is prohibited. DOE evaluated the risks of recreational use of the surface water, as discussed below.

Environmental Consequences

Commercial Use. Activities identified as potentially occurring under this scenario are similar to those currently ongoing at the site: light industrial, general office, and analytical laboratory. Potential hazards associated with these operations are the same as described below for the No Action Alternative. Activities identified in this scenario will not significantly increase the current air emissions and will not provide the groundwater for public use.

Personnel involved in construction or building modifications/demolitions would have the highest exposure to construction hazards and industrial accidents. The risks are similar to those at other construction sites. The potential for other workers on the facility and general public to be exposed to chemicals, toxic substances, radioactive substances, radioactive sources, tripping hazards, or industrial accidents would be low and typical for an office environment.

In the event that the ponds/wetlands are left intact under this scenario, the potential impacts to human health from contact with the ponds would be the same as that described below for the Open Space scenario.

Industrial Use. Activities identified in the Industrial Use scenario include mining the site for gravel resources. Potential hazards associated with these operations are consistent with gravel mining activities currently ongoing in active gravel pits near the site. Water generated from dewatering operations would be handled according to site-specific requirements established by the Colorado Department of Public Health and Environment. These requirements would minimize the risk to workers and the general public. Overall dust emissions would increase, but emissions of radionuclides would cease.

Gravel mining operations are regulated under Occupational Safety and Health Administration (OSHA) and industrial accidents would be about the same as for general construction or other earthmoving operations.

Mixed Use. Activities identified as potentially occurring under this scenario are similar to those currently ongoing at the site: light industrial, general office, analytical laboratory. Potential hazards associated with these operations are the same as described below for the No Action Alternative. Activities identified in this scenario will not significantly increase air emissions. Groundwater and surface water for use as drinking water would be prevented by institutional controls.

Risks to personnel involved in construction or building modifications/demolitions would be the same as those described above for the Commercial Use scenario. Similarly, the potential for other workers on the facility and general public to be exposed to chemicals, toxic substances, radioactive substances, radioactive sources, construction hazards, or industrial accidents would be low or about the same as that for a worker in a similar office environment.

In the event that the ponds are left intact under this scenario, the potential impacts to human health from contact with the ponds would be the same as that described below for the Open Space scenario.

Open Space. Under this scenario, all structures at the GJO facility would be demolished and the entire area would be used as an open space, park-like area. Since prior remedial actions have eliminated risks from the soils, the surface water bodies onsite would be the remaining potential sources of risk to human health. In order to quantify these potential risks, a human health risk assessment was conducted as a separate task from this EA. Various potential future uses of the GJO facility were analyzed, including potential uses of the surface water bodies onsite under the Open Space scenario. These potential uses include ecological viewing, teaching, field trips, and recreational fishing.

Two potential sources of risks are: (1) ingestion of fish from the ponds, and (2) unintentional ingestion of small quantities of pond water (incidental ingestion) during educational and recreational activities. From fish ingestion, manganese and uranium are contaminants that may pose risks to human health. Assuming consumption of fish at a conservative rate of 25 grams (approximately one ounce) per day, 365 days per year, the risks were determined to be unacceptable, according to EPA guidelines. It is therefore recommended that the taking of fish for human consumption not be permitted under all scenarios.

Incidental ingestion of surface water was assumed at the rate of 20 milliliters per day, 2 days a year for a child; and 20 milliliters per day, 7 days a year for an adult. No adverse human health effects would be expected to occur for these exposure assumptions. Sulfate, however, is present in the surface water above drinking water guideline levels. It is recommended that prohibitions on the use of surface water as a drinking water source remain in place.

No Action Alternative. As concluded in the June 1996 EA (DOE 1996a), current facility operations do not present a risk to the general public. Current air emissions are well below Federal and state standards. Potential health risks from contamination due to previous site uses would be mitigated via remediation. Contaminated groundwater would remain unavailable for public use. Also, institutional controls would restrict use of the surface water bodies at the GJO.

Workers involved in onsite activities would potentially be exposed to chemicals, toxic substances, and radioactive sources. All these personnel would be required to follow established operational, health, and safety procedures to reduce or eliminate their exposure to harmful elements. Additionally, standard operating procedures would require engineering or radiological controls to be implemented to reduce exposure limits.

4.8   ECOLOGICAL RESOURCES

Affected Environment

The existing environment is commercial and office use with residential style landscaping and maintenance. Developed areas on the property contain buildings, asphalt, concrete, gravel, roads, and lawns of low value as habitat for indigenous plants and wildlife. This habitat is marginal for most wildlife species that inhabit the adjacent native habitat. Areas closer to the Gunnison River include two small ponds (the North and South Ponds) and small patches of upland, wetland, and riparian vegetation (see Figure 1.2-2). Most of this open space was disturbed by environmental remediation activities between 1989 and 1994 and subsequently restored as natural habitats (DOE 1996a and 1999a). Lists of plant and wildlife species inhabiting the property and surrounding vicinity are included in the *Environmental Assessment of Facility Operations* completed in 1996 (DOE 1996a) and are copied in Appendix B. A list of bird species sighted on (or in the immediate vicinity of) the property by DOE-GJO employees is provided in Appendix C.

Specific ecological restoration activities completed on the property are described in Section 4.4 and depicted in Figure 1.2-2 (DOE 1998b). Vegetation in the restored wetland areas is presently dominated by reed canary grass (*Phalarus arundinacea*), hardstem bulrush (*Scirpus acutus*), common cattail (*Typha latifolia*), and willows (*Salix* sp.). Restored riparian vegetation includes seedlings of indigenous Fremont cottonwood (*Populus fremontii*), skunkbush sumac (*Rhus trilobata*), and buffaloberry (*Shepherdia argentea*). But it is dominated by naturalized exotic shrubs and trees such as saltcedar (*Tamarix ramosissima*) and Russian olive (*Elaeagnus angustifolia*) (DOE 1996a).

Vegetation in the restored wetland and riparian areas, and in undeveloped upland on the property, provides habitat for diversity of reptiles, amphibians, birds, and small mammals. Large mammals, such as coyotes (*Canis latrans*) and mule deer (*odocoileus hemionus*), may occasionally visit the undeveloped areas on the property but are likely discouraged by the urban surroundings. The surrounding area is too urban to provide habitat for bears. Mature trees adjacent to the property and along the Gunnison River provide roosting habitat for the bald eagle (*Haliaeetus leucocephalus*) and snowy egret (*Egretta thula*). The ponds also provide habitat for small fish (such as various minnows and shiners), but do not likely support sport fish. The Gunnison River supports a diverse fish population, including large mouth bass (*Micropterus salmoides*) and various species of trout.

A search of the Biological and Conservation Data system maintained by the Colorado Natural Heritage Program revealed six natural heritage resources that have been documented in the immediate vicinity of the property (Table 4.8-1) (Johnson 1999). Natural heritage resources include occurrences of significant natural communities and rare, threatened, or endangered plants or animals. Of the resources documented for the property, only the Colorado pikeminnow has a Federal or state status as threatened or endangered. The Colorado pikeminnow (or Colorado squawfish) migrates long distances in rivers and streams to spawn, using deep pools or eddies to rest and feed and riffles or shallow runs to mate. Although once inhabiting much of the Colorado and Gila River basins, its populations have declined due to changes in stream flow and temperature, loss of habitat from reservoir construction, blockage of migration routes, and the introduction of non-native fish (UNR 1999a; Arizona 1999).

The other species indicated by Colorado Natural Heritage Program have no special Federal status, but they are listed by the State of Colorado as "Special Concern." One, the roundtail chub, is another migratory fish inhabiting the Gunnison River. The other "Special Concern" species are amphibians, not likely to occur in developed areas, but they could potentially inhabit the wetlands, riparian lands, and other naturally vegetated areas adjoining the Gunnison River. The Grand Junction milkvetch and snowy egret lack Federal or state status but are considered somewhat rare in the state. Each could potentially occur in the naturally vegetated lands adjoining the Gunnison River.

In addition to the Colorado pikeminnow, a recent review of the area by the U.S. Fish and Wildlife Service noted that the Gunnison River may also provide habitat for three other federally endangered fish species: the humpback chub (*Gila cypha*), bonytail chub (*Gila elegans*), and razorback sucker (*Xyrauchen texanus*) (Moyer 1999). Each of these fish species have experienced population declines from the same causes as the Colorado pikeminnow (UNR 1999b, c, and d). The U.S. Fish and Wildlife Service review also noted that the riparian vegetation associated with the Gunnison River could provide habitat for the federally endangered willow flycatcher (*Empidonax traillii extimus*), and mature cottonwood trees along the river could provide roosting sites for the federally threatened bald eagle (*Haliaeetus leucocephalus*).

In a separate review, the Colorado Division of Wildlife noted the potential occurrence of the Colorado pikeminnow and razorback sucker in the Gunnison River, and the willow flycatcher in the associated riparian vegetation (Creeden 1999). The Colorado Division of Wildlife further emphasized that the riparian vegetation provides important habitat to a diversity of wildlife, including various hawk, eagle, and migratory songbird species. The Colorado Division of Wildlife review also noted that a kit fox (*Vulpes macrotis*) was sighted in the vicinity of Grand Junction but that it was unknown if it occurred on the DOE property. The preferred habitat for the kit fox is desert scrub and desert grassland (Southwest Wildlife 1999), which occurs in areas outside of the DOE property, but not inside. Extensive human activity has likely discouraged entry onto the property by this species.

Environmental Consequences

Commercial Use. The recipient of the property could convert all or part of the remaining 10 hectares (24.7 acres) of open space to commercial development, resulting in the permanent loss of up to approximately 0.8 hectares (2 acres) of wetland habitat, 1.7 hectares (4.2 acres) of riparian habitat, 4.4 hectares (11 acres) of upland habitat, and 1.2 hectares (3 acres) of shallow water habitat comprising the North and South Ponds. Most of the affected habitat is of recent origin, having been planted in 1995 and 1996 on exposed soils following an environmental remediation. Mature willow and cottonwood saplings in the adjacent areas provide roosting sites for the bald eagle or snowy egret. The habitat value of the riparian vegetation for the willow flycatcher is reduced by the predominance of invasive species such as saltcedar and Russian olive.

Development under the Commercial Use scenario would not disturb the existing dike and riverbanks and thus not likely affect the habitat value of the Gunnison River for the federally endangered fish species. Riparian

Table 4.8-1. *Natural Heritage Resources Documented for Immediate Area of DOE-GJO Site. Colorado Natural Heritage Program, Biological and Conservation Datasystem Township 1 South, Range 1 West, Sections 26 and 27.*

Scientific Name	Common Name	Taxon	Federal Status	State Status	Global Rank	State Rank	Typical Habitat
Astragalus linifolius	Grand Junction Milkvetch	Plant	None	None	G3Q	S3	Dry clayey slopes and gullies in pinyon-juniper woodlands and occasionally near cottonwoods.
Egretta thula	Snowy Egret	Bird	None	None	G5	S2B, SZN	Reservoirs, grassy marshes, wet meadows, and rivers. Nests in trees or shrubs adjacent to reservoirs and marshes.
Gila robusta	Roundtail Chub	Fish	None	SC	G2G3	S2	Slow moving water adjacent to faster water. Young in river eddies and irrigation ditches.
Hyla arenicolor	Canyon Treefrog	Amphibian	None	SC	G5	S2	Permanent pools or cottonwoods, especially in rocky canyons with pinyon-juniper cover on slopes.
Ptychocheilus lucius	Colorado Pikeminnow	Fish	LE	T	G1T?Q	S1	No information provided
Spea intermontana	Great Basin Spadefoot	Amphibian	None	SC	G5	S3	Pinyon-juniper woodland, sagebrush, semi-desert shrublands, usually in or near dry rocky slopes or canyons.

Federal Status: LE - Listed as Endangered under the Endangered Species Act

State Status: T - Threatened, SC - Special Concern

Global Ranks: G1- Critically imperiled, G2 - Imperiled, G3 - Rare or uncommon, G5 - Demonstrably secure
G2G3 - rank intermediate between G2 and G3
G1T?Q - Species is G1, variety or subspecies unranked, questionable taxonomy

State Ranks: S1 - Critically imperiled, S2 - Imperiled, S3 - Vulnerable
SZN - Non-breeding season imperilment of nonresident (migratory) species

Habitat descriptions based on draft descriptions under development by Colorado Natural Heritage Program using various scientific sources.

vegetation on the river shoreline would not be disturbed. The dike would prevent potential sedimentation of the river from construction activities. Stormwater and wastewater discharges from the commercial facilities would be directed to municipal sewers.

Industrial Use. The recipient of the property could convert all or part of the remaining 8 hectares (20 acres) of open space inside the dike to a gravel pit operation, resulting in the loss of up to approximately 0.8 hectares (2 acres) of wetland habitat, 1.7 hectares (4.2 acres) of riparian habitat, 4.4 hectares (11 acres) of upland habitat, and 1.2 hectares (3 acres) of shallow water habitat comprising the North and South Ponds. However, most of the affected habitat is of recent origin, having been planted in 1995 and 1996 on exposed soils following an environmental remediation. Mature willow and cottonwood saplings in the adjacent areas provide roosting sites for the bald eagle or snowy egret. The habitat value of the riparian vegetation for the willow flycatcher is reduced by the predominance of invasive species such as saltcedar and Russian olive.

Because the existing dike and riverbanks would not be disturbed, the gravel pit operation would not likely affect the habitat value of the Gunnison River or affect any of the federally endangered fish species potentially inhabiting the river. Riparian vegetation on the river shoreline would not be disturbed. The dike would prevent potential sedimentation from the gravel pit. The gravel pit operations would not likely discharge to the river, and any discharges would require a permit and have to meet applicable state and Federal water quality criteria.

There would be a future potential to restore riparian, wetland, and/or upland vegetation on areas of the gravel pit once extractive operations have been completed and the land reclaimed. The ability to restore such vegetation would depend upon future land use decisions for the site. Restoring natural vegetation to reclaimed mine sites is similar in practice to restoring vegetation to land disturbed by environmental remediation. Thus, re-establishment of habitats similar to those on the site at this time would be possible in the future.

Mixed Use. The recipient of the property could convert all or part of the remaining 8 hectares (20 acres) of open space inside the dike to various types of urban development, resulting in the permanent loss of up to approximately 0.8 hectares (2 acres) of wetland habitat, 1.7 hectares (4.2 acres) of riparian habitat, 4.4 hectares (11 acres) of upland habitat, and 1.2 hectares (3 acres) of shallow water habitat comprising the North and South Ponds. But, as explained for the Industrial and Commercial Use scenarios, most of the affected habitat is of recent origin, having been planted in 1995 and 1996 on exposed soils following an environmental remediation. Mature willow and cottonwood saplings in the adjacent areas provide roosting sites for the bald eagle or snowy egret. The habitat value of the riparian vegetation for the willow flycatcher is reduced by the predominance of invasive species such as saltcedar and Russian olive.

As for the Industrial and Commercial Use scenarios, the development under the Mixed Use scenario would not disturb the existing dike and riverbanks and thus not likely affect the habitat value of the Gunnison River for the federally endangered fish species. Riparian vegetation on the river shoreline would not be disturbed. The dike would prevent potential sedimentation of the river from construction activities. Stormwater and wastewater discharges from the new development would be directed to municipal sewers.

Open Space. The existing natural habitats on the property would not be disturbed under the Open Space scenario. Furthermore, these habitats would be complemented by additional upland habitats that establish, through restoration or natural succession, on formerly developed uplands on the property. Standard erosion control practices would be implemented during the demolition process to protect adjoining ponds, wetlands, and naturally vegetated areas. Departure of industrial activity from the site would likely make the existing natural habitats on the site more attractive to most wildlife. The entire site would likely be placed under an integrated wildlife habitat by the Colorado Department of Natural Resources or other state or Federal agency.

No Action Alternative. The existing natural habitats on the property would remain undisturbed under the No-Action Alternative. The DOE would continue to manage the open space as wildlife habitat.

4.9   CULTURAL RESOURCES

Affected Environment

Cultural resources are those aspects of the physical environment that relate to human culture and society, and those cultural institutions that hold communities together and link them to their surroundings. Cultural resources include expressions of human culture and history in the physical environment such as prehistoric or historic archaeological sites, buildings, structures, objects, districts, or other places including natural features and biota which are considered to be important to a culture, subculture, or community. Cultural resources also include traditional lifeways and practices, and community values and institutions.

The cultural resources present in western Colorado demonstrate the prehistoric use of the region for over 10,000 years; the ongoing tradition of the Utes and other Native American groups; EuroAmerican settlement, agriculture, ranching and mining; and the importance of the GJO in the history of uranium exploration, mining and processing activities for the Manhattan Project during World War II and the Cold War.

Cultural Resource Regulations. The identification of cultural resources and DOE responsibilities with regard to cultural resources are addressed by a number of laws, regulations, executive orders, programmatic agreements and other requirements. The principal Federal law addressing cultural resources is the *National Historic Preservation Act* of 1966, as amended (16 USC 470 et seq.), and implementing regulations (36 CFR 800) that describe the process for identification and evaluation of historic properties; assessment of the effects of Federal actions on historic properties; and consultation to avoid, reduce, or minimize adverse effects. The term "historic properties" refers to cultural resources that meet specific criteria for eligibility for listing on the National Register of Historic Places. This process does not require preservation of historic properties, but does ensure that DOE's decisions (as a Federal agency) concerning the treatment of these places result from meaningful considerations of cultural and historic values and of the options available to protect the properties.

The identification and evaluation of cultural resources for National Register of Historic Places eligibility is the responsibility of the DOE with the concurrence of the State Historic Preservation Officer. The Advisory Council on Historic Preservation, an independent Federal agency, administers the provisions of Section 106 of the *National Historic Preservation Act*, regarding cultural resources and has review and oversight responsibilities defined in 36 CFR 800.

Cultural Resources of the GJO. A literature review indicates that the GJO area has been extensively disturbed by past activities including development, environmental restoration, prior use as an ore processing facility, and floods. The potential for the existence and discovery of intact prehistoric or historic archaeological resources that would meet National Register of Historic Places eligibility requirements is considered very low. Likewise, no Native American or other traditional use areas or religious sites are known to be present on the GJO property. No Native American remains or artifacts of religious or cultural significance are known to exist or to have been removed from the GJO.

All buildings and structures on the GJO have been surveyed and evaluated for National Register of Historic Places eligibility. An historic district has been defined which encompasses the GJO area. The contributing elements to the district include 13 buildings (2, 12/12A, 19, 20, 26, 28, 29, 32, 40, 43, 810, 938, and 3022), an instrument calibration facility, and the protective dike (See Table 4.9-1). Twenty-seven buildings and structures within the boundaries are considered non-contributing elements and three buildings have been demolished since the survey was conducted. The district is considered significant for its association with the Manhattan Project during World War II, and the Cold War Federal programs for the exploration, mining and processing of uranium and vanadium. As an administrative center for Federal programs, the GJO was the focus of the uranium prospecting and mining boom of the 1950s and was associated with the development of technical processes that substantially advanced the exploration and processing of uranium ores. The proposed district includes buildings that appear to meet the criteria of "exceptional importance" required for listing properties that are less than 50 years old (Schweigert 1999a). In a Memorandum of Agreement dated August 14, 1998, the DOE-GJO agreed to consult with the State Historic Preservation Officer on the management of, and potential impacts to, the GJO Historic District.

Environmental Consequences

Impact Analysis Methods. Potential impacts on historic properties are assessed by applying the Criteria of Adverse Effect (as defined in 36 CFR 800.5a). An adverse effect is found when an action may alter the characteristics of a historic property that qualify it for inclusion in the National Register of Historic Places in a manner that would diminish the integrity of the property's location, design, setting, workmanship, feeling, or association. Adverse effects may include reasonably foreseeable effects caused by the action that may occur later in time, be farther removed in distance, or be cumulative.

Table 4.9-1. *Contributing Elements of the Grand Junction Project Office Historic District*

Building/Feature	Use/Function	Constructed	Integrity
2	Communications	1943	Fair
12/12A	Administration	Pre-1943/1948	Fair
19	Guard House	1948	Fair
20	Laboratory	1953/1957	Fair
26	Offices	1954	Good
28	Warehouse/Repair	1955	Fair
29	Truck Dispatch	1955	Good
32	Laboratories	1954	Good
43	Storage	Post-1967	Good
40	Utilities	1958	Excellent
810	Offices	1949/50/80	Good
938	Office/Auditorium	1954/55/63	Fair
3022	Laboratories/Offices	1953/55	Fair
	Calibration Facility	1950s	Good
	Earthen Dike	1957	Good

Source: Schweigert 1999.

Commercial Use. The lands proposed for transfer include the National Register of Historic Places-eligible GJO Historic District. Impacts to this historic property from the transfer itself would include the loss of Federal protection and responsibility for this resource if transferred to a non-Federal entity. The transfer, lease, or sale of historic properties out of Federal ownership or control without adequate and legally enforceable restrictions or conditions to ensure long-term preservation of the property's significance would be an adverse effect. When transferred, future consideration of this historic property under the *National Historic Preservation Act* and other Federal laws, regulations, guidelines, and executive orders would be limited. Transferred to a Federal entity would provide continued Federal protection and responsibility for the resources.

The All Commercial Land Use scenario represents a continuation of current land uses and expansion of similar site activities. Under this scenario, it is anticipated that many of the buildings in the GJO Historic District could be reused by the receiving entities. The continued use of historic buildings in a manner that does not diminish the integrity of the resource would be a positive impact and any abandonment leading to deterioration would be a negative impact.

New commercial construction is anticipated under this scenario. The Commercial Land Use scenario does not specifically call for the removal of any of the buildings or features that are contributing elements of the historic district, but removal by the receiving entities in the future is possible. Likewise, possible modifications by the receiving entities to enhance reuse of these historic buildings or features have not been defined, but such modifications could reasonably be expected to occur in the future. Modifications to historic buildings could negatively impact the integrity of the historic property. New development could also alter the setting of the historic district.

The proposed transfer of the facility would limit the effective options for management of the historic property unless it was transferred to another Federal entity. The DOE would not maintain an interest in the facility or control future uses. Any covenants or other restrictions on future owners would be unlikely to effectively preserve the historic appearance of the facility and would discourage reuse of the site. The long-term preservation of the GJO facility in its current form is practically impossible. Therefore, historic preservation actions by DOE must be undertaken before the property is divested.

The *Final Historic Structures Survey of the Department of Energy Grand Junction Office* recommends that the historical values of the facility can be preserved and made available for public appreciation by (a) completing Historic American Engineering Record documentation of the facility at a level determined in consultation with the National Park Service, (b) completion of a public information document that addresses the history and importance of the facility, and (c) the installation of commemorative signage at the site. The change in the proposed disposition of the facility and these mitigations will be reflected in a new Memorandum of Agreement between DOE and the State Historic Preservation Officer (Schweigert 1999a).

Industrial Use. Impacts of this scenario from the transfer itself would include the loss of Federal protection and responsibility for the GJO Historic District. In addition, most or all of the contributing features of the GJO District except the dike could be demolished under this land use scenario. The physical destruction of the historic property would be an adverse effect.

As described for the Commercial Use scenario, the proposed transfer of the facility would limit the effective options for management of the historic property and, therefore, historic preservation actions by DOE must be undertaken before the property is divested. The management recommendations of the *Final Historic Structures Survey of the Department of Energy Grand Junction Office* should be implemented prior to transfer.

Mixed Use. Impacts to the GJO Historic District and potential mitigations would be the same for the Mixed Use scenario as those described for the Commercial Use scenario.

Open Space. Impacts to the GJO Historic District and potential mitigations associated with the Open Space scenario would be similar to those described for the Industrial Use scenario. There would be a loss of Federal protection and responsibility for the resource and contributing elements of the district would be removed. The management recommendations of the *Final Historic Structures Survey of the Department of Energy Grand Junction Office* (Schweigert 1999b) should be implemented prior to transfer.

No Action Alternative. Under the No Action Alternative, the GJO would remain under the responsibility of the DOE and the treatment of the cultural resources present would continue to be subject to Federal laws, regulations, guidelines, and executive orders. The use of the historic structures for DOE and tenant activities would continue. Ongoing minor impacts from natural processes and aging on the physical integrity of the buildings would occur. The development of a Programmatic Agreement that addresses the potential effects to the GJO Historic District that may accrue from DOE-GJO's operation, remediation, divestiture, or other activities at the facility would continue in accordance with the Memorandum of Agreement dated August 14, 1998. In addition, the GJO would also develop a Cultural Resources Management Plan in consultation with the State Historic Preservation Officer and the National Park Service which will be referenced in the Programmatic Agreement. Management recommendations from the *Draft Historic Structures Survey of the Department of Energy Grand Junction Office*, which were predicated on the continued management of the GJO Historic District, would provide the basis for the provisions of the Cultural Resources Management Plan. These recommendations include Historic American

Table 4.10-1. *Federal Ambient Air Quality Standards*

Pollutant	Symbol	Averaging Time	Standard ppm	Standard $\mu g/m^3$	Violation Criteria
Ozone	O_3	1 Hour	0.12	235	If exceeded on more than 3 days in a 3-year period
		8 hours	0.08	160	If exceeded by the mean of annual 4th highest daily values for a 3-year period
Carbon Monoxide	CO	8 Hours	9	10,000	If exceeded on more than 1 day per year
		1 Hour	35	40,000	If exceeded on more than 1 day per year
Inhalable Particulate Matter	PM_{10}	Annual Arithmetic Mean	---	50	If exceeded as a 3-year single station average
		24 Hours	---	150	If exceeded by the mean of annual 99th percentile values over 3 years
Fine Particulate Matter	PM_{25}	Annual Arithmetic Mean	---	15	If exceeded as a 3-year spatial average of data from designated stations
		24 Hours	---	65	If exceeded by the mean of annual 98th percentile values over 3 years
Nitrogen Dioxide	NO_2	Annual Average	0.053	100	If exceeded
Sulfur Dioxide	SO_2	Annual Average	0.03	80	If exceeded
		24 Hours	0.14	365	If exceeded on more than 1 day per year
		3 Hours	0.5	1,300	If exceeded on more than 1 day per year
Lead Particles	Pb	Calendar Quarter	---	1.5	If exceeded

Source: 40 CFR 50, 53, and 58

Environmental Consequences

Commercial Use. Demolition of structures would have temporary short-term effects similar to those described below for the industrial scenario but at a smaller scale. Construction of additional commercial space in developable parcels would result in fugitive dust emissions from soils disturbance and vehicle exhaust emissions from construction equipment. Site grading in particular has the potential for creating localized dust nuisance conditions. These conditions would be temporary in nature and, if necessary, could be reduced using standard dust control measures, such as watering.

The Commercial Use scenario would result in the continuation of many of the same uses of the site as under existing conditions, though the number of vehicle trips projected under this scenario would be less than under baseline conditions. Operation of the Analytical Laboratory by a private entity would be required to comply to the same standards and permitting requirements as under DOE operation. Air quality conditions would be similar to those under the No Action alternative.

Industrial Use. Demolition of existing structures would result in temporary short-term emissions from construction equipment exhaust, from site disturbance, and from demolition of the buildings themselves. Demolition activities could introduce asbestos and lead particles into the air if present in the structures, creating a potentially hazardous situation for workers. If asbestos and lead-based paint are present, demolition activities should be undertaken by personnel certified by the OSHA to handle hazardous materials and wastes.

Engineering Record documentation of the site and contributing elements, maintenance of property boundaries, maintenance of visual associations among contributing elements where possible, preservation of exterior appearances of contributing elements, and encouragement of adaptive reuse of contributing buildings (Schweigert 1999b).

4.10 AIR QUALITY

Affected Environment

Regional Air Quality. The Federal *Clean Air Act* (42 USC 7401, et seq.), as amended, authorizes the EPA to establish national ambient air quality standards to protect public health and welfare. Federal ambient air quality standards have been adopted for the following six criteria pollutants: ozone, carbon monoxide, nitrogen dioxide, sulfur dioxide, inhalable and fine particulate matter (PM_{10} and $PM_{2.5}$), and lead. National ambient air quality standards for these pollutants are presented in Table 4.10-1. Colorado has adopted the national ambient air quality standards as the state air quality standards, with the exception of sulfur dioxide.

Areas that violate Federal air quality standards are designated as Federal nonattainment areas for the relevant pollutants. Nonattainment areas are sometimes further classified by degree (marginal, moderate, serious, severe, and extreme). Areas that comply with air quality standards are designated as attainment areas for the relevant pollutants. Areas for which monitoring data are lacking are designated as unclassified. Unclassified areas are treated as attainment areas for most regulatory purposes. Mesa County, Colorado, where the GJO Site is located, is unclassified for all criteria pollutants.

Air Quality Emission Sources. Sources of criteria air pollutants associated with GJO facility and tenant operations include vehicle traffic, building heating, painting activities, and small amounts of fugitive dust.

Two radon emission sources and two radioparticulate emission point sources are located at the GJO facility (DOE 1996a). Radon is emitted from instrument calibration facilities and radon calibration chambers, and radioparticulates are emitted from the Analytical Laboratory and Baghouse. Radon emissions released from the GJO facility do not affect atmospheric radon concentrations at the facility boundary (DOE 1996a). Radioparticulate emission dose modeling indicates that the total dose to off-site receptors is well below EPA and DOE standards (DOE 1996a).

DOE maintains an air permit from the Colorado Department of Public Health and Environment for the Analytical Laboratory; all other stationary sources are exempt from permit requirements.

Regulatory Considerations. Section 176(c) of the *Clean Air Act* (42 USC 7401, et seq.) requires Federal agencies to ensure that their actions are consistent with the *Clean Air Act* (42 USC 7401, et seq.) and with applicable air quality management plans (state implementation plans). Agencies are required to evaluate their proposed actions to make sure they will not violate or contribute to new violations of any Federal ambient air quality standards, will not increase the frequency or severity of any existing violations of Federal ambient air quality standards, and will not delay the timely attainment of Federal ambient air quality standards.

The EPA has promulgated separate rules that establish conformity analysis procedures for transportation-related actions and for other (general) Federal agency actions. The EPA general conformity rule requires a formal conformity determination document for Federal actions occurring in nonattainment areas or in certain designated maintenance areas when the total direct and indirect emissions of nonattainment pollutants (or their precursors) exceed specified thresholds. The *Clean Air Act* (42 USC 7401, et seq.) conformity guidelines do not apply to Federal actions at the GJO Site since it is not in a nonattainment area.

Operation of the site as a gravel pit would be subject to state permitting requirements for this type of operation. Gravel pits can be substantial sources of particulate emissions from crushing and loading operations and may require emissions controls to reduce dust generation.

Mixed Use. Construction in developable areas would have temporary short-term effects similar to those described for the Commercial Use scenario.

The Mixed Use scenario would result in the continuation of many of the same uses of the site as under existing conditions, though more light industrial uses are projected than under the Commercial Use scenario. Use of newly developed areas would result in minor increases in air pollutant emissions similar in type to current nonradiological pollutants emitted by existing users. Operation of the Analytical Laboratory by a private entity would be required to comply to the same standards and permitting requirements as under DOE operation. The number of vehicle trips projected under this scenario would be less than under baseline conditions and overall air quality conditions would be similar to those under the No Action Alternative.

Open Space. Demolition of existing structures to restore the site to open space would have the same temporary short-term effects as described for the Industrial Use scenario. Configuring the site for open space uses, such as parkland or a wildlife preserve, would result in fugitive dust emissions from site grading and in minor emissions from construction vehicle exhaust; these emissions also would be temporary and short-term.

Operation of the site as parkland or as another public use area would result in emissions from employee and visitor vehicle trips to the site; these vehicle trips likely would be less than the numbers of vehicle trips under baseline conditions, though use times could be concentrated more on weekends and evenings during spring, summer and fall months.

No Action Alternative. Under the No Action Alternative, air pollutants would continue to be emitted at current rates. Because current emissions comply with permitting regulations, conform to DOE and EPA standards for radioparticulates, and do not result in a violation of air quality standards, no adverse effects to air quality are predicted. Because the GJO Site is not in a nonattainment area, it is not subject to the requirements of the *Clean Air Act* (42 USC 7401, et seq.) general conformity rule.

4.11   NOISE

Affected Environment

Sound travels through the air as waves of minute air pressure fluctuations caused by some type of vibration. Sound level meters measure pressure fluctuations from sound waves, with separate measurements made for different sound frequency ranges. These measurements are reported in a logarithmic decibel (dB) scale. Because the human ear is not equally sensitive to all frequencies, the "A-weighted" decibel scale (dBA) is used to weight the meter's response to approximate that of the human ear.

Average noise exposure over a 24-hour period often is presented as a day-night equivalent noise level. Equivalent noise level values are calculated from 24-hour averages, with the values for the nighttime period (10 PM to 7 AM) increased by 10 dB. The weighting of nighttime noise levels reflects the greater disturbance potential from nighttime noises.

Existing Noise Conditions

Noise Receptors. Sensitive receptors are land uses, such as residences, schools, libraries, and hospitals, that are considered to be sensitive to noise. There are no sensitive receptors on-site. Off-site receptors include a cemetery and residences across the river within one-half mile of the GJO Site.

Noise Sources. The primary noise sources at the GJO Site are vehicle traffic and light industrial activities. Temporary sources of noise are construction and cleanup activities. Off-site noise sources include a police firing range located about 200 yards east of the site and the railroad. Use of the firing range results in intermittent periods of sudden, high noise. The railroad is an intermittent source of noise; trains run by the site from twice a day to several times a day at five to ten miles per hour. The crossings do not have gates; therefore, the train engineers use the locomotive horns to warn motorists and pedestrians.

Regulatory Guidelines. The Federal *Noise Control Act of 1972* (42 USC 4901, et seq.) established a requirement that all Federal agencies must comply with applicable Federal, state, interstate, and local noise control regulations. Federal agencies also were directed to administer their programs in a manner that promotes an environment free from noise that jeopardizes public health or welfare.

The U.S. Department of Housing and Urban Development is the lead Federal agency setting standards for interior and exterior noise for housing. These standards, outlined in 24 CFR 51, establish site acceptability standards based on day-night equivalent sound levels. The standards are used to designate noise levels as acceptable, normally unacceptable, or unacceptable. The acceptable exterior noise level for residential housing is 65 dB or less, the normally unacceptable noise level is 65 dB to 75 dBA, and the unacceptable noise level is above 75 dBA.

The OSHA, Occupational Noise Exposure guidelines, codified at 29 CFR 1910.95, set an action level of 85 dBA as the maximum acceptable noise level for the workplace.

Environmental Consequences

Commercial Use. The Commercial Use scenario would result in the continuation of similar site uses at a similar level of activity as under current conditions. Demolition and construction activities could result in temporary noise disturbances to adjacent lands. Construction noise would be greatest in the immediate vicinity of construction equipment. Given the commercial nature of the surrounding parcels, noise effects on existing land uses would be minor. New commercial development would be compatible with existing land uses; new noise-sensitive land uses may not be compatible with existing commercial uses or with the off-site firing range.

Industrial Use. This scenario would result in a much more industrial use of the site than under current operating conditions. Demolition and construction would result in temporary noise disturbances that would be greatest in the immediate vicinity of the construction equipment; given the commercial nature of the surrounding parcels, construction noise effects on existing land uses would be minor. Because the site would be completely redeveloped for use as a gravel pit, noise levels under this scenario would be greater than under current conditions, both from mining operations and from hauling operations. This land use may not be compatible with some surrounding land uses, such as nearby residences and the cemetery. Restrictions on operations may be required to lessen the effects of noise; restrictions could include limiting the time of day or days of week of noise-generating operations or placing conditions on haul routes.

Mixed Use. The Mixed Use scenario would result in the same noise effects as described for the Commercial Use scenario since this scenario is also a continuation of similar site uses at a similar level of activity.

Open Space. The Open Space scenario would result in a lower intensity of use than under existing conditions. Demolition and construction would result in temporary noise disturbances that would be greatest in the immediate vicinity of the construction equipment; as the site would be vacant and given the nature of the surrounding parcels, noise effects on existing land uses would be minor. The primary source of noise from use of the site would be employee and visitor vehicles trips; however, the number of vehicle trips likely would be similar or less in number than under current operation conditions. Noise generated from the adjacent firing range could make the site less attractive for some uses, such as parkland.

No Action Alternative. Under the No Action Alternative, there would be little to no change in existing noise levels or noise patterns; therefore, no adverse noise effects would occur.

4.12   VISUAL RESOURCES

Affected Environment

Visual resources are those aspects of the environment that determine the physical character of an area and the manner in which it is viewed by people. Visual resources include scenery in the near, middle, and distant landscape and include cultural modifications, landforms, water surfaces and vegetation. This analysis inventories the existing visual resources, assesses any changes that could result from the alternatives, and determines the impact of those changes on the viewsheds observed by the public.

Visual Resources of the GJO. The visual character of the GJO Site reflects its past and current uses as a uranium milling site, an administrative center, and commercial/industrial center. As such, the site has been subject to extensive disturbance that has altered the original landscape through grading, construction, environmental restoration, and active use as an industrial site. The developed areas include pavement, gravel road base, fencing, some grassy areas and older utilitarian buildings in good to fair condition. The building development pattern does not represent a unified architectural style or campus arrangement. There are several temporary and modular buildings and many buildings have been altered to respond to changing needs. The site is adjacent to the Gunnison River on the north, south and west side but a dike limits views of the river from the site. There are approximately 8.1 hectares (20 acres) of open space with vegetation and manmade ponds and wetlands. There is a large mesa west of the site with some homes visible on top. On the east side there is a parking area, and railroad tracks. An escarpment that includes the city cemetery overlooks the site on the east side. The scenic quality of the GJO would be considered low.

The site is primarily viewed by workers and tenants. Views are limited as the single road that connects the highway with the GJO is used primarily by workers accessing the site. Overall views are observed by residents living in the Little Park Road area and by visitors to the city cemetery.

Environmental Consequences

Impact Analysis Methods. Potential impacts to visual resources are assessed by determining whether changes could result from the alternatives that would noticeably increase visual contrast and reduce scenic quality from current conditions; would block or disrupt existing views or reduce public opportunities to view scenic resources; or would conflict with regulations governing aesthetics.

Commercial Use. The Commercial Use scenario anticipates the reuse of many of the existing buildings by the receiving entities and the construction of new commercial properties on the site. The reuse of existing buildings would not likely result in any changes to visual resources. The construction of new facilities and possible replacement of aging structures and temporary buildings would maintain or improve visual resources through planned development. If current open space or wetland areas are developed, there may be some loss in quality of visual resources. If development is extensive, there may be some loss in quality of views from the Little Park Road area and the cemetery.

Industrial Use. The Industrial Use scenario would remove the current buildings and revert the site back to its pre-war use as a gravel pit. The removal of buildings and use of the site for a single purpose would provide more visual unity to viewers, but the industrial use would maintain or decrease overall visual quality. The loss of current open space and wetland areas would reduce current quality of visual resources. Dust associated with gravel mining activities could also reduce visibility and access to current views.

Mixed Use. The Mixed Use scenario anticipates the reuse of many of the existing buildings by the receiving entities and the construction of new commercial or industrial properties on the site. The impacts to visual resources of the Mixed Use scenario would be similar to those of the Commercial Use scenario.

Open Space. The Open Space scenario would remove the current buildings and restore the site to open space and recreational use. The removal of buildings and other manmade features would provide more visual unity to site viewers and would represent an aesthetic improvement over current conditions. The possible expansion and enhancement of wetland areas would also improve the visual quality of the site. The potential development of a walkway/bikeway along the riverfront would also improve the visual quality of the tract. Opening these areas to the public would increase opportunities to view scenic resources.

No Action Alternative. Under the No Action Alternative, the visual resources of the GJO would remain the same.

4.13   SOLID AND HAZARDOUS WASTE MANAGEMENT

Affected Environment

Hazardous wastes are generated at the GJO in typical day-to-day activities, although in small quantities. Hazardous waste is regulated under the Federal *Resource Conservation and Recovery Act* (RCRA) (42 USC 6901, et seq.) and State of Colorado equivalent regulations. The Analytical Chemistry Laboratory (Building 20) is the primary user of hazardous materials and hence generates the majority of hazardous wastes. The hazardous materials used are mainly various solvents and calibration standards. These wastes and quantities are typical of those generated during normal day-to-day operations at facilities such as GJO. Wastes are stored in accordance with the RCRA (42 USC 6901 et seq.) in three modular hazardous waste storage units (Buildings. 61A, B, and C). They are shipped offsite for treatment and disposal at facilities that operate under RCRA (42 USC 6901, et seq.) permits.

The facility typically operates as a conditionally exempt small quantity generator under the RCRA (42 USC 6901, et seq.). However, the GJO occasionally moves into small quantity generator status and has been a large quantity generator once or twice in the past, primarily from generating regulated wastes during remedial actions. In order to accommodate the possibility of future generation of large quantities of waste, GJO maintains full compliance with all of the requirements of the RCRA (42 USC 6901, et seq.) for large quantity generators.

In addition to the RCRA (42 USC 6901, et seq.) regulated wastes, the GJO also generates waste regulated under the *Toxic Substances Control Act* (15 USC 2601, et seq.) – PCBs and asbestos. The rate of generation of these wastes at GJO is low and is generated primarily from replacement and removal of PCB-containing light ballasts. Asbestos waste is generated from the removal of asbestos-containing materials such as ceiling insulation, damper material, and linoleum. The PCB waste generated is stored on site in Building 42 for later disposal at offsite facilities (within the mandated 9-month time period).

Because the GJO was the site of uranium processing, residual radioactive material still exists on the site, and both asbestos and PCB waste present on the GJO (such as light ballasts) may be radioactively contaminated with residual radioactive material. The GJO stores radioactive PCB wastes in Building 42 in compliance with 40 CFR 761.65, *Facilities Compliance Agreement on the Storage of Polychlorinated Biphenyls*. Approximately 204 kilograms (450 pounds) of this material is currently in storage. This waste will be shipped offsite for treatment and disposal.

Non-radioactive asbestos that is removed from buildings is disposed of at the Mesa County Landfill in compliance with local, state, and Federal regulations; radioactive asbestos is disposed of at the Cheney Disposal Cell. Small amounts of asbestos containing materials will be produced as remediation efforts continue and buildings are remodeled or demolished. All asbestos abatement work has been and will continue to be performed by a licensed subcontractor in accordance with Colorado Regulation 8, The Control of Hazardous Air Pollutants. The site also generates non-hazardous, non-radioactive solid wastes, including sanitary wastes and building debris. This waste is hauled to the Mesa County Landfill for disposal.

Environmental Consequences

Commercial Use. Under this scenario, future users of the site will likely be small quantity generators similar to the current situation. In the event that they generate sufficient quantities to require reporting status, they would likely qualify as conditionally exempt small quantity generators. Users would be expected to comply with the temporary storage provisions under the RCRA (42 USC 6901, et seq.). Under this scenario, similar quantities of solid nonhazardous waste would be generated at the site and disposed at the Mesa County Landfill. Minor increases in demolition material may occur as older buildings are demolished and replaced with either new construction or open space.

Industrial Use. Minor quantities of hazardous waste and hazardous materials would likely be handled at the site under this scenario. As is the case with the Commercial Use scenario, quantities would likely be small and not require reporting. Fuel would be stored at the site for the gravel pit equipment and solvents and degreasers would be used for vehicle maintenance. Large quantities of demolition debris (15,000-25,000 cubic meters [60,000-100,000 cubic yards]) would be generated in clearing the site for gravel pit operations. It is anticipated that some small percentage of this material would be classified as asbestos containing material and would need to be disposed of in accordance with the State of Colorado solid waste regulations, Title 6, Code of Colorado Regulations Part 1007-2. In addition, previous surveys have indicated the presence of lead-based paint on most of the buildings. Future demolition activities will need to conform with Regulation No. 19 of the Colorado Air Quality Control Commission. The remainder of the solid waste would be taken to the Mesa County Landfill. The Mesa County Landfill currently has a life-expectancy of 50 years based on a design capacity of 11,250,000 cubic meters (15,000,000 cubic yards); removal of this debris would decrease the landfill life expectancy by approximately 0.07-0.1 years.

Mixed Use. Under this scenario, impacts to solid and hazardous waste management would be similar to the Commercial Use scenario with the possibility that slightly larger quantities of materials and wastes would be generated. Quantities would likely be small and not require reporting. Minor demolition could occur at the site but debris volume would be considerably smaller than for the Industrial Use scenario.

Open Space. Under the Open Space scenario, all buildings on site would be demolished and debris would be transported to an appropriate landfill. Quantities would be similar to the Industrial Use scenario. It is anticipated that a small percentage of the material would be classified as asbestos-containing material and would need to be disposed of in accordance with State of Colorado solid waste regulations, Title 6, Code of Colorado Regulations Part 1007-2. In addition, previous surveys have indicated the presence of lead-based paint on most of the buildings. Future demolition activities will need to conform with Regulation No. 19 of the Colorado Air Quality Control Commission. The remainder of the solid waste would need to be taken to the Mesa County Landfill. The Mesa County Landfill currently has a life expectancy of 50 years based on a design capacity of 11,250,000 cubic meters (15,000,000 cubic yards); removal of this debris would decrease the landfill life expectancy by approximately 0.07-0.1 years.

No Action Alternative. Under the No Action Alternative, the GJO would continue to operate as a conditionally exempt small quantity generator utilizing Buildings 61A-C and 42 as hazardous waste storage areas. Hazardous and toxic waste would continue to be shipped offsite for treatment and disposal. Though not a large quantity generator, the site would continue to maintain full compliance with all of the requirements of the RCRA (42 USC 6901, et seq.) for large quantity generators. The site would continue to generate a similar volume of non-regulated solid waste and contract with a commercial vendor to collect and transport the waste to the Mesa County Landfill.

4.14   TRANSPORTATION

Affected Environment

Daily traffic to and from the GJO facility primarily consists of 300-330 vehicle trips per day by employees and about 50 vehicle trips per day by service vehicles driven by subcontractors or delivery personnel. The only ingress to and egress from the GJO facility is a two-lane, city-maintained road (B ¾ Road) about 0.8 kilometers (0.5 miles) in length. This road connects the GJO facility to U.S. Highway 50, one of the major transportation routes through Grand Junction and across southern Colorado. Within the city limits, U.S. Highway 50 has four lanes and numerous traffic lights. Outside the city limits it has two lanes and crosses sparsely populated desert rangelands. Other major transportation routes in the vicinity of the GJO are U.S. Interstate 70, which is part of a major east-west transcontinental trucking route; Colorado State Highway 141, which provides access to the south along with U.S. Highway 50; and the Union Pacific Railroad, which borders the east side of the facility.

Walker Field Airport, nine miles northeast of the GJO, provides scheduled commercial airline, air cargo and general aviation services. It is also used by military and fire fighting aircraft. It is outside of the area affected by the proposed transfer.

Environmental Consequences

Commercial Use. Under this scenario, the number of vehicle trips per day to the site would likely be similar to or slightly less than under the No Action Alternative. Vehicular emissions and the potential for vehicle-related accidents would also be similar.

Industrial Use. Under this scenario, the number of vehicle trips to the site would be greatly reduced, based on the reduced number of workers. There would be more truck traffic related to the gravel pit, but overall vehicular emissions and the potential for vehicle-related accidents would be reduced from the baseline. Increased heavy truck traffic would pass by residences on the road to State Highway 50.

Mixed Use. Under this scenario, the number of vehicle trips per day to the site would likely be similar to or slightly less than under the No Action Alternative. Vehicular emissions and the potential for vehicle-related accidents would also be similar.

Open Space. Under this scenario, there would be little or no employment or development at the site. The number of vehicle trips to the site would be reduced, but vehicle trips could be more concentrated on weekends and evenings. Overall, vehicular emissions and the potential for vehicle-related accidents would be reduced from the baseline.

No Action Alternative. Under the No Action Alternative, there would be no change from the baseline level of vehicle trips, vehicular emissions, or the potential for accidents involving vehicles. At the baseline level of activity, traffic volume is considered to be within the existing transportation infrastructure's capacity and therefore the potential for accidents is considered acceptable. Vehicle emissions at the baseline level have no adverse effects on air quality in the area.

4.15   SOCIOECONOMICS

Affected Environment

This section provides an overview of the current socioeconomic conditions within the Grand Junction Region of Influence. The Region of Influence for this analysis is Mesa County, Colorado.

Employment and Income. The Region of Influence has historically been dependent on the wholesale and retail trade and service sectors for employment. These sectors have become increasingly important in recent years as farming and mining employ a smaller percentage of the workforce, as shown in Table 4.15-1. In 1997, the service sector provided almost 32 percent of the regional employment while wholesale and resale trade provided almost 25 percent of the employment (BEA 1999).

The unemployment rate in the Region of Influence has averaged much higher than the unemployment rate in Colorado, as shown in Table 4.15-2. The 1998 unemployment rate was 5.0 percent in the Region of Influence, but only 3.8 percent in Colorado (BLS 1999). Employment in the Region of Influence totaled 55,779 in 1998, while the labor force totaled 58,691.

The per capita income in the Region of Influence was $20,593 in 1997, significantly lower than the state average of $27,015. The Region of Influence per capita income increased 35 percent from the 1990 level of $15,280, while the state per capita income increased more than 40 percent from the 1990 level of $19,290 (BEA 1999).

Population and Housing. The Region of Influence population grew steadily between 1980 and 1998, increasing an average of 1.3 percent annually, the same rate of increase as the state population. Region of Influence population totaled 112,891 in 1998, and is projected to reach 163,602 by 2020 (Census 1995, Census 1999). Historic and projected populations for the Region of Influence and Colorado are shown in Table 4.15-3.

In 1990, there were 39,208 housing units in the Region of Influence, 36,250 of which were occupied. The majority of these were single family, detached houses. The owner-occupied vacancy rates in the Region of Influence was 2.2 percent and the rental vacancy rate was 5.9 percent (Census 1992). Region of Influence housing characteristics are shown in Table 4.15-4.

Community Services. There are five hospitals in the Region of Influence with a total of 785 beds. All of the hospitals operate well below capacity (AHA 1995). In addition, there are 215 physicians in the Region of Influence (AMA 1995).

The Region of Influence encompasses three school districts with 42 schools, and approximately 19,750 students and 1,100 teachers. The student/teacher ratios range from 5.6 in the DeBeque School District to 14.0 in the Plateau Valley School District (CDE 1999). Mesa State College in Grand Junction is the only post-secondary school in the Region of Influence (HPI 1999).

There are six law enforcement agencies in the Region of Influence with approximately 300 officers (HPI 1999). The Grand Junction Police Department and Mesa County Sheriffs Department are the largest departments in the Region of Influence with 107 and 174 employees, respectively.

Environmental Justice. Executive Order 12898, "Federal Actions to Address Environmental Justice in Minority Populations and Low-Income Populations" (59 Federal Register 7629, February 16, 1994), requires Federal agencies to identify and address any disproportionately high and adverse human health or environmental impacts on minority or low-income populations from Federal actions. In the Region of Influence, almost 95 percent of the population was identified as white, compared to 88.2 percent of the population in Colorado, as shown in Table 4.15-5. Over 15 percent of the Region of Influence population was identified as living in poverty, compared to 11.7 percent of the state population. Minority and low-income populations are distributed throughout the county and are not concentrated in any one area.

Environmental Consequences

Commercial Use. Under this scenario, employment at the site would be similar to the baseline employment. Due to variations in potential workforce, there could be either a slight increase or decrease in the Region of Influence employment. Variations either way would represent less than 1 percent of the labor force.

Table 4.15-1. *Employment by Sector in the Region of Influence.*

Sector	Percentage of ROI Employment		
	1980	1990	1997
Services	23.7	31.3	31.9
Wholesale and Retail Trade	22.2	23.4	24.7
Government and Government Enterprise	12.5	13.2	12.2
Construction	8.8	5.7	7.8
Manufacturing	6.4	7.3	6.8
Finance, Insurance, and Real Estate	9.7	7.9	6.5
Transportation and Public Utilities	5.7	4.9	5.3
Farm	4.3	3.6	2.5
Agriculture Service, Forestry, Fishing, and other	0.6	1.1	1.3
Mining	5.9	1.7	0.9

ROI = Region of Influence; Source: BEA 1999

Table 4.15-2. *Unemployment in the Region of Influence and Colorado.*

Area	1990	1995	1998
ROI	5.9	5.5	5.0
Colorado	5.0	4.2	3.8

ROI = Region of Influence; Source: BLS 1999

Table 4.15-3. *Historic and Projected Population for the Region of Influence and Colorado.*

Area	1980	1990	1998	2000	2005	2010	2015	2020
ROI	81,530	93,145	112,891	117,317	128,201	139,624	151,321	163,602
Colorado	2,889,964	3,294,394	3,970,971	4,175,003	4,542,169	4,892,567	5,230,705	5,547,647

ROI = Region of Influence; Source: Census 1999, Census 1995, CDLA 1998

Table 4.15-4. *Region of Influence Housing Characteristics.*

Area	Total Number of Housing Units	Number of Owner-Occupied Units	Owner-Occupied Vacancy Rate	Median Value	Number of Occupied Rental Units	Rental Vacancy Rates	Median Monthly Contract Rent
ROI	39,208	23,534	2.2%	$62,700	12,716	5.9%	$275

ROI = Region of Influence; Source: Census 1992

Population in the Region of Influence could be affected if the site workforce decreased. Some workers and their families may out-migrate from the Region of Influence. This would result in a less than 1 percent decrease in the Region of Influence population. Some currently occupied housing units would become vacant or the housing construction rate would decrease as a result of the out-migration. If the site workforce increases over the baseline

Table 4.15-5. *Race, Ethnicity, and Poverty for the Region of Influence and Colorado.*

	ROI	Colorado
White	94.7	88.2
Black	0.4	4.0
American Indian, Eskimo, or Aleut	0.7	0.8
Asian or Pacific Islander	0.7	1.8
Other	3.5	5.1
Hispanic[a]	8.1	12.9
Living in Poverty	15.1	11.7

[a] Note: Persons of Hispanic ethnicity may be of any race.
ROI = Region of Influence; Source: Census 1992

level, there would not likely be any change in the Region of Influence population or housing markets. The current Region of Influence labor force would be sufficient to fill any additional employment requirements.

Industrial Use. Under this scenario, employment at the site would decrease significantly from the baseline level. No more than 10 employees would be involved in the gravel pit operation. Total employment generated by the site (including both direct employment at the site and indirect employment in local suppliers within the Region of Influence) would be much less than the baseline level. The decrease in total employment would represent approximately 1 percent of the Region of Influence labor force. Total income in the Region of Influence would also decrease.

There could be some change in Region of Influence population and housing as a result of the change in workforce requirements. Some workers and their families may out-migrate from the Region of Influence. This would result in a less than 1 percent decrease in the Region of Influence population. Some currently occupied housing units would become vacant or the housing construction rate would decrease as a result of the out-migration.

Mixed Use. Under this scenario, employment at the site would be similar to the baseline employment. Due to variations in potential workforce, there could be either a slight increase or decrease in Region of Influence employment. Variations either way would represent less than 1 percent of the labor force.

Population in the Region of Influence could be affected if the site workforce decreased. Some workers and their families may out-migrate from the Region of Influence. This would result in a less than 1 percent decrease in the Region of Influence population. Some currently occupied housing units would become vacant or the housing construction rate would decrease as a result of the out-migration. If the site workforce increases over the baseline level, there would not likely be any change in the Region of Influence population or housing markets. The current Region of Influence labor force would be sufficient to fill any additional employment requirements.

Open Space. Under this scenario, there would be no employment at the site. The decrease in total employment would represent approximately 1 percent of the Region of Influence labor force. Total income in the Region of Influence would also decrease.

There could be some change in Region of Influence population and housing as a result of the change in workforce requirements. Some workers and their families may out-migrate from the Region of Influence. This would result in a less than 1 percent decrease in the Region of Influence population. Some currently occupied housing units would become vacant or the housing construction rate would decrease as a result of the out-migration.

No Action Alternative. Under the No Action Alternative, there would be no change from the baseline level of employment. There would be no change in the Region of Influence employment, income, population, housing, or community services.

Environmental Justice. As shown in the other environmental impacts sections, there would be no significant adverse impact from implementing either of the alternatives. Therefore, there would be no disproportionately high or adverse impacts to minority or low-income populations in the area.

4.16 CUMULATIVE EFFECTS

The Council on Environmental Quality regulations implementing NEPA (40 CFR 1500-1508) define cumulative effects as "the impact on the environment which results from the incremental impact of the action when added to other past, present, and reasonably foreseeable future actions regardless of what agency (Federal or non-federal) or person undertakes such other actions (40 CFR 1508.7)." The regulations further explain that "cumulative effects can result from individually minor but collectively significant actions taking place over a period of time." The cumulative effects section presented is based on the potential effects of transfer of the GJO property on resources also affected by other past, present, and reasonably foreseeable future actions in the Region of Influence.

The DOE assessed cumulative effects by examining potential future activities at the GJO Site after its transfer and other activities in the Grand Junction area. Potential activities that may occur after the transfer of the GJO property have been presented in Chapter Three, Description of Proposed Action and No Action Alternatives. It is important to note that the DOE will not control future land uses under the Proposed Action, and there is no assurance that any specific scenario or activity will in fact take place. For purposes of this analysis, the All Industrial Land Use scenario will be used to examine cumulative effects. It provides a reasonable upper limit of impacts when combined with the cumulative effects from the Region of Influence.

Anticipated activities in the local region include ongoing activities as well as reasonably foreseeable future activities. The most significant construction activity is the ongoing widening of U.S. Highway 50, outside of the city and several miles southeast of the site. The highway is being widened from a mostly two lane road to four lanes between Grand Junction and Delta, Colorado. This activity represents a source of added heavy truck traffic and the noise and fugitive dust emissions associated with such activity. It also temporarily contributes to a reduced level of service on this important commercial and tourist route. Long-term effects will be beneficial because the widening will contribute to an improved level of service rating upon completion of the project.

The existing gravel pit operation, approximately one mile (1610 meters) south of the site, contributes to the noise exposure experienced in the Region of Influence. In addition, it is a source of heavy truck traffic on local surface streets leading to U.S. Highway 50 north of the site. Mining operations also contribute minor fugitive dust emissions to the local ambient air conditions along with particulate emissions from crushing and loading operations. The gravel pit is a permitted operation and is required to control dust emissions.

The police firing range, located approximately 200 yards (183 meters) east of the GJO site, is a source of intermittent periods of sudden, high noise. Noise is considered to be a nuisance when there are sensitive local receptors that would be affected by the intrusion of noise. The receptors considered to be sensitive in the local area are visitors to the cemetery, adjacent to both GJO and the firing range, and residents of the Little Park Road neighborhood across the river west of the site.

Another source of noise at the site and in the surrounding area is the Union Pacific Railroad running along the eastern boundary of the GJO Site. Trains run by the site from twice to several times a day at five to ten miles per hour and are a short-term minor source of noise. Noise levels are such that speech is interrupted, but only for brief periods of time.

As a result of analyzing the combined effects of the industrial land use scenario and the previously described current and future activities in the Region of Influence, four resource areas warranted further evaluation. Noise, transportation, air quality, and surface water quality were evaluated for cumulative effects and are described below.

Four activities have been identified in the Region of Influence that currently contribute to the noise environment. The widening of U.S. Highway 50, the Union Pacific Railroad, continued operation of the firing range, and continued operation of the gravel pit south of the site all contribute to the noise environment around GJO. All noise is of an intermittent nature and would be mostly noticeable to receptors within approximately one-fourth of a mile (402 meters) from the sources. Noise contributions from operation of the GJO Site as a gravel pit would also be localized and affect receptors very close to the site. Restrictions on operations at the site may be required to lessen the effects of noise; restrictions could include limiting the time of day or days of the week of noise generating operations or placing conditions on haul routes.

The widening of U.S. Highway 50 and the continued operation of the gravel pit are contributing elements to the heavy vehicle traffic in the local Region of Influence. Daily traffic to and from the GJO consists of 300-330 passenger vehicle trips by employees and about 50 service vehicle trips per day. Under the All Industrial Land Use scenario, the number of daily vehicle trips to the site would be greatly reduced and replaced by periodic truck traffic hauling gravel to various job sites throughout the region. When combined with activities in the local region, there will be a slight increase in heavy truck traffic, though service levels on local transportation routes are not expected to be affected. Overall vehicle trips will be significantly reduced, potentially providing for an increase in the level of service on local transportation routes.

Air quality in the region could also receive cumulative effects from two of the identified activities. The widening of U.S. Highway 50 and continued operation of the gravel pit south of the site are contributing minor sources of fugitive dust emissions and particulate emissions. In addition, operation of the heavy equipment is a minor source of vehicle exhaust emissions. It is expected that both activities are required to use standard dust control measures, such as water sprays. In addition, any future operator of a gravel pit mining operation at the former GJO Site would be subject to state permitting requirements for this type of operation. It is thus expected that dust emissions would not cause adverse impacts in the area. Overall vehicle emissions would decrease due to reduced numbers of personnel working at the site and the subsequent reduction in the number of vehicle trips to and from the site on a daily basis. The decrease may be offset to some extent by the widening of U.S. Highway 50 and increased residential development south of Grand Junction, resulting in more vehicle trips from the Delta and white water areas.

As discussed in the surface water and groundwater sections, there is a potential for excavation activities below the water table to increase the suspended solids load of the shallow groundwater. Because this groundwater is hydraulically connected to the Gunnison River, it is possible that the suspended solids load of the river could be increased under this scenario. If this is occurring at the gravel pit located one mile (1610 meters) southeast of the GJO Site, there is potential for a cumulative effect on the Gunnison River. There is currently insufficient data available to assess the potential cumulative effect. However, this issue would be considered by the Colorado Department of Public Health and Environment during the permitting process for the gravel mining operation at the GJO Site.

Cumulative effects on noise, transportation, and air quality associated with the proposed transfer of the GJO facility to a non-DOE entity and local activities in the Region of Influence are minor in nature. In the case of transportation and air quality, there may be beneficial effects due to a reduction in vehicular trips. The potential cumulative effects on the Gunnison River from two gravel mining operations located within one mile (1610 meters) of each other have yet to be determined. However, if this scenario were to occur, the Colorado Department of Public Health and Environment would consider the potential cumulative effects on the Gunnison River during the permitting process for the gravel mining operation at the GJO Site.

No past, present, or reasonably foreseeable future projects have been identified in the local region which, when added to the effects of the proposed action, would result in a significant impact.

5.0   LIST OF AGENCIES AND PERSONS CONSULTED

Laine Johnson
Colorado Natural Heritage Program
Colorado State University
College of Natural Resources
254 General Services Building
Fort Collins, Colorado 80523

Paul J. Creeden
Colorado Division of Wildlife
West Region Service Center
711 Independent
Grand Junction, Colorado 81505

Susan T. Moyer
U.S. Fish & Wildlife Service, Ecological Services
764 Horizon Drive, Building B
Grand Junction, Colorado 81506-3946

Jill Parker
U.S. Fish & Wildlife Service, Division of Endangered Species
PO Box 25486
Denver Federal Center
Denver, Colorado 80225

Ms. Georgianna Contiguglia
State Historic Preservation Officer
Office of Archaeology and Historic Preservation
Colorado Historical Society
Denver, Colorado

Mr. Kurt P. Schweigert
Associated Cultural Resource Experts
Englewood, Colorado

6.0 REFERENCES

10 CFR 1022	U.S. Department of Energy, "Compliance with Floodplains/Wetlands Environmental Review Requirements."
24 CFR 51	U.S. Department of Housing and Urban Development, Office of Assistant Secretary for Equal Opportunity, "Environmental Criteria and Standards."
29 CFR 1910	U.S. Department of Labor, Occupational Safety and Health Administration, "Occupational Safety and Health Standards."
36 CFR 800	Advisory Council on Historic Preservation, "Protection of Historic Properties."
40 CFR 50	U.S. Environmental Protection Agency, "National Primary and Secondary Ambient Air Quality Standards."
40 CFR 53	U.S. Environmental Protection Agency, "Ambient Air Monitoring Reference and Equivalent Methods."
40 CFR 58	U.S. Environmental Protection Agency, "Ambient Air Quality Surveillance."
40 CFR 192	U.S. Environmental Protection Agency, "Health and Environmental Standards for Uranium and Thorium Mill Tailings."
40 CFR 261	U.S. Environmental Protection Agency, "Identification and Listing of Hazardous Wastes."
40 CFR 403	U.S. Environmental Protection Agency, "General Pretreatment Regulations for Existing and New Sources of Pollution."
40 CFR 761	U.S. Environmental Protection Agency, "Polychlorinated Biphenyls Manufacturing, Processing, Distribution in Commerce, and Use Prohibitions."
40 CFR 1508	Council on Environmental Quality, "Terminology and Index."
Executive Order 11988	*Floodplain Management*, 42 FR 26951, May 24, 1977.
Executive Order 11990	*Protection of Wetlands*, 42 FR 26961, May 24, 1977.
Executive Order 12898	*Federal Actions to Address Environmental Justice*, 59 FR 7629, February 16, 1994.
AHA 1995	American Hospital Association, 1995. *The AHA Guide to the Health Care Field.*
AMA 1995	American Medical Association, 1995. *Physician Characteristics and Distribution in the U.S.*
Arizona 1999	Arizona Game and Fish Department, 1999. Colorado River Squawfish (Ptychocheilus lucius). http://www.gf.state.az.us/frames/fishwild/idx_fish.htm.
BEA 1999	Bureau of Economic Analysis, 1999. *Regional Economic Information System, 1969-1997.*

BLS 1999   Bureau of Labor Statistics, 1999. *Bureau of Labor Statistics Data.*
                      http://146.142.4.24/cgi-bin/dsrv.

CDE 1999   Colorado Department of Education, 1999. *Pupil Membership and Classroom Teacher
                      data- Fall 1997.* http://cde.state.co.us/cdemgmt/rvpupmem.htm.

CDLA 1998   Colorado Department of Local Affairs, 1998. *County and State Population
                      Projections.* http://www.dlg.oem2.state.co.us/demog/project.htm.

Census 1992   U.S. Bureau of the Census, 1992. *General Profile for Mesa County, CO.*
                      http://govinfo.library.orst.edu/cgi-bin/buildit2?filenam=la-077.coc.

Census 1995   U.S. Bureau of the Census, 1995. *Colorado Population of Counties by Decennial
                      Census: 1900 to 1990.* http://www.census.gov/population/cencounts/co190090.txt.

Census 1999   U.S. Bureau of the Census, 1999. *County Population Estimates for July, 1998.*
                      http://www.census.gov/population/estimates/county/co-98-2/98c2_08.txt.

COE 1989   Department of the Army, Corps of Engineers, Permit No. 10040 Issued by U.S.
                      Army COE, Sacramento District to U.S. Department of Energy.

Creeden 1999   September 29, 1999 letter from Colorado Division of Wildlife, to J. Peyton Doub,
                      Tetra Tech NUS.

DOE 1986   U.S. Department of Energy (DOE), 1986. *Final Environmental Impact Statement.
                      Remedial Actions at the Former Climax Uranium Company Uranium Mill Site,*
                      Grand Junction Project Office Site, 2597 B ¾ Road, Grand Junction, Colorado.

DOE 1989   U.S. Department of Energy, 1989. Final RI/FS for the GJORAP Program, Grand
                      Junction, Colorado.

DOE 1990a   *Declaration for the Record of Decision and Record of Decision Summary*–Grand
                      Junction Projects Office Remedial Action Project, Idaho Operations Office, U.S.
                      Department of EnergyGrand Junction Projects Office, Grand Junction, Colorado.

DOE 1994   National Environmental Policy Act Documentation for the Proposed Action to
                      Perform Remedial and Demolition Type Activities on Radiologically Contaminated
                      Buildings at the U.S. Department of Energy Grand Junction Projects Office (GJP-94-
                      031).   Administrative Record for Compliance with DOE NEPA Process.
                      Albuquerque Field Office.   Grand Junction Projects Office, Grand Junction,
                      Colorado, August 15, 1994.

DOE 1995a   Final Report of the Decontamination and Decommissioning of the Exterior Land
                      Areas at the Grand Junction Projects Office Facility, GJPO-GJ-13, prepared by Rust
                      Geotech for U.S. Department of Energy Grand Junction Projects Office, Grand
                      Junction, Colorado.

DOE 1995b   "Survey Plan for Releasing the Buildings at the Grand Junction Projects Office for
                      Unrestricted Use", Grand Junction, Colorado.

DOE 1996a   Environmental Assessment of Facility Operations at the U.S. Department of Energy Grand Junction Projects Office, DOE/EA-0930, Grand Junction, Colorado.

DOE 1996b   Grand Junction Projects Office Remedial Action Project: Building 2 Public Dose Evaluation, DOE/ID/12584-265, Grand Junction, Colorado.

DOE 1998a   "*Facility Condition Assessment*", Prepared by WASTREN, Grand Junction, Colorado.

DOE 1998b   Fourth Annual Monitoring Reports for the U.S. Departments of Energy Grand Junction Projects Office Wetland Mitigation Project, prepared by WASTREN for U.S. Department of Energy, Grand Junction Projects Office, Grand Junction, Colorado.

DOE 1998c   U.S. Department of Energy, 1998. *Strengths, Weaknesses, Opportunities, and Barriers Analysis*, Department of Energy, Grand Junction Office, Prepared by Transitions to Tomorrow, Inc., Grand Junction, Colorado.

DOE 1999a   *Site Environmental Report for Calendar Year 1998*, Prepared by WASTREN, Grand Junction, Colorado.

DOE 1999b   U.S. Department of Energy, 1999a, *U.S. Department of Energy Grand Junction Office Site Transition Plan*, prepared by unknown, Grand Junction, Colorado.

DOE 1999c   U.S. Department of Energy, 1999. *Title Search on Grand Junction Project Office Site*, prepared by Western Colorado Title Company, Grand Junction, Colorado.

FEMA 1992   Federal Emergency Management Agency Flood Insurance Rate Map, Mesa County, Colorado (Unincorporated Areas). Panel 470 of 1000. Community Panel Number 080115 0470 B. Map Revised July 15, 1992.

HPI 1999   Harden Political Info Systems, 1999. State, County and Municipal Courts in Mesa County, Colorado. http://hpi.www.com/colaw/08077.html.

ICBO 1997   International Conference of Building Officials, 1997. *Uniform Building Code*, Whittier, California.

Johnson 1999   Johnson, L. 1999. Letter dated September 9, 1999 from Laine Johnson of the Colorado Natural Heritage Program to Peyton Doub of Tetra Tech NUS, Gaithersburg, Maryland. Subject: Natural Heritage Resource Review for Township 1 South, Range 1 West, Sections 26 and 27.

Mesa 1996   Mesa County, 1996, *Mesa County Wide Land Use Plan; From Issues to Action*, Prepared by Design Studios West, Inc. and Frerlich, Lectner & Carlisle, for Mesa County Long Range Planning Division, Grand Junction, Colorado.

Moyer 1999   Letter dated October 6, 1999 from S. Moyer of the USFWS to Peyton Doub of Tetra Tech, NUS.

Schweigert 1999a   Draft Historic Structures Survey of the Department of Energy Grand Junction Office.
                          Prepared for the U.S. DOE, Grand Junction Office, Grand Junction, Colorado. May
                          1999.

Schweigert 1999b   *Final Historic Structure Survey of the Department of Energy Grand Junction Office.*
                          Prepared for the U.S. Department of Energy, Grand Junction Office, Grand Junction,
                          Colorado. October 1999.

Southwest Wildlife 1999   Southwest Wildlife Fact Sheets: Kit Fox.
                          http://www.extremezone.com/SWRGF/factsheets/Kitfox.html.

USAR 1999   U.S. Army Reserves, 1999, "Environmental Baseline Survey". Prepared by J. M.
                          Waller Associates, Inc., Salt Lake City, Utah.

UNR 1999a   University of Nevada – Reno, 1999, *Ptychocheitus lucius:* Colorado Squawfish
                          Biological Resources Research Center.
                          http://www.brcc.unr.edu/data/fish/ptycluci.html.

UNR 1999b   University of Nevada – Reno, 1999, *Gila cypha*: Humpback Chub. Biological
                          Resources Research Center. http://www.brrc.unr.edu/data/fish/gilacyph.html.

UNR 1999c   University of Nevada – Reno, 1999, *Gila elegans*: Bonytail Chub. Biological
                          Resources Research Center. http://www.brrc.unr.edu/data/fish/gilaeleg.html.

UNR 1999d   University of Nevada – Reno, 1999, *Xyrauchen texanus*: Razorback Sucker.
                          Biological Resources Research Center.
                          http://www.brrc.unr.edu/data/fish/xyratexa.html.

7.0   LIST OF PREPARERS

Name: Amy Cordle
Affiliation: Tetra Tech, Inc.
Education: B.S., Civil Engineering
Technical Experience: 6 years of experience in air quality, noise analyses, and NEPA documentation
GJO Transfer Responsibility: Air and Noise Resources

Name: Peyton Doub
Affiliation: Tetra Tech, Inc.
Education: B.S., Plant Sciences, Cornell University
                    M.S., Botany, University of California at Davis
Technical Experience: 17 years of experience preparing NEPA documents, remedial investigations, environmental baseline surveys, wetland delineations, mitigation plans
GJO Transfer Responsibility: Ecological Risk Assessment and Ecological Resources; Floodplains and Wetlands Assessment

Name: Kevin Doyle
Affiliation: Tetra Tech, Inc.
Education: B.A., Sociology, University of California, Santa Barbara
Technical Experience: 13 years of experience in archaeology, cultural resources management, and NEPA documentation
GJO Transfer Responsibility: Cultural and Visual Resources

Name: David Flynn
Affiliation: Tetra Tech, Inc.
Education: B.S., Geology, Southern Illinois University
Technical Experience: 19 years of experience in geology and 10 years in western geology and environmental investigations
GJO Transfer Responsibility: Human Health and Ecological Risk Assessment

Name: Clifford J. Jarman
Affiliation: Tetra Tech, Inc.
Education: B.S., Geology, University of New Mexico
                    M.S., Geophysics, New Mexico Institute of Mining and Technology
Technical Experience: 13 years of experience in preparation of NEPA documents, NEPA compliance, management of environmental programs, and seismic risk assessments
GJO Transfer Responsibility: Third-Party Document Review

Name: John W. Lynch, P.E.
Affiliation: Tetra Tech, Inc.
Education: B.S., Civil Engineering, University of Notre Dame
                    M.S., Civil Engineering, University of Notre Dame
Technical Experience: 17 years of experience in preparation of NEPA documents, NEPA compliance, and management of environmental programs
GJO Transfer Responsibility: GJO Transfer EA Project Manager, DOPAA preparation, Land Use and Infrastructure, Solid and Hazardous Waste Management

Name: Sara McQueen
Affiliation: Tetra Tech, Inc.
Education: B.A., Economics, Wittenberg University

Technical Experience: 4 years of experience with socioeconomic impacts and environmental justice analysis for NEPA documentation
GJO Transfer Responsibility: Socioeconomics/Environmental Justice/Transportation

Name: John Nash
Affiliation: Tetra Tech, Inc.
Education: B.A., Political Science, LaSalle University
Technical Experience: 5 years of experience in technical editing, including NEPA regulatory compliance
GJO Transfer Responsibility: Technical Editor

Name: Kevin Taylor
Affiliation: Tetra Tech, Inc.
Education: B.S., Physics, Clemson University
M.S., Nuclear Engineering, Georgia Institute of Technology
Technical Experience: 5 years experience in the preparation of radiological risk modeling, environmental assessments, and statistical analysis
GJO Transfer Responsibility: Human Health Risk Assessment

Name: Scott Truesdale, P.G.
Affiliation: Tetra Tech, Inc.
Education: B.A., Environmental Science, University of Virginia
Technical Experience: 12 years of experience in site characterization, environmental programs, geology, groundwater, and NEPA analysis
GJO Transfer Responsibility: Geology/Topography/Surface and Groundwater

APPENDIX A: CONSULTATION LETTERS

Colorado Natural Heritage Program, 254 General Services Building · CSU, Fort Collins, CO 80523

9/09/1999

J. Peyton Doub
Tetra Tech NUS, Inc.
910 Clopper Road, Suite 400
Gaithersburg, MD 20878-1399

Dear J. Peyton Doub:

The Colorado Natural Heritage Program (CNHP) is in receipt of your request for information regarding the Grand Junction project. In response, CNHP has searched its Biological and Conservation Datasystem (BCD) for natural heritage resources (occurrences of significant natural communities and rare, threatened or endangered plants and animals) documented from the immediate area of T1S R1W S26, 27.

We have enclosed two reports from BCD. One describes natural heritage resources known from the area and gives location (by Township, Range, and Section), precision of the locational information, and the date of last observation at that location. Please note that "precision" reflects the resolution of original data. For example, an herbarium record from "4 miles east of Colorado Springs" provides much less spatial information than a topographic map showing the exact location of the occurrence. "Precision" codes of Seconds, Minutes, and General are defined in the report footer.

You may notice that some occurrences do not have sections listed. Those species have been designated as "sensitive" due to their rarity and threats by humans. Peregrine falcons, for example, are susceptible to human breeders removing falcon eggs from their nests. For these species, CNHP does not provide locational information beyond township and range. Please contact us should you require more detailed information for sensitive occurrences.

A second report outlines the status of the known elements. We have included status according to Natural Heritage Program methodology and legal status under state and federal statutes. Natural Heritage ranks are standardized across the Heritage Program network, and are assigned for global and state levels of rarity. They range from "1" for critically imperiled or extremely rare elements, to "5" for those that are demonstrably secure. For your convenience we have also included habitat descriptions. However, please be aware that these descriptions are in draft form and have not been edited for content. Please do not cite CNHP habitat information; instead, cite the original source of the habitat information as indicated.

The Colorado Division of Wildlife has legal authority over wildlife in the state. CDOW would therefore be responsible for the evaluation of and final decisions regarding any potential effects a proposed project may have on wildlife. If you would like more specific information regarding

Colorado Natural Heritage Program, 254 General Services Building – CSU, Fort Collins, CO 80523

these or other vertebrate species in the vicinity of the area of interest, please contact the Colorado Division of Wildlife.

The information contained herein represents the results of a search of Colorado Natural Heritage Program's (CNHP) Biological and Conservation Data System (BCD). However, the absence of data for a particular area, species or habitat does not necessarily mean that these natural heritage resources do not occur on or adjacent to the project site, rather that our files do not currently contain information to document their presence.

The information provided can be used as a flag to anticipate possible impacts or to identify areas of interest. If impacts to wildlife habitat are possible, these data should not be considered a substitute for on-the-ground biological surveys.

Although every attempt is made to provide the most current and precise information possible, please be aware that some of our sources provide a higher level of accuracy than others, and some interpretation may be required. CNHP's data system is constantly updated and revised. Please contact CNHP for an update or assistance with interpretation of this natural heritage information.

Sincerely,

Laine Johnson
Acting Environmental Review Coordinator

enc.

United States Department of the Interior

FISH AND WILDLIFE SERVICE
Ecological Services
764 Horizon Drive, Building B
Grand Junction, Colorado 81506-3946

IN REPLY REFER TO:
ES/CO:DOE
MS 65412 GJ

October 6, 1999

J. Peyton Doub, Environmental Scientist
Tetra Tech Nus, Inc.
910 Clopper Road, Suite 400
Gaithersburg, Maryland 20878-1399

Dear Mr. Doub:

The Fish and Wildlife Service has reviewed your September 9, 1999, letter requesting a species list for the Grand Junction, Colorado, Department of Energy site. The 56.4 acre site is located in sections 26 and 27, T. 1 S., R. 1 W., Ute P.M. along the Gunnison River.

Federally listed species that occur or may occur in the area include the threatened bald eagle (*Haliaeetus leucocephalus*) and the endangered southwestern willow flycatcher (*Empidonax traillii extimus*), Colorado pikeminnow[1] (*Ptychocheilus lucius*), bonytail (*Gila elegans*), humpback chub (*Gila cypha*), and razorback sucker (*Xyrauchen texanus*). The bald eagle may occasionally roost in the cottonwood trees in the area especially in the winter. The southwestern willow flycatcher may occur in shrubby riparian vegetation. Any activities on the site should avoid impacting habitat for the bald eagle and southwestern willow flycatcher. The Colorado pikeminnow and the razorback sucker are known to occur in the area. They, along with the other two endangered fishes, could be impacted by water depletions or contamination of water as a result of activities on the site.

If the site does not contain contaminants at levels of concern to the endangered fishes, the Service would be particularly interested in cooperating with the DOE to use and modify habitat on the site to benefit the endangered fishes. Section 7(a)(1) of the Endangered Species Act directs all Federal agencies to use their authorities to further recovery of threatened and endangered species and the Service believes this would be an excellent opportunity for the DOE to exercise that authority. Habitat modifications may include removal of dikes to allow water onto the historic floodplain and to flow through existing ponds, creation of seasonally ponded backwater areas, and/or creation of side channels. These modifications will provide spawning, resting, feeding, and nursery sites for the endangered fishes. Additionally, the Service may be interested in using the existing ponds, with some reconfiguration, for raising young endangered fishes.

---

[1]formerly squawfish

2

The Service would appreciate a response back from the DOE with their interest in cooperating on recovery activities for the endangered fishes. If the Service can be of further assistance, please contact Terry Ireland at the letterhead address or (970) 243-2778.

Sincerely,

Susan T. Moyer
Assistant Colorado Field Supervisor

pc: FWS/ES, Lakewood
     CDOW, Grand Junction

TIreland:DOEClosing.ltr:100699

STATE OF COLORADO
Bill Owens, Governor
DEPARTMENT OF NATURAL RESOURCES

DIVISION OF WILDLIFE

AN EQUAL OPPORTUNITY EMPLOYER

John W. Mumma, Director
6060 Broadway
Denver, Colorado 80216
Telephone: (303) 297-1192

*For Wildlife-
For People*

Mr. J. Peyton Doub, CEP
Tera Tech NUS
910 Clopper Rd., Suite 400
Gaithersburg, MD 20878-1399

RECEIVED
NOV 1 1999

29 September 1999

Dear Mr. Doub,

I have examined our database for information on the possible occurrence of threatened, endangered, and special-status species in the vicinity of the DOE-GJO site along the Gunnison River in Mesa County, Colorado. This report is only an estimate of the possible occurrence of rare species in that area, and is not a substitute for a thorough biological inventory of that location.

The Gunnison River corridor with its native cottonwood-willow riparian habitat is critically important to many species of wildlife, including hawks, owls, eagles and migratory songbirds. The DOE-GJO property is located within bald eagle winter range and a winter roost is located near the site. The mature cottonwoods located on-site serve as potential roost sites or hunting perches and the preservation of those trees is of critical importance. The river corridor on and adjacent to the site is considered potential suitable habitat for the southwest willow flycatcher. Disturbance of the willow riparian habitat would negatively impact its value to that species.

The Gunnison River itself is considered occupied habitat for the endangered fishes of the Colorado River drainage, including the Colorado pikeminnow and the razorback sucker. Any impacts development of the site would have on the Gunnison River could potentially impact those species.

We have a recorded sighting of a kit fox in T 1S, R1W, Sec. 36, but it is unknown if kit fox occur on the DOE-GJO site.

Thank you for your interest in Colorado's wildlife. Don't hesitate to call if you have any further questions. I can be reached at (970) 255-6112.

Sincerely,

Paul J. Creeden
Wildlife Manager
Glade Park District

xc: Yamashita

APPENDIX B: PLANT AND WILDLIFE SPECIES OBSERVED ON DOE-GJO SITE

APPENDIX B

Table B-1. Plant and Wildlife Species Observed at DOE-GJO Site

Scientific Name	Common Name	Scientific Name	Common Name
Invertebrates		*Mephitis mephiti*	striped skunk
Cambarus Spp.	crayfish [a]	*Mustela frenata*	long-tailed weasel
		Myotis leibil	small-footed myotis [a]
Fish		*Odocolleus hemionus*	mule deer [a]
Catostomus commersoni x Catostomus discobolus	white sucker x bluehead sucker [b]	*Ondatra zibethicus*	muskrat [a]
Catostomus latipinnis	flannelmouth sucker	*Peromyscus maniculatus*	deer mouse
Catostomus latipinnis x C. commersoni	white x flannelmouth sucker	*Pipistrellus hesperus*	western pipistrelle [a]
Catostomus latipinnis x Xyrauchen texanus	flannelmouth x razorback sucker	*Piecotus towwnsendil*	Townsend's big-eared bat [a]
Cyprinus carpio	carp [a]	*Procyon lotor*	raccoon
Gila cypha	humpback chub	*Rattus norvegicus*	Norway rat
Gila robusta	roundtail chub	*Spilogale gracilis*	spotted skunk
Ictalurus melas	black bullhead	*Sylvilagus audubonii*	desert cottontail [a]
Lactalurus punctatis	channel catfish	*Tamiasciurus hudsonicus*	red squirrel [a]
Lepomis cyanellus	green sunfish	*Taxidea taxus*	badger
Lepomis machrochirus	bluegill	*Urocyon cinereoargenteus*	gray fox
Micropterus salmoides	largemouth bass	*Vulpes fulva*	red fox
Notropis lutrensis	red shiner		
Notropis stramineus	Sand shiner	**Amphibians**	
Pimephales promelas	flathead minnow [a]	*Ambystoma tigrinum*	tiger salamander
Prychocheilus lucius	Colorado squawfish	*Bufo woodhousei*	woodhouse toad [a]
Pomoxis nigromaculatus	black crappie	*Rana catesbeiana*	bullfrog
Phinichthys osculus	speckled dace	*Rana pipiens*	leopard frog [a]
Salmo clarki	cutthroat trout		
Salmo gairdneri	rainbow trout	**Reptiles**	
Salmo trutta	brown trout	*Chelydra serpentina*	snapping turtle
		Chrysemys picta	painted turtle
Mammals		*Cnemidophorus velox*	plateau whiptail
Antilocapra americana	pronghorn antelope	*Crotalus viridis concolor Woodbury*	midget faded rattlesnake
Canis latrans	coyote	*Pituophis melanoleucus*	bullsnake
Castor canadensis	beaver	*Phrynosoma douglassii*	short horned lizard
Cynomys leucurus	white-tailed prairie dog	*Sceloporus graciosus*	sagebrush lizard
Felis cattus	feral cat [a]	*Sceloporus undulatus*	eastern fence lizard
Lepus californicus	blacktail jackrabbit	*Thamnophis elegans*	wandering garter snake [a]

Table B-1. *Plant and Wildlife Species Observed at DOE-GJO Site (continued)*

Scientific Name	Common Name	Scientific Name	Common Name
Birds			
Accipiter cooperii	Cooper's hawk	*Corvus corax*	common raven
Aix sponsa	wood duck	*Dendroica petechia*	yellow warbler
Alectoris chukar	chukar	*Eremophila alpestris*	horned lark
Amphispiza Belli	sage sparrow[a]	*Euphagus cyanocephalus*	Brewer's blackbird[a]
Anas acuta	northern pintail	*Falco sparverius*	American kestrel[a]
Anas cyanoptera	cinnamon teal	*Hellaeetus leucocephalus*	bald eagle[a]
Anas discors	blue-winged teal	*Hirundo pyrrhonota*	cliff swallow[b]
Anas formosa	green-winged teal	*Hirundo rustica*	barn swallow[b]
Anas platyrhynchos	mallard[a]	*Junco hyemalis*	dark-eyed junco[a]
Aquila chrysaetos	golden eagle[a]	*Meleagris gallopavo*	wild turkey
Ardea herodias	great blue heron[a]	*Melospiza melodia*	song sparrow[a]
Asio otus	long eared owl	*Molothrus ater*	brown-headed cowbird
Aythya valisineria	canvasback	*Nycticorax nycticoras*	black-crowned night heron
Branta canadensis	Canada goose	*Passerina amoena*	lazuli bunting
Bucephala albeola	bufflehead	*Phalaenoptilus nuttallii*	whippoorwill
Bucephala ciangula	common goldeneye	*Phasianus colchicus*	ring-necked pheasant
Buteo jamaicensis	red-tailed hawk	*Pica pica*	black-billed magpie[a]
Calamospiza melanocorys	lark bunting	*Picoides pubescens*	downy woodpecker[a]
Callipepla gambelii	Gambel's quail	*Pipilo chlorurus*	green-tailed towhee
Carpodacus mexicanus	house finch[a]	*Selasphorus platycercus*	broad-tailed hummingbird[a]
Cathartes aura	turkey vulture	*Sialis currucoides*	mountain bluebird
Charadrius vociferus	killdeer	*Spizella breweri*	Brewer's sparrow[b]
Chen caerulescens	snow goose	*Sturnella neglecta*	western meadowlark
Chondestes grammacus	lark sparrow[b]	*Sturnus vulgaris*	European starling[a]
Chordeiles minor	common nighthawk[a]	*Tachycineta thallassina*	violet-green swallow[b]
Colaptes auratus	northern flicker[a]	*Turdus migratorius*	American robin[a]
Columba livia	rockdove[a]	*Tyrannus verticalis*	western kingbird
Contopus sordidulus	western wood pewee	*Zenaida macroura*	mourning dove[a]
Corvus brachyrhynchos	American crow[a]		

[a] Species observed at the GPO facility.
[b] "x" indicates a hybrid between the two species listed.
[c] Animals possibly observed at the GJO facility, but which were only identified in general terms (i.e., bat, swallow, toad, etc.).

APPENDIX C: BIRD SPECIES OBSERVED ON DOE-GJO SITE

APPENDIX C

Table C-1. *Bird Species Observed on DOE-GJO Site* [a]

Scientific Name[b]	Common Name[c]
Anseriformes: Anatidea[d]	
Aix sponsa	Wood Duck
Anas acuta	Northern Pintail
Anas cyanoptera	Cinnamon Teal
Anas platyrhynchos	Mallard Duck
Aythya collaris	Ring-necked Duck
Branta canadensis	Canada Goose
Bucephala clangula	Common Goldeneye
Mergus mersanger	Common Mersanger
Apodiformes: Apodidae	
Aeronautes saxatalis	White-throated Swift
Apodiformes: Trochilidae	
Archilochus alexandri	Black-Chinned Hummingbird
Selasphorus platycercus	Broad-tailed Hummingbird
Selasphorus rufus	Rufous Hummingbird
Caprimulgiformes: Caprimulgidae	
Chordeiles minor	Common Nighthawk
Charadriiformes: Charadriidae	
Charadrius vociferus	Killdeer
Ciconiiformes: Ardeidae	
Ardea herodias	Great Blue Heron
Nycticorax nycticorax	Black-crowned Night Heron
Columbiformes: Columbidae	
Zenaida macroura	Mourning Dove
Coraciiformes: Alcedinidae	
Ceryle alcyon	Belted Kingfisher
Falconiformes: Accipitridae	
Accipiter cooperi	Cooper's Hawk
Accipiter striatus	Sharp-shinned Hawk
Aquila chrysaetos	Golden Eagle
Buteo jamaicensis	Red-tailed Hawk
Circus cyaneus	Northern Harrier
Haliaeetus leucocephalus	Bald Eagle
Pandion haliaetus	Osprey
Falconiformes: Cathartidae	
Cathartes aura	Turkey Vulture
Falconiformes: Falconidae	
Falco peregrinus or	Peregrine Falcon or
Falco mexicanus	Prairie Falcon
Falco sparverius	American Kestrel
Galliformes: Phasianidae	
Callipepla gambelii	Gambel's Quail
Phasianus colchicus	Ring-necked Pheasant
Gruiformes: Rallidae	
Fulica americana	American Coot
Passeriformes: Aegithalidae	
Psaltriparus minimus	Bushtit
Passeriformes: Bombycillidae	
Bombycilla cedrorum	Cedar Waxwing
Bombycilla garrulus	Bohemian Waxwing

Table C-1. *Bird Species Observed on DOE-GJO Site* [a]

Scientific Name[b]	Common Name[c]	*(continued)*
Passeriformes: Corvidae		
Corvus brachyrhynchos	American Crow	
Corvus corax	Common Raven	
Pica pica	Black-billed Magpie	
Passeriformes: Emberizidae		
Agelaius phoeniceus	Red-winged Blackbird	
Amphispiza bilineata	Black-throated Sparrow	
Chondestes grammacus	Lark Sparrow	
Dendroica coronata	Yellow-rumped Warbler	
Euphagus cyanocephalus	Brewer's Blackbird	
Guiraca caerulea	Blue Grosbeak	
Icteria virens	Yellow-breasted Chat	
Icterus galbula	Northern Oriole	
Junco hyemalis	Dark-eyed Junco	
Melospiza melodia	Song Sparrow	
Molothrus ater	Brown-headed Cowbird	
Passerina amoena	Lazuli Bunting	
Pipilo erythrophthalmus	Rufous-sided Towhee	
Piranga ludoviciana	Western Tanager	
Pheucticus melanocephalus	Black-headed Grosbeak	
Quiscalus quiscula	Common Grackel	
Sturnella neglecta	Western Meadowlark	
Vermivora celata	Orange-crowned Warbler	
Zonotrichia leucophrys	White-crowned Sparrow	
Passeriformes: Fringillidae		
Carduelis psaltria	Lesser Goldfinch	
Carduelis tristis	American Goldfinch	
Carpodacus cassinii	Cassin's Finch	
Carpodacus mexicanus	House Finch	
Passeriformes: Hirundinidae		
Hirundo rustica	Barn Swallow	
Tachycineta thalassina	Violet-green Swallow	
Passeriformes: Muscicapidae		
Myadestes townsendi	Townsend Solitaire	
Regulus calendula	Ruby-crowned kinglet	
Sialia mexicana	Western Bluebird	
Sialia currucoides	Mountain Bluebird	
Turdus migratorius	American Robin	
Passeriformes: Paridae		
Parus atricapillus	Black-capped Chickadee	
Parus gambeli	Mountain Chickadee	
Parus inornatus	Juniper Titmouse	
Passeriformes: Passeridae		
Passer domesticus	House Sparrow	
Passeriformes: Sturnidae		
Sturnus vulgaris	European Starling	
Passeriformes: Troglodytidae		
Catherpes mexicanus	Canyon Wren	
Salpinctes obsoletus	Rock Wren	
Thryomanes bewickii	Bewick's Wren	
Troglodytes aedon	House Wren	

Table C-1. *Bird Species Observed on DOE-GJO Site* [a]

Scientific Name[b]	Common Name[c]
Passeriformes: Tyrannidae	*(continued)*
Contopus sordidulus	Western Wood-Peewee
Tyrannus verticalis	Western Kingbird
Piciformes: Picidae	
Colaptes auratus	Red-shafted Northern Flicker
Picoides pubescens	Downy Woodpecker
Sphyrapicus nuchalis	Red-naped Sapsucker
Podicipediformes: Podicipedidae	
Podilymbus podiceps	Pied-billed Grebe

[a] Source: Personal observations of Larry Arnold of DOE-GJO, as communicated to J. Peyton Doub of Tetra Tech, Inc.
[b] Scientific names and taxonomy are based on *Colorado Birds* by Robert Andrews and Robert Righter, 1992, published by Denver Museum of Natural History.
[c] Other birds sighted but for which species data is not available include certain flycatchers, owls, swallows, swifts, terns, woodpeckers, wrens, and warblers.
[d] The first term refers to Order; the second refers to Family.

INDEX